Intelligent Mars III:

Aum and the Architect

Arthur Raymond Beaubien

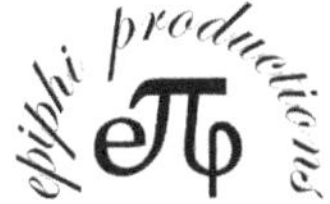

Intelligent Mars III: Aum and the Architect

ISBN 9780994032126

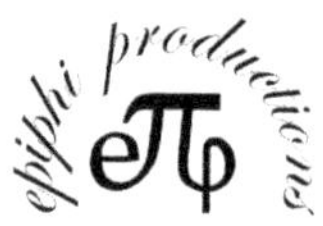

epiphi productions
Ottawa, Ontario
Canada

www.epiphiproductions.ca

Library and Archives Canada Cataloguing in Publication

Title: Intelligent Mars III : Aum and the architect / Arthur Raymond Beaubien.
Names: Beaubien, Arthur Raymond, 1942- author.
Description: Includes bibliographical references.
Identifiers: Canadiana 2020022719X | ISBN 9780994032126 (softcover)
Subjects: LCSH: Martian craters. | LCSH: Mars (Planet)—Surface. | LCSH: Life on other planets.
Classification: LCC QB641 .B434 2020 | DDC 559.9/23—dc23

Dedication

*To those who have chosen the path
of spiritual evolution*

Acknowledgements

The author wishes to express his gratitude to Gayle Peterson and their daughter Kery for giving an abundance of meaning to his life. He also wishes to thank George A. Neville for his invaluable enduring friendship, suggestions and support, and to Liana Manole for her generous gift of time and many helpful comments in her review of this manuscript.

Contents

Preface

In order to properly understand Mars, it is paramount that we get beyond our indoctrination that Mars has never been home to intelligent life. We accept this view as dogma without realizing that it has never been proven. Another barrier is that we think in terrestrial terms, totally ignorant of the ancient Martian mindset. In other words, we lack a technological language to interpret the data that we do have. To give an example, NASA informs us that the mean orbital distance of Deimos from the centre of Mars is 23,459 km and that of Phobos is 9,378 km. These are rather uninteresting numbers because they are given in terrestrial units. The kilometer is very close to 1/10,000 of the distance from the Equator to the North Pole of the planet Earth and has nothing to do with Mars. However, if we translate these distances into Martian kilometers (Mkm, see *Intelligent Mars I*[1], chapter 12) we find that Deimos is very close to 20,000√2 Mkm and Phobos, 8,000√2 Mkm, to the centre of Mars. In *Intelligent Mars I*, I defined a Martian kilometer as the equatorial radius of Mars divided by 2^{12}. The Martian kilometer becomes a very credible unit of measurement when you are informed that it is almost exactly the length of 1000 Megalithic yards. A Megalithic yard is the unit of measure that Alexander Thom determined to have been used by the architects of the megalithic sites in the British Isles and Brittany, and it must have been brought to Earth from Mars. When formulated in terms of Mkm, we see immediately that there is something very unnatural in the distances to the Martian moons. Mother Nature does not create distances in units of 1000 except by extreme fluke, but it cannot be a fluke when it happens for both moons. Nor is it credible to assume that the orbits are natural when the ratio of the 2 distances works out to 5/2. The ratio of 5/2 can be re-written as (5/4) x 2 which is the formula for the major 3rd interval in the second octave of the chromatic musical scale. Note also the use of the irrational number √2. The use of irrational numbers in the distances between sites on Mars is very common because it seems that the Martian architects loved to incorporate sacred geometry into their designs. They often included one or two of the special irrational numbers of π, φ, e, √2, √3 and √5 in distance values as well as the planetary radius to create what I have termed sacred formulae. I use the word "sacred" because these numbers are very important in sacred geometry. For instance, the value of √2 is a factor in the diagonal of a square and π is a factor in the circumference of a circle. Both of these are important shapes used in sacred geometry.

Another example of the need for a technological language to interpret Mars is the location of a small mountain, Issedon Tholus. Inside its 2 km wide caldera is a structure shaped like the tip of an arrow. I measured the longitude of the tip to be 265.1717° E. Not exactly an interesting number. However, when you reference it to one of the ancient Martian prime meridians which I discovered (*Intelligent Mars II* [2]), the longitude becomes 36.0007° W. Now that's much more interesting since not only is it very close to a pure integer, but that integer happens to be the number of degrees in the star point of a pentagram. Even more intriguing is that the latitude of the tip of the caldera arrow is exactly 36.0000° N! Once again we see that the Martian architects had a passion for sacred geometry which modern science and technology tends to ignore.

A third example reported in *Intelligent Mars II* is Ulysses Tholus and its 2 large craters. The northern crater has 2 peaks in its central region. The more northern peak has the coordinates of 238.6047° E 3.5343° N, both rather uninteresting numbers. However, when you reference the longitude value to one of the ancient Martin prime meridians, it becomes 50°°° W. Why did I put 3 degree signs after 50 instead of just a single one? That is because I discovered that the ancient Martians used a 288 degree system as well as a 360 degree system. Each degree in that system is 1.25 of the degrees we are used to. I use the triple degree symbol to distinguish between the 2 types of degrees. Of course, to have the number 50 appear in a coordinate value of an important site is very unlikely to happen under the random conditions of an impact from outer space, but you will not see this clue to intelligent engineering unless you are privy to both the ancient prime meridian and the knowledge of the 288 degree system. The latitude of 3.5343° N becomes π°° N when you convert it to yet another degree system used on Mars. This system measures out the planet in 320 degrees instead of 360 degrees and its degrees are equal to 1.125 of the regular degrees. I use the double degree sign to distinguish it from the other 2 systems. Now if you measure the distance between the northern central peak of the northern crater of Ulysses Tholus and the centre of the large crater on its southeast side, you get a value of 59.3 km. This is exactly the distance covered by 1 degree of latitude! So here you have a marker, visible from outer space, of the size of 1 degree of latitude. Would you expect Mother Nature to provide such a convenient calibration marker by accident? Thus Ulysses Tholus is a large billboard telling us that engineering has taken place on Mars, but we are blind to it unless we have the necessary technical background to understand it.

Another very important component of the Martian mindset that escapes our perception is what I call *participatory* sacred geometry (see *Intelligent Mars II*). The Martian architects don't always provide complete images of the concepts which they wish to portray. Instead, they often give only the basic starting information, requiring the observer to complete the picture. If you are not aware of this, you will miss the big picture entirely and see only what seem to be disconnected pieces. A good example of this is to be found with the 4 giant mountains and the Pentagram Pyramid. The observer has to construct the Vitruvian Martian from this basic information or else it remains invisible. As another example, polygonal-shaped craters often only provide part of the information which the observer requires in order to complete the geometric shapes that fit them. In other words, the observer has to be a co-creator. A prime example of participatory sacred geometry in this book will be the Martian portrayal of the concept of Aum.

Some of the most glaring signs of artificiality seem to have escaped the scientific community despite being in full view in the quadrangular MOLA (Mars Orbiter Laser Altimeter) maps of Mars which have been around for more than a decade. Hence, besides NASA's negative indoctrination, our lack of necessary technical concepts and our ignorance of the Martian mindset, there is a fourth reason why we are unaware of the intelligent life that has occupied Mars in the past and possibly still does. This is what is known as inattentional blindness. We tend not to perceive things that we do not expect. Hence, we expect craters to be round and do not take notice when they are shaped like an arrowhead or a polygon. We do not expect to find approximately kilometer-wide straight grooves on the surface of Mars so we don't notice them even when they travel intermittently at a constant bearing angle for thousands of kilometers. We do not expect mountains to be placed in geometric patterns so we never bother to examine their interrelationships.

Once we are able to finally free ourselves of the tethers that bind our mindset to the concept that Mars is a lifeless cratered spheroidal hunk of rock with mountains, we can open up to the reality that this is a very special planet whose topography has been intelligently designed by a highly advanced civilization to honour sacred geometry. The *Intelligent Mars* series presents the results of a decade-long study of the surface of Mars. My findings are based on thousands of careful measurements made mostly from the quadrangular MOLA maps of Mars which I have found to be the most faithful to the original data. *Intelligent Mars I* deals mainly with how the major mountains of Mars are deliberately arranged in patterns reflecting sacred geometry. One of these patterns depicts the Vitruvian Martian in a virtual form with the aid of an adjacent pyramid that has the shape of a pentagram. The pyramid has been carefully positioned to enable an observer to sketch out the circle, equilateral triangle and pentagram that fit the humanoid shape. *Intelligent Mars II* demonstrates that many craters are the result of engineering as opposed to arising from impacts. Artificiality was detected in the shape and size of a number of craters, with some of them having an overall contour of a regular polygon (square, pentagon, hexagon or octagon). Craters in the shape of a regular polygon were found to be sized according to the interval sizes of the chromatic scale, betraying an advanced knowledge of music by the ancient Martian architects. Other craters have very sophisticated functions such as mimicking an eye which looks over the mountain architecture from a huge distance. *Intelligent Mars II* also presents the discovery of coordinate systems and prime meridians used by the ancient

Martian civilization. These are absolutely critical to the detection and understanding of much of the artificiality of Martian topography so I have included a summary in this current book.

So why a third book? What else is there on Mars other than mountains and craters? Well, to begin with, there is a massive rift in the crust called the Valles Marineris that is analogous to the Grand Canyon here on Earth, but is about a full order of magnitude longer and deeper. Then there are the huge impact basins of Hellas and Argyre. This current book does indeed examine these enormous formations, but only after first spending a large part of its resources on a single, seemingly insignificant, crater about 70 km in diameter. I alluded to this crater in the final chapter of *Intelligent Mars II* but did not include it in that book as it would have taken up too much space to adequately explain it. It took a huge amount of effort to decipher the purpose and meaning of this crater. After many false starts, I can now say with considerable confidence that I have finally arrived at the fundamental concept of what the crater is all about and that this crater provides a powerful unifying force to the layout of the topography of a large portion of the surface of the planet. Besides being linked to sites far beyond its perimeter, this crater provides a crucial insight into the very deep spirituality which the early Martian civilization must have possessed, a spirituality so profound that they based the construction of most, if not all, of their living environment upon it.

The title of this book also contains the word "architect". This not only alludes to the hypothetical master craftsman who designed the artificial surface topography, but also to an actual replica of a person embedded in the planetary landscape who just might be this person - the chief architectural genius who drew up the blueprints and provided the oversight while the construction took place.

The discovery of planetary engineering on such a massive scale of course has massive implications. But once again, it's not just the engineering but the motivation behind the engineering that has the most important implications for humanity. We live in a rather materialistic, godless, "rational" mental space which is really more of a prison than true enlightenment. The *Intelligent Mars* series brings solid evidence of a race so technologically advanced that it humbles our own achievements down to a relative stone age. It also humbles our current boastful view of the spiritual dimension as being an unscientific delusion of pre-Renaissance thinking. Just perhaps we need to rethink all of this. What are our premises? Where do they come from? What are the logical endpoints? Do they serve humanity and the planet or do they serve the originators and maintainers of the modern mental space. Are we evolving or devolving? A species that destroys its own habitat can hardly be looked upon as advanced, and is in great danger of relegating itself to becoming no more than an archaeological curiosity to the species that will eventually replace it.

It is my hope that readers of the *Intelligent Mars* series will start to question the delusion that we humans now call "reality". This delusion is so engrained that those who do catch a glimpse of what is actually going on are ridiculed as being either stupid, misguided or even insane. Our present delusional state results from an authoritarian control of our "education", our "history", our "media" and our "science". We are discouraged at a very early age from independent thinking and are not given proper access to the vital information required to construct a proper assessment of our place in the universe. In fact, we are deliberately fed with a lot of misinformation in all aspects of our "education". Objective science has now been mostly removed from the public domain and has become the proprietary tool of large corporations, the military, intelligence agencies and secretive governments. The public is increasingly being fed a version of "science" that is little more than propaganda serving the interest of a ruling elite.

I believe that a new golden era awaits humanity if we can just seize the opportunity. But it will never be arrived at simply through revolution. "Revolution" is a process that members of the ruling elite use to give humanity the false impression of real change. A lot of lives are sacrificed and a huge amount of destruction takes place. The intermediaries and/or the governmental model may be changed, but in the end, the ruling elite remain in control. True history has proven this again and again. No, our next step needs to be an evolution rather than a revolution. And the evolution needs to be on the spiritual plane rather than the physical. Real change can take place only if we open our hearts to rise to a resonance with the infinite Divine Spirit. Only then will our relationship to others and to the planet move to the higher plane necessary for our survival and well-being. There is much to be learned from these ancient Martians in this respect, and I sincerely hope that this book will provide at least the first glimpse of what they have to offer. It is obvious to me that they understood the concept of Aum at a level far more profound than our own understanding of it. Opening ourselves to their tutelage may provide important insights to help us on our path to a true union with the Divine. It may also help us advance our understanding of fundamental physics. This is definitely an opportunity that we would be wise to pursue.

Arthur R. Beaubien

1. *Intelligent Mars I: Sacred Geometry of the Mountains. Did Da Vinci Know? Arthur Raymond Beaubien. Epiphi Productions, Ottawa, Ontario, Canada. 2015.*

2. *Intelligent Mars II: Code of the Craters. Arthur Raymond Beaubien. Epiphi Productions, Ottawa, Ontario, Canada. 2019.*

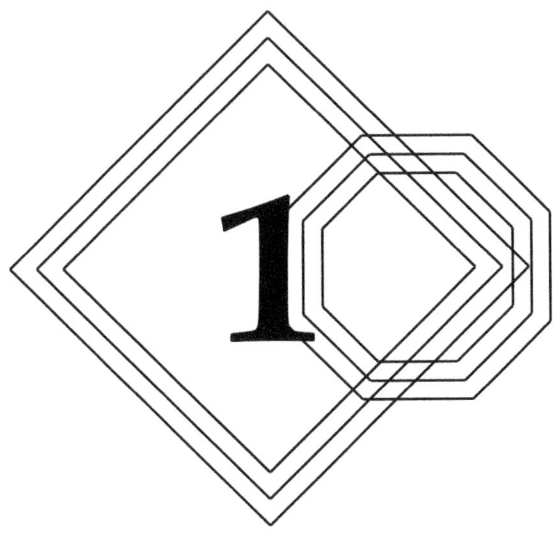

Introduction

Before I begin with the subject matter of this book, there are certain concepts that I wish to convey to readers who are either unfamiliar with the material in the previous 2 books of the *Intelligent Mars* series or else need to refresh their memories. Otherwise, the material which is presented in this third book might not be well understood. In the first 2 books of this series, I presented the finding that many of the mountains and craters on Mars are not randomly arranged as they would have been if they had arisen from natural forces alone. Rather, they have been deliberately created and positioned according to the principles of sacred geometry by a very ancient and highly advanced civilization which was capable of both space travel and mega-engineering projects. In the course of my study, I discovered that the Martians used several prime meridians (i.e., reference longitudes) instead of a single prime meridian as we are accustomed to using on the planet Earth. The route to the discovery of these prime meridians was a combination both of good luck and a careful study of the Martian terrain as is outlined in *Intelligent Mars II*. I found that the Martian prime meridians (other than perhaps one of them) were less likely to have been used for navigational purposes than for generating meaningful numbers for the longitude coordinates of significant sites and/or of some of their architectural components. It was also discovered that a second reference latitude was used in addition to the equator for the same purpose with regard to latitudes.

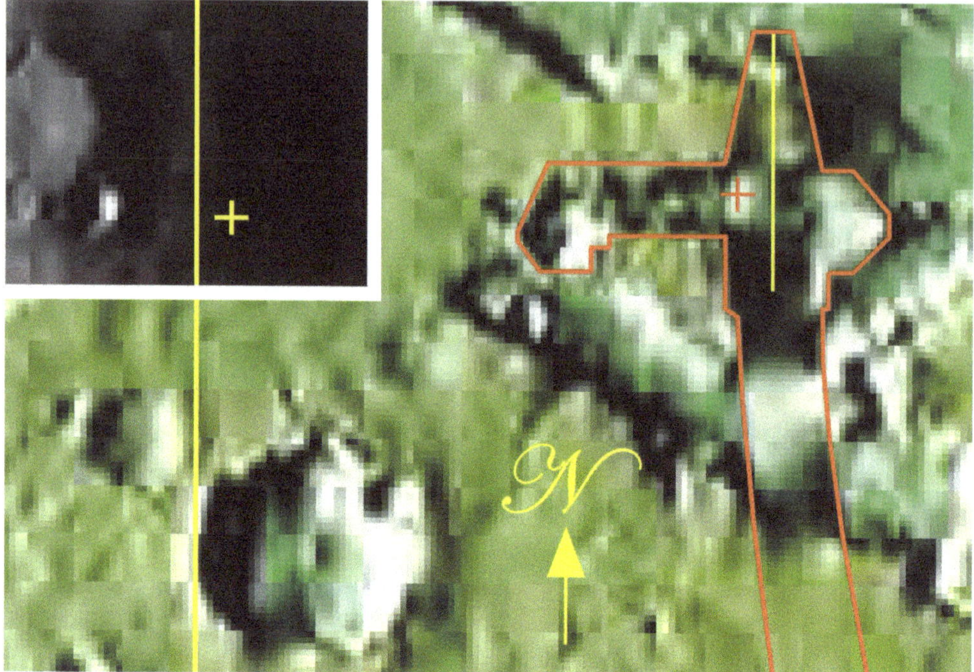

Fig. 1.1: *The Crater Edge Prime Meridian (vertical yellow line at bottom left) lines up with the western edge of a crater just west of a dagger-shaped region. Inset at top left shows the Elysium Mons survey site (yellow cross) which marks the Elysium Mons Prime Meridian. Within the hilt portion of the dagger-shaped area are the Dagger Peak Prime Meridian marked by a red cross, and the Dagger Midline Prime Meridian marked by a vertical yellow line to the right of the red cross. USGS Astrogeology.*

Prime Meridians & Prime Latitudes of Mars

A total of 8 prime meridians were identified. The west-most prime meridian (PM) occurs at 147.1047° E (NASA longitude) and is marked by the western edge of an unnamed crater (see vertical yellow line at bottom left in Fig. 1.1). The survey centre of Elysium Mons is about 4 minutes of a degree to the east of this and marks the second prime meridian (yellow cross at top left). A second pair of prime meridians occurs exactly 1 degree east of the first pair. These are marked by a dagger-shaped image carved out in the Martian landscape just south of the equator. A peak (red cross at top right) in the dagger's hilt marks the longitude of the western member of the pair and the midline (yellow line at top right) of the dagger hilt marks the eastern member of the pair. The peak and midline are about 4 minutes of longitude apart and are shown in the right half of Fig. 1.1.

A third pair of prime meridians occurs on the top of Pavonis Mons (Fig. 1.2) 100 degrees east of the first pair. The west-most member is

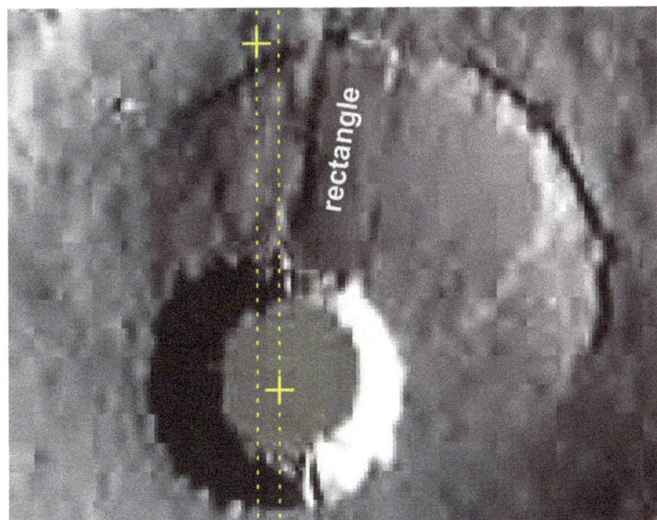

Fig. 1.2: *Location of Pavonis Mons prime meridians. The top yellow cross marks the Pavonis Mons PM at the mountain survey centre, and the yellow cross at the bottom marks the Pavonis Caldera PM. Note the rectangular area extending from the northern side of the caldera at a clockwise bearing angle of about 9 degrees. The length of the rectangle is approximately π times its width. USGS Astrogeology.*

labeled by a yellow cross at the top of Fig. 1.2 which marks the survey centre of Pavonis Mons. The eastern member of the pair is located at the centre of the caldera of Pavonis Mons (bottom yellow cross). These 2 prime meridians are also about 4 minutes of longitude apart.

The final pair of prime meridians is associated with the Sharonov Crater (Fig. 1.3). These prime meridians are located 54 degrees east of the Pavonis Mons pair. The west-most member of the pair is marked by a triangular trench which occurs just outside the northern perimeter of the Sharonov Crater. A red triangle is superimposed on the course of the trench in Fig. 1.3. The prime meridian is located at the exact point where the triangle touches the white circle which outlines the outer perimeter of the Sharonov Crater. The eastern member of the pair is located at the white-coloured spot identified as the "Tower" in Fig. 1.3. Like the previous 3 pairs of prime meridians, this pair is approximately 4 minutes of longitude apart.

In *Intelligent Mars I*, I found that a large number of sites seemed to be an integer or an integer + 1/2 number of degrees from the latitude of the Arsia Mons survey centre (8.0996° S). As a result, I declared this latitude to be prime in addition to the normally used equator. A large number of craters on the mountains were found in *Intelligent Mars II* to have meaningful latitude displacements from the latitude of Arsia Mons thus validating it as a prime latitude. This book will use the terminology of Arsia Mons Prime Latitude (AMPL) to denote latitudes using this reference. All latitude values will be considered to be from the equator unless specifically stated to be with reference to the Arsia Mons Prime Latitude.

To better orient the reader towards the location of the various prime meridians and prime latitudes, I have constructed Fig. 1.4 which points

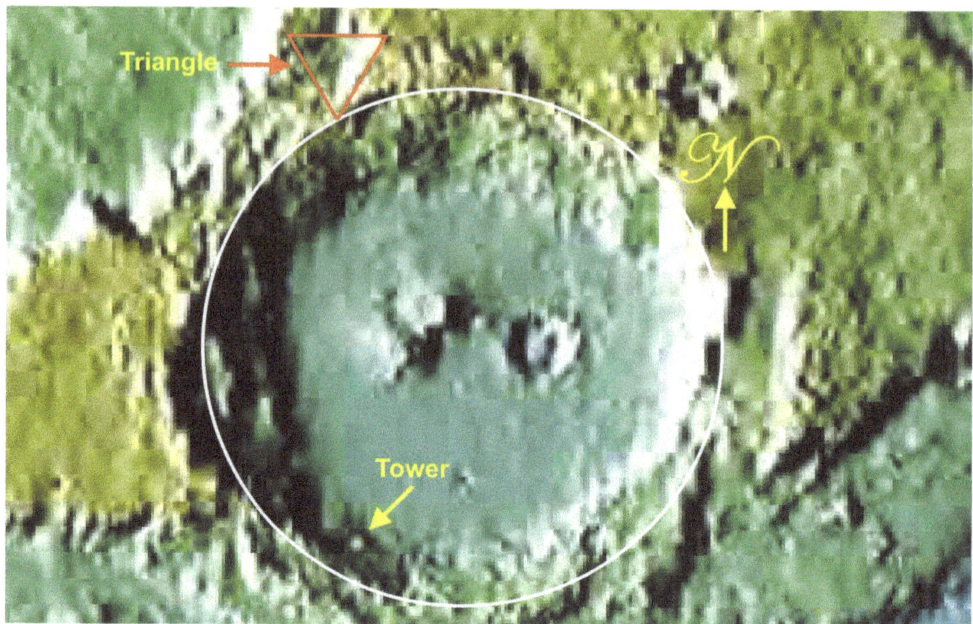

Fig. 1.3: *Picture of the Sharonov Crater showing the locations of the Sharonov Triangle and the Sharonov Tower prime meridians. The large white circle is fitted to the outer perimeter of the crater. USGS Astrogeology.*

out the placement of these coordinate reference lines on a map showing about half of the planet between the latitudes of 30° S and 30° N. Since the resolution is very low at this magnification, I had to use single lines to indicate the locations of each of the 4 pairs of prime meridians: the Crater Edge and Elysium Mons pair, the dagger pair, the Pavonis Mons pair, and the Sharonov Crater pair.

The various prime meridians are summarized in Table 1.1. Throughout this book I will refer to the 8 prime meridians as: CEPM (Crater Edge Prime Meridian), EMPM (Elysium Mons Prime Meridian), DPPM (Dagger Peak Prime Meridian), DMPM (Dagger Midline Prime Meridian), PMPM (Pavonis Mons Prime Meridian), PCPM (Pavonis Caldera Prime Meridian), STrPM (Sharonov Triangle Prime Meridian) and SToPM (Sharonov Tower Prime Meridian). Since the decimal fractions of the CEPM, DPPM, PMPM and STrPM were all within 15 seconds of a degree of each other, I assumed that the prime meridians were actually integer numbers of degrees apart. I therefore used the average of their decimal part (0.1047°) when calculating longitude coordinates from them. The EMPM, DMPM, PCPM and SToPM are also extremely close to integer numbers of degrees apart and have an average decimal fraction of 0.1724°. The difference between the average decimal

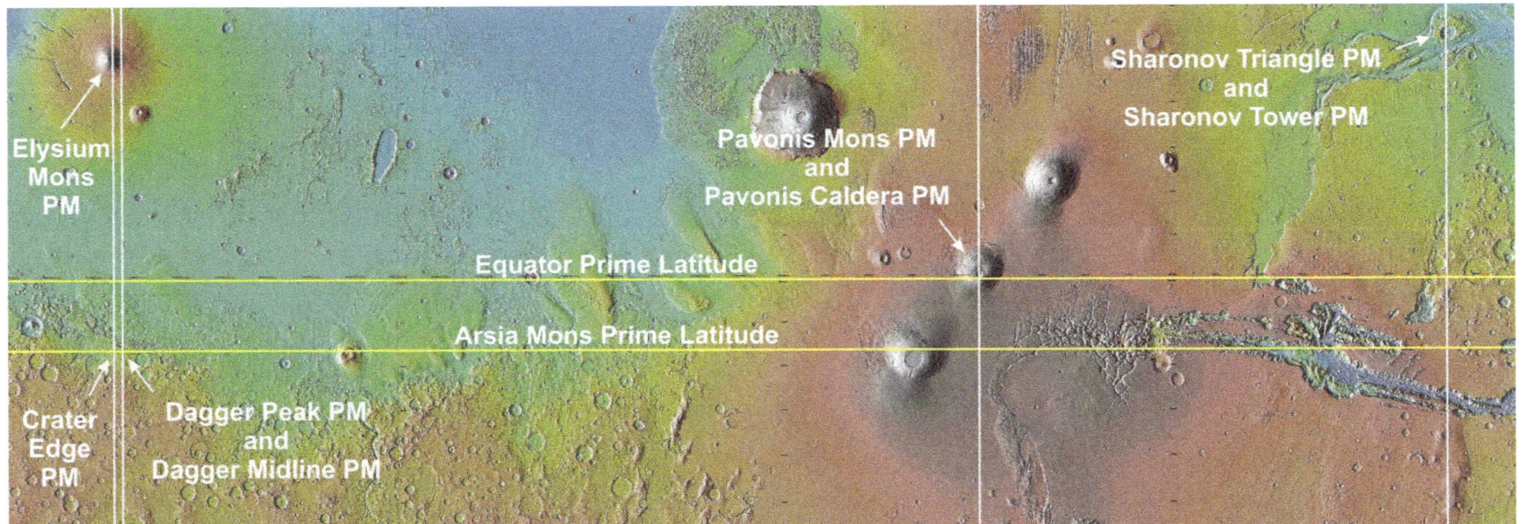

Fig. 1.4: *Placement of the 8 prime meridians and 2 prime latitudes on the planetary surface of Mars. Arrows point to the location of the markers for the specific prime meridians which are indicated by vertical white lines. Since the individual lines of each pair of prime meridians are too close to be distinguished at this magnification, single lines were used for the Crater Edge and Elysium Mons pair, the dagger pair, the Pavonis Mons pair, and the Sharonov Crater pair. Horizontal yellow lines indicate the Equator and Arsia Mons prime latitudes. The Arsia Mons Prime Latitude passes through the survey centre for Arsia Mons at 8.0996° S. USGS Astrogeology.*

Table 1.1: *Summary of the 8 prime meridians likely to have been used on Mars.*

Prime Meridan	Abbreviation	Longitude (°E)		Longitude (°E)	
		Measured	Mean decimal fraction	CEPM	EMPM
Crater Edge	CEPM	147.1047	147.1047	0	
Elysium Mons	**EMPM**	**147.1701**	**147.1724**		**0**
Dagger Peak	DPPM	148.1052	148.1047	1	
Dagger Midline	**DMPM**	**148.1727**	**148.1724**		**1**
Pavonis Mons	PMPM	247.1037	247.1047	100	
Pavonis Caldera	**PCPM**	**247.1724**	**247.1724**		**100**
Sharonov Triangle	STrPM	301.1052	301.1047	154	
Sharonov Tower	**SToPM**	**301.1743**	**301.1724**		**154**

fraction for each of the 2 sets of prime meridians amounts to only 4.06 minutes of a degree. This is very close to exactly 4 minutes of a degree which may have been the intended difference.

Coordinate Systems Used on Mars

Another very important concept in understanding the artificial nature of Martian topography is the use of several different coordinate systems by the architects. It would appear that they used a 720 degree system as well as our terrestrial 360 degree system. Even more interesting are a 320 degree system and a 288 degree system. The former uses degrees (which I call big degrees) equal to 1.125 ordinary degrees, and the later uses degrees (which I call sacred degrees) equal to 1.25 ordinary degrees. The existence of coordinate systems using big and sacred degrees has been well verified in *Intelligent Mars II* by the discovery of their use in the positioning of important craters. The most likely degree systems used on Mars are listed in Table 1.2. Besides the double degree systems of 720, 640 and 576 degrees, I have also included the quadruple degree systems of 1440, 1280 and 1152 degrees for which there is evidence in the coordinates of polygon crater centres and vertices. To distinguish between the degree systems, my custom has been to use a double degree sign for big degrees and a triple degree sign for sacred degrees. Thus 56 big degrees in the 320 degree system is written as 56°° and 30 sacred degrees in the 288 degree

Table 1.2: Degree systems likely used on Mars

Degree System	Type	Degrees Symbol	Size (°)
360	**Regular**	°	**1.00000**
720			0.50000
1440			0.25000
320	**Big**	°°	**1.12500**
640			0.56250
1280			0.28125
288	**Sacred**	°°°	**1.25000**
576			0.62500
1152			0.31250

system is written as 30°°°. To minimize confusion, I have avoided using double and quadruple degree systems, and have only used the 360, 320 and 288 degree systems to express coordinate values in this book.

Sacred Formulae

I have found that the Martian architects overwhelmingly used formulae reflecting sacred geometry in the distances between structures in their layout of the Martian topography. The formulae always included the value of either the equatorial radius (R = 3396.19 km) or the northern polar radius (R' = 3376.20 km). The formulae also often included integer values and the irrational numbers of φ, π, e, √2, √3, or √5. The term "sacred distance formula" used in this book refers to sacred geometry formulae of the forms:

(i/j)mnr
(i/j)(1/m)(1/n)r
or **(i/j)(m/n)r**

where i and j are integers, r is the equatorial (R) or northern polar (R′) radius of Mars, and m and n have the value of 1, φ, π, e, √2, √3, or √5. The simplest formulae are either R or R′ in which all the other variables are set equal to 1. Distances are given in units of kilometers in this book. However, the Martians probably expressed them either as radians by setting R to 1 and R′ to 0.994114, or as Martian kilometers (Mkm) by setting R to 4096 Mkm and R′ to 4071.0909 Mkm. [In *Intelligent Mars I*, it was determined that 1 Mkm = (R km)/(2^{12}).] The Martian architects also often used sacred geometry formulae for latitude and longitude coordinate values. When referring to a coordinate value, the term "sacred formula" refers to the same formulae forms as used for distances but without the 'r' variable. Coordinates are in degrees or radians.

Bearing Angles

The term "bearing angle" refers to the clockwise or counterclockwise angle from due north of a straight line on the surface of Mars. In this book, clockwise bearing angles are assigned negative values and counter-clockwise bearing angles are given positive values.

Planetocentric vs. Planetographic

Martian coordinates can be expressed as either planetocentric or planetographic values. Planetocentric coordinates assume the planet to be a sphere of radius R = 3396.19 km, the equatorial radius of Mars. Planetographic latitudes take into account the fact that Mars is an oblate spheroid (or ellipsoid) with the polar radius smaller than the equatorial radius. These latitudes are measured from the angle made with the equatorial plane by a line perpendicular to the ellipsoidal surface. When NASA longitudes are specified in this book, they are given in degrees east (planetocentric) with respect to the Airy-0 Crater Prime Meridian. The Airy-0 Crater is a 0.5 km diameter crater that lies within the wider Airy Crater in the Sinus Meridiani region of Mars. When the ancient Martian prime meridians are used in this book, longitude values are expressed either in degrees east or degrees west depending on which is closest to the prime meridian used. Latitude values can be expressed either in terms of planetocentric or planetographic coordinates. NASA latitudes given in this book are planetocentric. I found that the Martian architects used planetocentric latitudes most commonly. However, there are important exceptions to this, such as the latitude of the centre of the Pentagram Pyramid which is exactly $\pi°$ N in planetographic coordinates.

Rhumb vs. Great Circle

Spherical geometry offers 2 main choices for measuring the distance and bearing angle of a line joining any 2 points on a spherical surface: either great circles or rhumb lines. A great circle is the intersection of the surface of a sphere with a plane passing through its centre. Any great circle divides the surface of the sphere into 2 equal halves. Examples of great circles are the Earth's equator and meridian lines (shortest distance lines running along the surface of the sphere which connect the North Pole with the South Pole and mark out the longitude coordinates of locations on a planet). A great circle that passes through any 2 different points on a planet provides the shortest possible path between them. However, the bearing angle of this path varies over its length unless the 2 points lie on the equator or on the same meridian line. A rhumb line on the other hand is the line formed by a path maintaining a constant bearing between the 2 points. Rhumb lines were often used by navigators to simplify the process of maintaining course on ocean journeys despite the expense of taking a longer path. They also needed to be used prior to the invention of clocks accurate enough to enable longitude to be calculated at sea. I have found that the Martian architects used rhumb lines rather than great circle lines between 2 sites since they wanted to produce lines of constant bearing angles set to meaningful sacred geometry values.

Rather than following the true surface of the planet for measuring distances, site-to-site distances in the *Intelligent Mars* series are measured with rhumb line methodology assuming the planet to be a perfect sphere set to the equatorial radius of Mars. Thus all of the distances presented in this book are actually virtual distances which suppose that all the sites are situated at the elevation of a virtual sphere. This assumes that the mega-architecture of the planet was intended by the creators to be viewed from high altitudes rather than from the ground. This approach receives much support from the many meaningful distance values which it yields.

Latitude Degrees and Radians

The Martian architects often measured out dimensions of geometric shapes such as regular polygons in terms of degrees. Since longitude degrees vary in length depending on latitude, they used the length of a degree of planetocentric latitude (or longitude measured at the equator) since it is constant independent of latitude. I use the term 'latitude degree' to refer to this unit of measurement in the book. It is equal to 59.2747 km. They also used the length of 1 radian (rad) measured at the equator as a unit of measurement where 1 rad = R km or 57.2958 latitude degrees. As well, 1 big rad = 1.125 rad and 1 sacred rad = 1.25 rad.

Maps for Distance and Coordinate Measurements

For making careful mountain and crater coordinate measurements, I used a high resolution (1:5 million-scale) series of 30 coloured MOLA maps which divided the surface of Mars into 30 separate regions. These maps are colour coded according to elevation. I downloaded them from the

Internet (Courtesy of the USGS Astrogeology Research Program, http://planetarynames.wr.usgs.gov/Page/mars1to5mMOLA). They are no longer available at this site but can be downloaded from the Wikipedia site at https://en.wikipedia.org/wiki/List_of_quadrangles_on_Mars. The MOLA maps were produced from the data obtained from the Mars Orbiter Laser Altimeter (hence, the abbreviation MOLA), an instrument aboard the Mars Global Surveyor (MGS) spacecraft which was launched on Nov. 7, 1996. It achieved research orbit on Sept 12, 1997, and stopped functioning at the end of June, 2001. During that time, the MOLA instrument made approximately 1 billion elevation measurements of the planet's surface. Each measurement was made by transmitting an infrared laser pulse to the planet's surface and measuring the time it took to receive the reflected pulse by the MGS spacecraft. The resolution of the coloured MOLA maps ranges from ±0.33 to ±0.47 km since the side length of one of the tiny square pixels composing the map is 0.67 km at the equator, giving a pixel diagonal size of 0.94 km. The absolute accuracy (the closeness to the true value) of the MOLA measurements upon which the map pixels are based is of the order of 100 meters horizontally and 1 meter vertically.

Orientation Maps

Figures 1.5 and 1.6 have been constructed to help orient the reader with regard to the location of various important sites on Mars. In Fig. 1.5, the major mountains, the Aum Crater (see Chapter 2), the west part of the Valles Marineris and the site of the bust of the Architect (see Chapter 11) are identified. In Fig. 1.6, the entire surface of Mars is shown between the latitudes of 70° S and 70° N. The left side of the map starts at 0° using the Crater Edge Prime Meridian (CEPM) instead of the Airy-0 prime meridian used by NASA. In this way, the reader can get a sense of how the planet was probably mapped in ancient times. The equator runs through the lower half of Pavonis Mons. The map used for both Figs. 1.5 and 1.6 is a NASA MOLA map downloaded from: https://planetary.s3.amazonaws.com/assets/images/4-mars/2008/mola_mercat.jpg.

Survey Centres of Mountains

The term "survey centre" is used throughout this book to refer to the centre of a mountain surveyed from special survey craters which were created by the Martian architects to locate the centre of a mountain. The survey centre coordinates of the major mountains were determined in *Intelligent Mars I* by locating the unique point on a given mountain that was exactly at a meaningful distance [e.g., $R/(2\pi)$ km] from 2 or more survey craters.

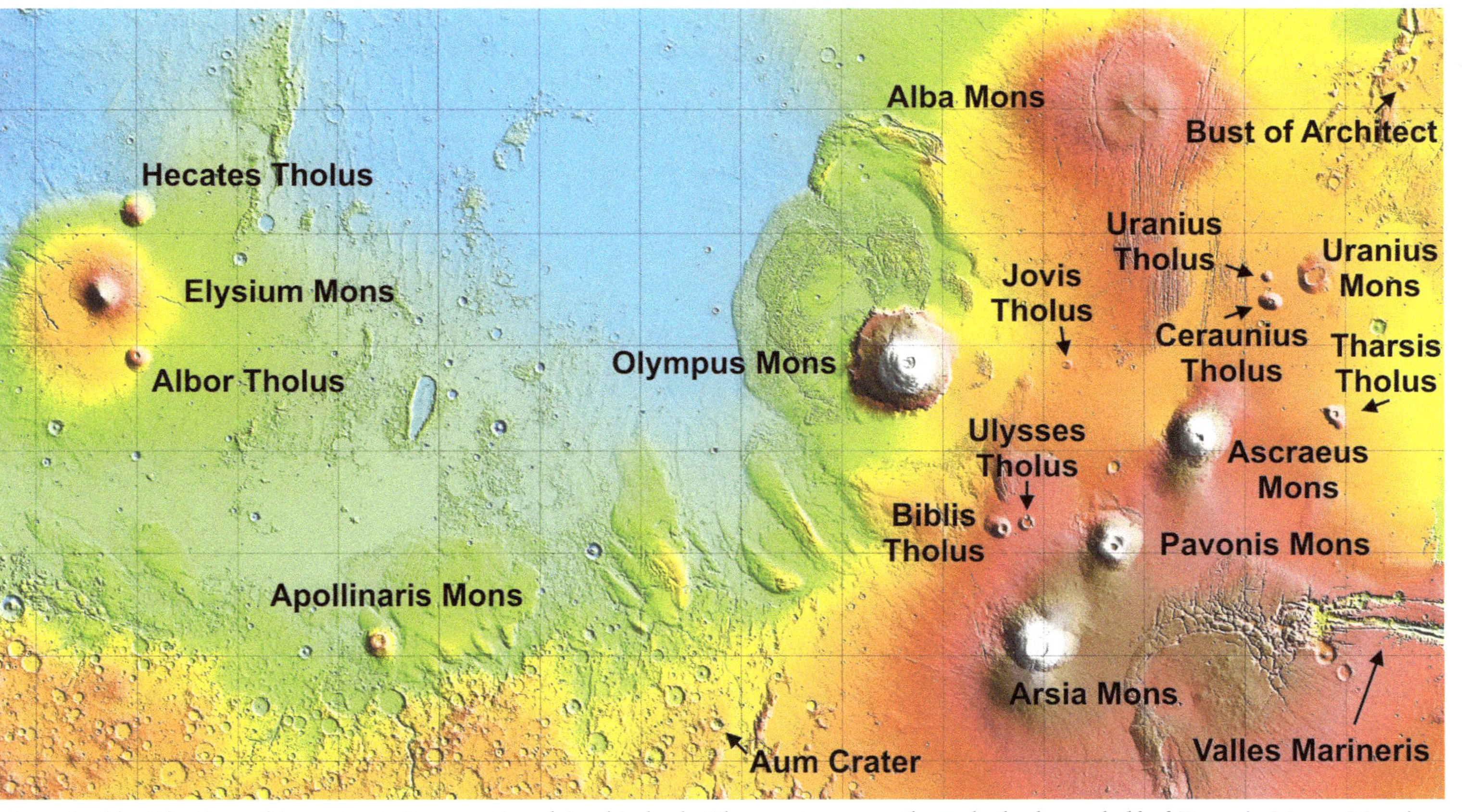

Fig. 1.5: *Identification of important sites mentioned in this book. The equator runs through the lower half of Pavonis Mons. Map is a portion of a rearranged NASA MOLA map.*

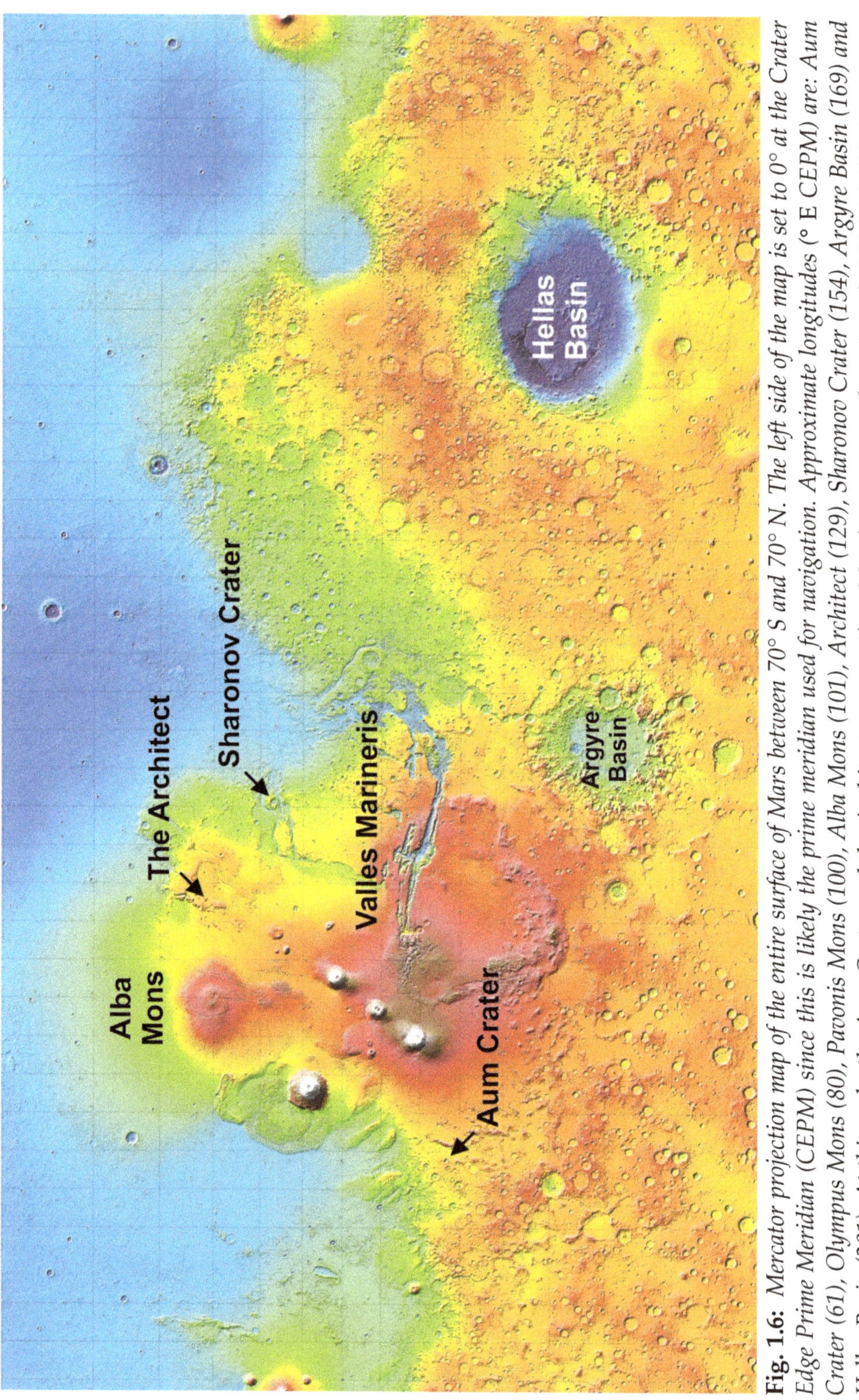

Fig. 1.6: *Mercator projection map of the entire surface of Mars between 70° S and 70° N. The left side of the map is set to 0° at the Crater Edge Prime Meridian (CEPM) since this is likely the prime meridian used for navigation. Approximate longitudes (° E CEPM) are: Aum Crater (61), Olympus Mons (80), Pavonis Mons (100), Alba Mons (101), Architect (129), Sharonov Crater (154), Argyre Basin (169) and Hellas Basin (281). At this scale, the Aum Crater and the Architect are very tiny. Map is a rearranged version of a NASA MOLA map.*

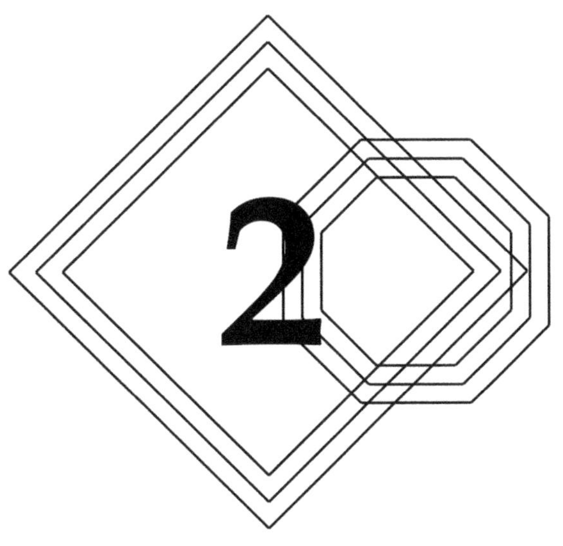

A Mysterious Crater

An extremely interesting crater lies to the southwest of Olympus Mons, and it is mysterious in that it does not seem to simply inhabit the part of the Martian real estate that it sits on. Rather, many sites around the crater appear to have been systematically positioned in relationship to this crater, thereby extending its zone of influence far beyond the borders of its perimeter. What could be so important about this crater that other landmarks seem to pay homage to it in their own construction? Although it is approximately 70 km in diameter, there are many more sizeable craters than this which you would think would attract the greatest attention. The special nature of this crater will gradually become apparent as I take you through the process of acquiring more and more data on its zone of influence. You will find that not only is this seemingly insignificant crater special, but it is one of the most important structures on the planet along with the mighty Olympus Mons, the Valles Marineris and the Hellas Basin.

I first came across this crater when I was analyzing polygon craters for my second book. It was one of several which had a square shape rather than the round shape that natural craters should have. It was also one of the thousands of craters which do not yet have an official name assigned to them. During my many initial attempts to make sense of this crater, I started to call it by several different names. Firstly I called it the 'Infinity Crater' since its zone of influence appeared to expand outwards indefinitely. Next I decided to call it the 'Web Crater' since it seemed to spawn a large number of concentric squares which resembled a sort of web when viewed together. I eventually decided against using either of these 2 names in favour of a third name which seemed the most appropriate. But for now, I will use the term "Mystery Crater" until the rationale for the new name becomes apparent.

The Fitting of a Square

When you first look at the Mystery Crater (Fig. 2.1), you cannot help but notice that it has an underlying square shape even though its corners are somewhat rounded. Within the eastern half of the crater is another relatively large crater which appears to have 6 straight sides and would best be fit by a regular hexagon. Appearances can be deceiving, however, and this interior crater is no exception. A more rigorous analysis reveals that it is actually intended to be a template for a regular octagon, i.e., an 8-sided polygon whose sides are equal. It will be discussed in great detail

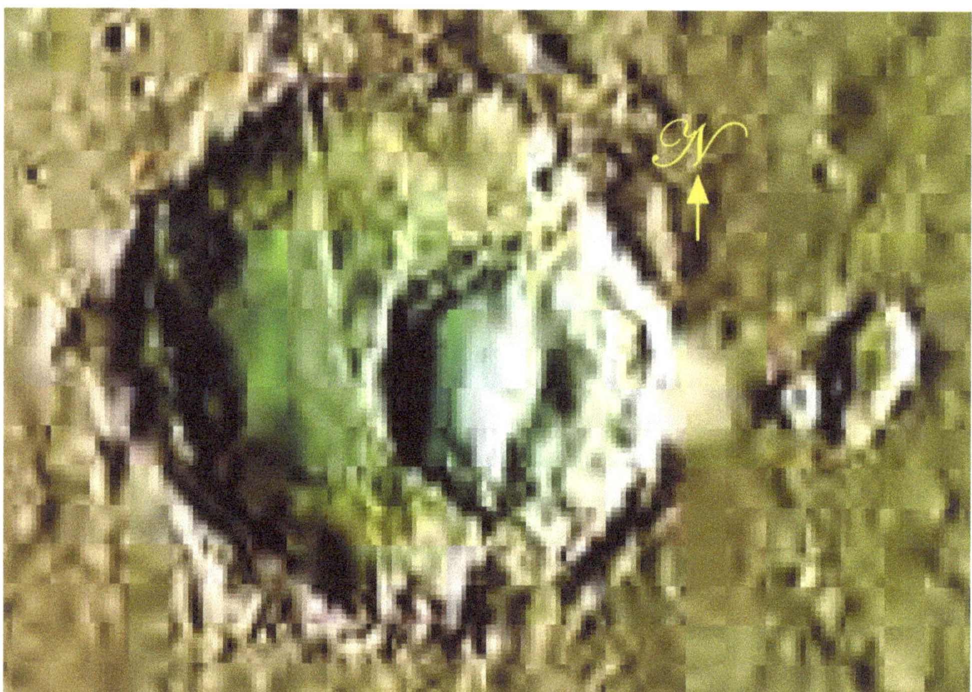

Fig. 2.1: *The Mystery Crater has an overall square-like appearance with its sides tilted at 45° with regard to the north-south direction. Within the east side of the Mystery Crater there is a second large crater which has linear sides. Just to the west of the interior crater lies a straight line ridge which has a bearing angle of 7.5° in the counterclockwise direction. USGS Astrogeology.*

in the following chapter. Just to the west of the 8-sided crater is a ridge with a bearing angle of 7.5° in the counterclockwise direction. It is one of the groove/ridge lines that will be examined in Chapter 6. It changes from a ridge to a trench beyond the boundaries of the crater to the north.

The bearing angles of linear sections of the sides of the square-shaped perimeter of the Mystery Crater are ±45°. This would suggest that the crater is rotated 45° from having its sides parallel to lines of longitude and latitude, and would allow the crater to be used as a compass with the 4 vertices pointing to north, east, south and west. However, since the vertices are rounded and since the perimeter of the east vertex is not very distinct, only a rough approximation of direction would be possible unless a virtual square could be fit and superimposed onto the crater.

My initial attempts to fit a square to the Mystery Crater were quite frustrating. The problem that I faced was that it was impossible to find a square which included the entire perimeter on all 4 sides while fitting linear lengths of the perimeter on at least 3 sides simultaneously. The best fit that I could come up with is shown in Fig. 2.2. This square fits linear sections of the crater perimeter over a substantial distance on the southeast, the southwest and the northwest sides. However, parts of the perimeter go beyond the limits of the square on all 3 of these sides, and on the northeast side the entire perimeter falls short of the square. The longitude of the centre of the square is 53.0000°° E of the Dagger Midline Prime Meridian (DMPM) and 83.0000°° W of the Sharonov Tower Prime Meridian (SToPM). Notice that these longitude coordinates are in terms of big degrees and are meaningful since they are pure integers. However, the latitude coordinate of 16.5648° S could not be expressed exactly as a meaningful sacred formula. Because of the shortcomings of this square in fitting the crater perimeter and because I could not find any other square which fit the boundaries of the Mystery Crater satisfactorily, I did not include this crater in the database of square craters in *Intelligent Mars II.*

During my attempts to fit a square to the Mystery Crater I noticed that large squares which exceeded the crater perimeter on one side or another often fit features of other craters in the immediate vicinity. This led me to try even larger squares, squares which exceeded the crater perimeter on all sides, and to my astonishment they also tended to fit features of craters and other landmarks at great distances from the crater perimeter. This went well beyond the concept of auxiliary craters which were found in *Intelligent Mars II* to help define the boundaries of polygonal shapes which fit various aspects of some craters. I therefore invested countless hours placing various squares at different central locations and then expanding them to see how well enlarged squares fit distant sites. If these sites were really synchronized to the Mystery crater, there should be a

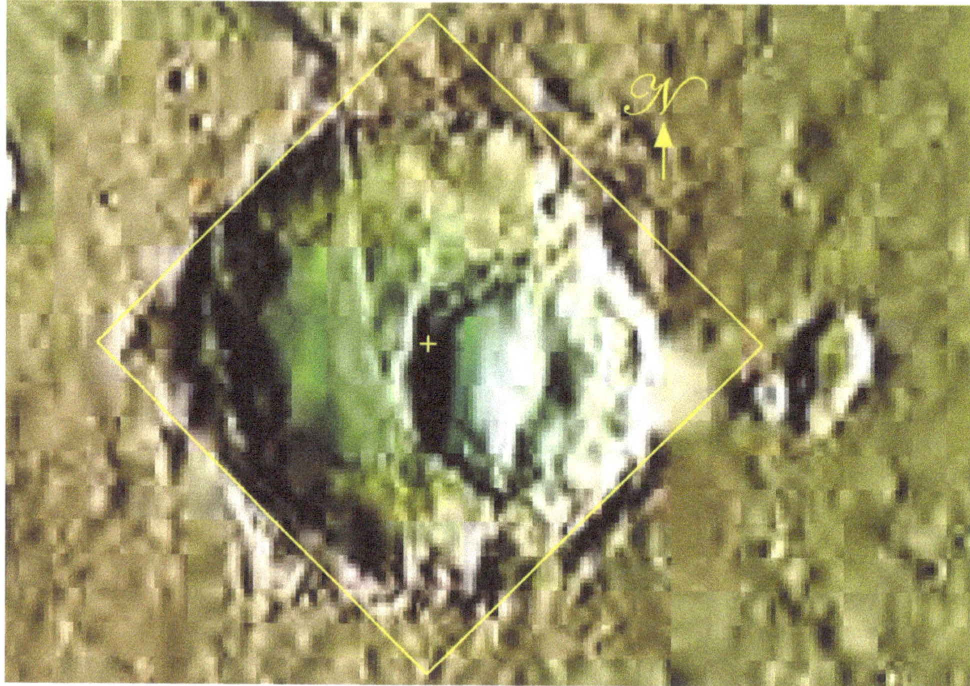

Fig. 2.2: *The Mystery Crater fit to a square of diagonal size 1.6667 latitude degrees with sides tilted at 45° to the north-south direction. The fit to the crater perimeter is best on the southeast, southwest and northwest sides while the northeast side lies beyond the crater perimeter. Small regions of the crater perimeter extend beyond the 3 best fit sides. The centre of the square (207.7974° E 16.5648° S) is marked by a cross. USGS Astrogeology.*

centrally located square which would fit the sites optimally when expanded outwards. The central locations that I used were based on latitudes and longitudes that were either an integer, an integer plus 1/2, 1/4 or 3/4, or could be expressed as a sacred formula. These could be in any one of the 3 basic degree systems used on Mars. I followed this procedure since I discovered in my previous experience with "square" craters (*Intelligent Mars II*) that the centres of the squares which fit them generally had meaningful coordinates.

Template Square for the Mystery Crater

The square that I finally decided to use as the primary fundamental for the centring and expansion factors of all larger squares is shown in Fig. 2.3. It should be noted that I will express the size of all squares in terms of their diagonal lengths in latitude degrees. This is the parameter which I found in *Intelligent Mars II* to best describe the size of "square" craters. The diagonal sizes of "square" craters were discovered to follow

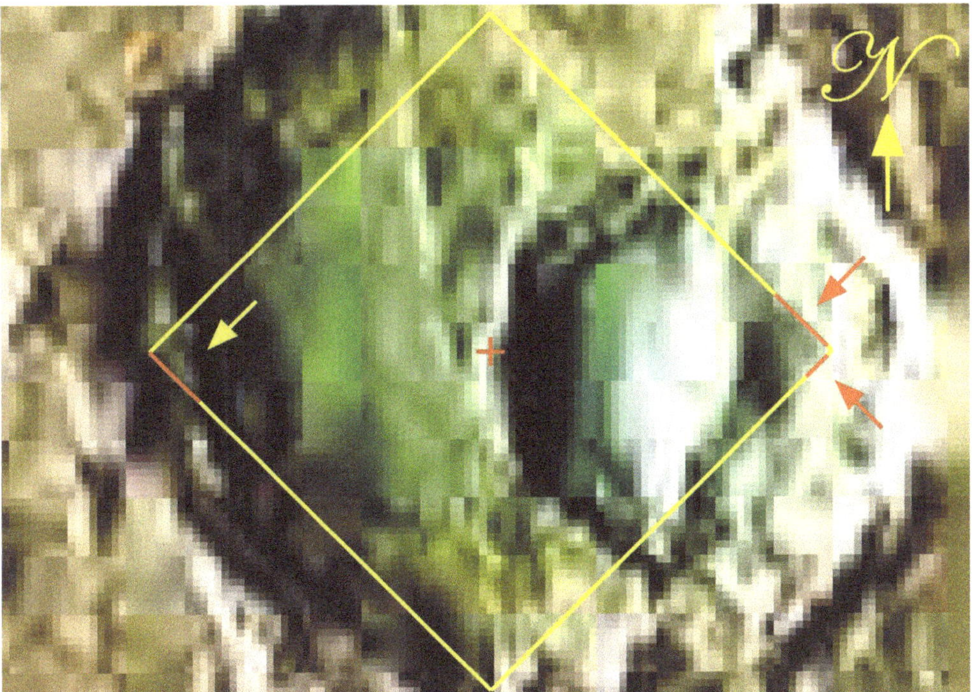

Fig. 2.3: *The fundamental square with a diagonal size of 1 latitude degree. The centre is marked by a cross and is located at 207.7297° E 16.6254° S. The square's east corner runs along 2 of the linear edges of a grey region shown by red lines superimposed on the square. The southwest side of the square aligns to a linear edge of a light grey structure interior to the crater near the west vertex. USGS Astrogeology.*

chromatic scale music intervals. Latitude degrees were used by the Martian architects as a unit of measurement as will be seen in Chapter 7. These degrees are constant over all latitudes whereas longitude degrees vary in size with latitude. Each latitude degree is equal to 59.2747 km which is the same size as a longitude degree at the equator.

The size of the square (diagonal length = 1 latitude degree) in Fig. 2.3 is well below the size of the crater and therefore the square does not touch the crater perimeter anywhere. However, in the region of the east vertex, the sides of the square follow 2 of the linear edges of a dark grey region which suggests the presence of a cavity in the crater. The southwest edge of the square aligns to a linear edge of a barely visible structure lying within the Mystery Crater near its western corner. No other significant alignments occur for this square. The coordinates of the centre of the square (207.7297° E 16.6254° S) are quite meaningful. Its longitude is 53.0000°° E of the Dagger Peak Prime Meridian (DPPM) and 83.0000°° W of the Sharonov Triangle Prime Meridian (STrPM). Note that the longitude values are exactly the same numerically as for the best fit square in Fig.

2.2, but that the prime meridians are different: DPPM vs DMPM and STrPM vs SToPM. Its latitude is $2(e^2)^{\circ\circ}$ S rather than the value of 16.5648° S for the previous square. Note that both the longitude and latitude are in terms of big degrees. My reason for choosing this size and location for the base square will become evident as the fitting of larger squares to features of the Mystery Crater and to other sites beyond the crater limits proceeds. Since the latitude of the centre of this square is meaningful, when it is expanded so that its northern vertex lies right on the equator, it produces a square of very special dimensions (see below).

Larger Squares Concentric to the Fundamental Square

I noticed that if you expand the fundamental square by factors which are the ratios of pure integers and keep the centre the same, the resulting squares align with structures within the crater's interior as well as with the crater perimeter and structures beyond the crater perimeter. However, as you will notice, the ratios of the diagonal lengths of the expanded squares to the diagonal length of the fundamental square very often do not correspond to standard chromatic scale music intervals as was the case for squares which were found to fit "square" craters in *Intelligent Mars II*. The first expanded square to show alignments was a square which enlarged the fundamental square by a factor of 35:32 to give a diagonal size of 1.0938 latitude degrees (Fig. 2.4). The southeast side of the square aligns with the linear edge of the same dark area which the fundamental square aligns to, but in a different region. The southwest side does not align to any linear edge. It only touches the corner of a structure near the crater wall in the interior of the crater.

A second expanded square with an expansion factor of 37:32 and a diagonal size of 1.1563 latitude degrees is shown in Fig. 2.5. The northwest side of this square aligns with the linear interface between a light coloured area outside the square and a dark coloured area inside the square (upper yellow arrow). The southwest side of the square aligns with the northeast linear edge of the same structure as with the previous square (lower yellow arrow).

The diagonal of the next expanded square is 1.2344 latitude degrees which corresponds to an integer ratio of 79:64 above the fundamental square (Fig. 2.6). The southeast side makes 2 substantial alignments with the linear interfaces between grey and light areas (see red lines superimposed on the outline of the square). The southwest side aligns with the southwest edge of the same structure inside the crater perimeter that aligned to the 2 previous squares. It also aligns to a lengthy linear part of the crater perimeter near the south vertex.

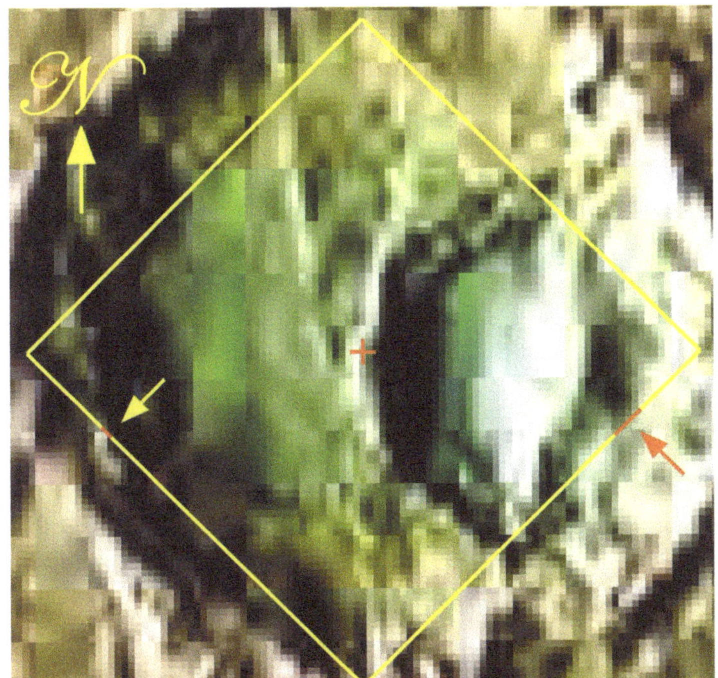

Fig. 2.4: *Fundamental square in the previous figure enlarged by a factor of 35:32 = 1.09375. The red arrow points to an alignment with the linear edge of a dark region inside the crater. The yellow arrow points to where the southwest side of the square touches the northeast corner of a structure within the crater's interior. USGS Astrogeology.*

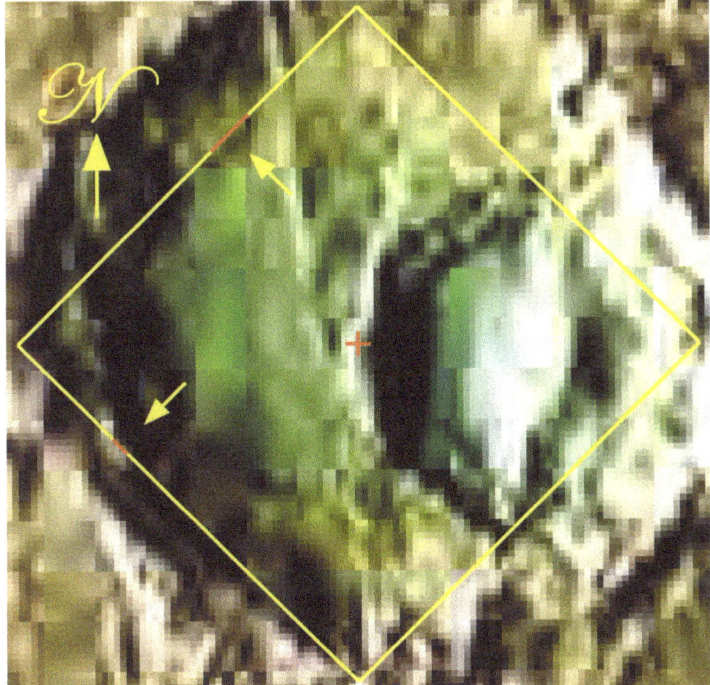

Fig. 2.5: *Fundamental square enlarged by a factor of 37:32 = 1.1563. Upper arrow points to an alignment with the linear interface between a light region outside the square and a dark region inside the square. The lower arrow points to an alignment with the northeast edge of a structure within the crater's interior. USGS Astrogeology.*

The next size of square to make major alignments is a square with a diagonal size of 41:32 or 1.2813 latitude degrees (Fig. 2.7). Here the

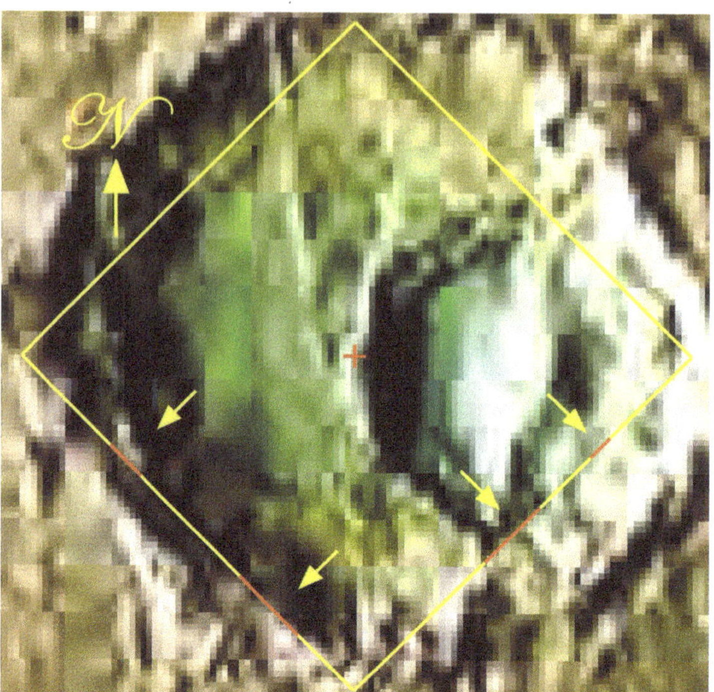

Fig. 2.6: *Square with a diagonal size of 79:64 or 1.2344 latitude degrees. The southeast side aligns to the linear interfaces between grey and light coloured areas. The southwest side aligns to a linear section of the crater perimeter near the south vertex, and to the southwest side of a structure in the interior of the crater. See arrows and short red lines for regions of alignment. USGS Astrogeology.*

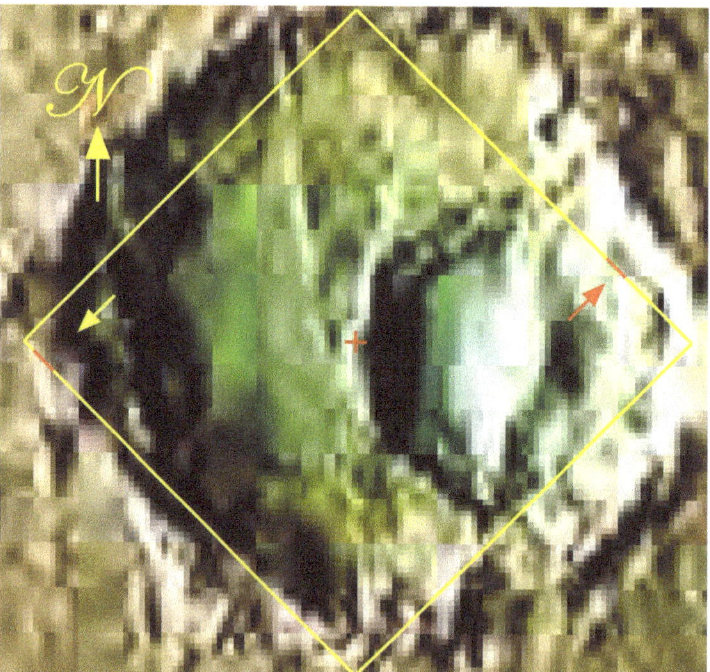

Fig. 2.7: *Square with a diagonal size increased to 41:32 or 1.2813 latitude degrees. The northeast side makes a major alignment with the southwest edge of a grey area. The southwest side of the square aligns to the crater perimeter near the west vertex. USGS Astrogeology.*

northeast side of the square aligns with the southwest edge of a grey area. This grey area (likely a depression in the landscape) makes alignments

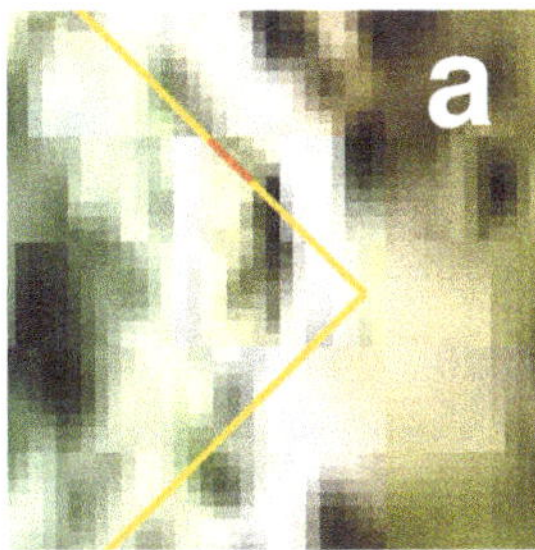

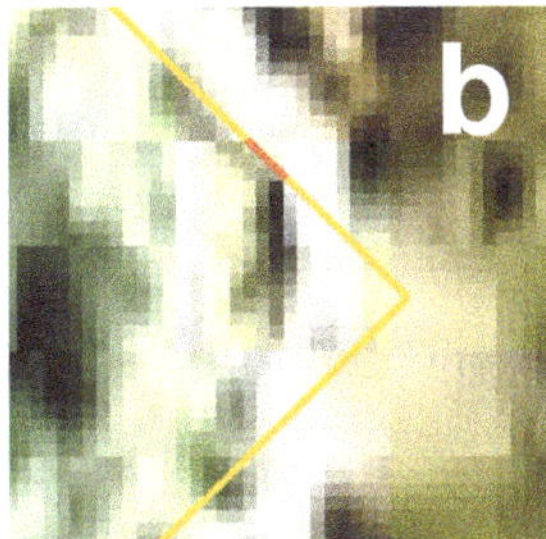

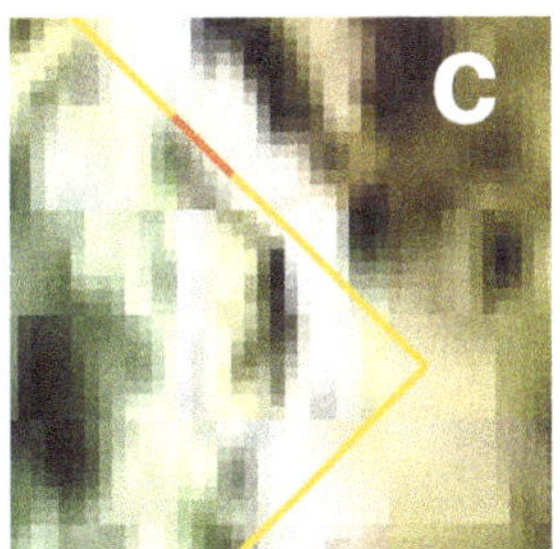

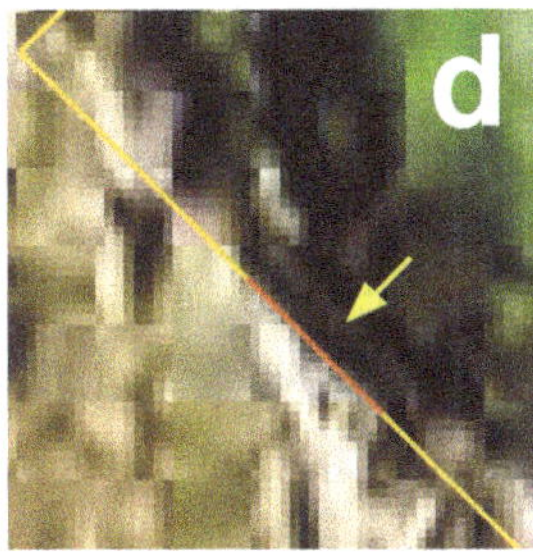

Fig. 2.8: *Expanded squares with diagonal sizes of (a) 21:16 = 1.3125, (b) 11:8 = 1.3750 and (c) and (d) 45:32 = 1.4063 latitude degrees. The northeast sides of the squares align with different linear edges of the same grey area which aligned to the square in Fig. 2.7. The largest square (c) also aligns to a long linear portion of the Mystery Crater's southwest perimeter (d). The top side of each figure is north as is the case for all figures up to Fig. 2.29 except Fig. 2.27. USGS Astrogeology.*

with the next 3 squares. The southwest side of the square is aligned to a short length of the crater perimeter near the west vertex.

Enlarging the square yet again to 21:16 or 1.3125 latitude degrees (Fig. 2.8a) causes the northeast side of the square to align to the southwest side of the same grey area as before but now the alignment is with a region southeast of the previous alignment site. No other alignments were noted for this square. Similarly, by enlarging the diagonal of the square to 11:8 or 1.3750 latitude degrees (Fig. 2.8b), an alignment with the northeast side of the square is created with the northeast side of the same grey region as for the previous square. The final square to align with the eastern grey region has a diagonal size of 45:32 or 1.4063 latitude degrees (Figs. 2.8c and 2.8d). This is one of the ratios used for the tritone interval which lies midway in the chromatic scale. The northeast side of this square aligns with the most northerly region of the eastern grey area (Fig. 2.8c). The southwest side of the square aligns with a long linear stretch of the southwestern perimeter of the Mystery Crater (Fig. 2.8d, yellow arrow).

The next pair of expanded squares have diagonal sizes of 3:2 = 1.5000 and 203:128 = 1.5859 latitude degrees (Fig. 2.9). The northeast side of the smaller square aligns with the southwest edge of a flask-shaped indentation in the crater wall (see red arrow, Fig. 2.9). The northeast side of the larger square aligns with a short length of the northeast edge of the crater perimeter. No other alignments with linear edges of structures could be found for these squares.

The next size of square (13:8 or 1.625 lateral degrees) takes us close to the outermost edge of the southeast crater perimeter (Fig. 2.10). Here the

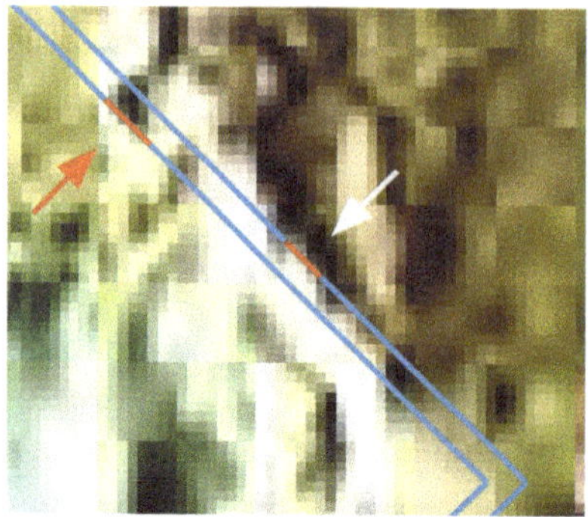

Fig. 2.9: *Expanded squares with diagonal sizes of 3:2 = 1.5000 and 203:128 = 1.5859 latitude degrees. The northeast side of the smaller square aligns with the southwest side of an indentation of the crater perimeter (red arrow). The northeast side of the larger square aligns with a short linear portion of the crater perimeter. USGS Astrogeology.*

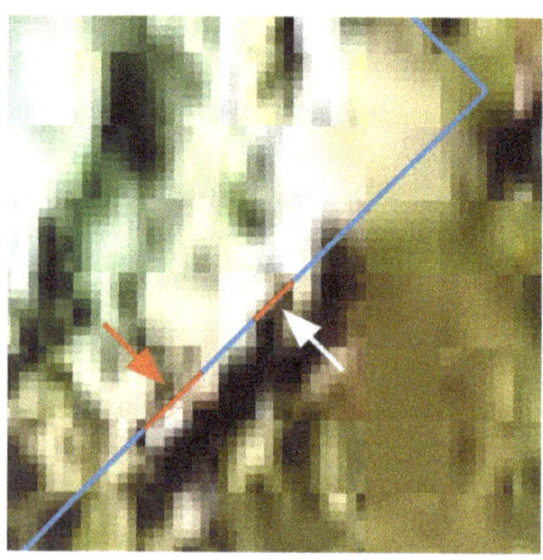

Fig. 2.10: *Square with a diagonal size of 13:8 or 1.625 latitude degrees. The southeast side aligns to a short length of the crater perimeter (white arrow) and also to the linear southeast edge of a grey region in the crater wall (red arrow). The grey region is likely a depression in the crater wall. USGS Astrogeology.*

southeast side of the square aligns with a short linear segment of the crater perimeter (white arrow) and also with the southeast linear edge of a depression (grey area) in the crater wall which is connected to the crater interior (red arrow). The northwest side of the square makes contact with the western perimeter of the crater in several locations (not shown in Fig. 2.10) but each of these sites only touch the square at a single point rather than follow the course of the square over a linear distance.

Expanding the square 2 more times produces the pair of squares in Fig. 2.11. These have diagonal sizes of 439:256 = 1.7148 and 241:128 = 1.8828 latitude degrees. The 2 squares bound the southwest and northeast sides of a structure lying outside the northeast crater perimeter (upper pair of red and yellow arrows). They also bound the outside edges of a wide linear trench which is adjacent to the southeast crater perimeter (lower groups of 3 red and 3 yellow arrows). With the larger of these 2 squares we are now beyond the limits of the Mystery Crater on all 4 sides.

The final expansion of the square in the first octave above the square with a diagonal of 1 latitude degree has a diagonal size of 499:256 = 1.9492 latitude degrees. The southeast side of this square (Fig. 2.12) aligns with a linear segment of the outside edge of the apron of the long trench lying to the southeast of the Mystery Crater. It also passes through the

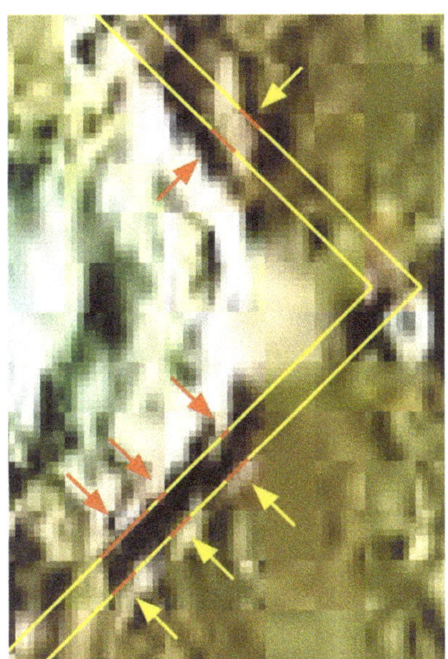

Fig. 2.11: *Expanded squares with diagonal sizes of 439:256 = 1.7148 and 241:128 = 1.8828 latitude degrees. The northeast sides of the squares align with linear edges of a structure lying beyond the northeast crater perimeter. The southeast sides align to short lengths of the outside edges of a wide linear trench which bounds the crater perimeter on its southeast side. USGS Astrogeology.*

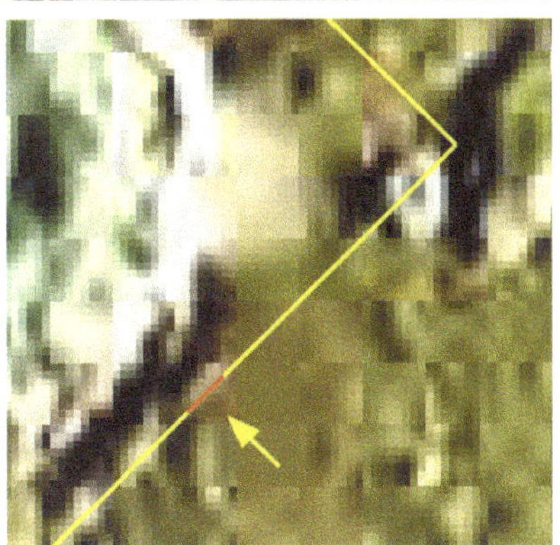

Fig. 2.12: *Largest expanded square in the first octave. It has a diagonal size of 499:256 = 1.9492 latitude degrees. The southeast side of the square aligns to the southeast linear edge of material bounding the edge of the wide linear trench shown in the previous figure. The square also passes through the centre of a small grey region near the eastern vertex. This grey region appears to be a depression in an elevated plateau in the southwest corner of a crater to the east of the Mystery Crater. USGS Astrogeology.*

centre of a small oblong grey area near the eastern vertex of the square.

Summary of First Octave of Concentric Squares

Together with the square having a diagonal size of 1 latitude degree we now have a total of 14 intervals in 1 octave of square expansions. Most of these squares fit linear edges of cavities or of structures inside and surrounding the Mystery Crater rather than the crater perimeter itself. It is the linear nature of these alignment sites that give rise to the high probability that this is not simply an accident of nature. It almost seems that these objects form some sort of a non-uniform coded grid for the alignment of geometric shapes, in this instance, the square.

Except for the fundamental (1:1), the tritone (45:32) and the Perfect 5th (3:2) intervals, the intervals that were found to produce squares which fit linear edges are not those found in the standard chromatic scale. This is in marked contrast to the discoveries presented in *Intelligent Mars II* in which the diagonals of "square" craters were most often found to be sized according to standard chromatic interval ratios. A big exception to this was found for the Janssen Crater in which most of the squares fitting the crater were sized according to the side length rather than the diagonal length. There, the side length was most often equal to the irrational number of e = 2.7183 latitude degrees, and the latitude of the centre of the squares was e° N. So obviously another concept is required for the squares of the Mystery Crater.

All of the integer ratios that give the size of a diagonal of a fitted square in the first octave are harmonics of the fundamental square size. A harmonic interval is one whose frequency is an integer multiple of the fundamental frequency which is the first harmonic. The numerator of the ratio is the harmonic number, and the denominator is a number that can be expressed as an integer power of 2. The denominator is simply a scaling factor to bring the interval into the range of the current octave. Thus the harmonics represented by the squares in the first octave range from 1 to 499.

The squares centred on the Mystery Crater do not stop at one octave. It is now time to examine the squares which fit landforms well beyond the Mystery Crater's perimeter.

Second Octave Squares

I could not locate any landforms having linear edges which aligned to the fundamental square for the second octave of expanded squares (diagonal size of 2.0000 latitude degrees). The first set of 5 expanded squares in the second octave of squares making alignments do so with a crater lying immediately to the east of the Mystery Crater (Fig. 2.13). This crater has a very odd shape being almost rectangular with a large bulge protruding from its southwest corner. Within the bulge there is a white area which probably represents an elevated region. In the centre of this region there is a grey structure which may represent a depression.

The smallest square (leftmost) in Fig. 2.13 aligns with the southeast border of the elevated area in the southwest corner of the crater. It has a diagonal size of 2.1250 latitude degrees and it makes an interval ratio of 17:16 with respect to the fundamental square for the second octave. All interval ratios given in this chapter are with respect to the fundamental square of the current octave. The next size of square has a diagonal size of 2.2500 latitude degrees and has an interval ratio of 9:8 in the second octave of squares. The ratio of 9:8 is the ratio for a Major 2nd in the

Fig. 2.13: *Expanded squares in the second octave making alignments with the crater east of the Mystery Crater. From left to right, the squares have diagonal sizes of 2.1250, 2.2500, 2.3828, 2.4375 and 2.5391 latitude degrees. All alignments are marked with short red lines. Alignments are with the southeast edge of an elevated area within the west side of the crater, with the bottom of the crater wall, and with the crater perimeter and its apron. USGS Astrogeology.*

chromatic scale. This square aligns with the bottom edge of the southeast wall of the eastern crater. The third square has an interval ratio of 305:256 and a diagonal size of 2.3828 latitude degrees. It aligns with a linear section of the southeast perimeter of the crater. The fourth square has a diagonal size of 2.4375 latitude degrees and has an interval size of 39:32 in the second octave. It also aligns with the southeast perimeter of the eastern crater, but in a different region. The fifth square in Fig. 2.13 makes an alignment with the northeast perimeter of the eastern crater and has a diagonal size of 2.5391 latitude degrees. It lies at an interval of 325:256 in the second octave of squares. The southeast side of this square aligns with the southeast edge of a section of the dark coloured apron of the crater.

The next set of expanded squares in the second octave contains 6 squares which make alignments with 4 craters having arrow-shaped perimeters and 1 crater with an arrow shape in its interior (Fig. 2.14). The southeast side of the square in image a aligns with a small arrow-shaped crater pointing due east. This square makes an interval ratio of 91:64 with the fundamental square of the second octave, and has a diagonal size of 2.8438 latitude degrees. The square in image b has an interval ratio of 193:128 and a diagonal size of 3.0156 latitude degrees. It passes through the centre of what appears to be a depressed area central to an arrow shape inside a crater north of the Mystery Crater. The northern side of the interior arrow is aligned in image c to a square having an interval ratio of 25:16 and a diagonal size of 3.1250 latitude degrees. The west pointing arrow shape of the crater in image d is more covert since it is quite blunted. Its northwest side aligns to the northwest side of a square having

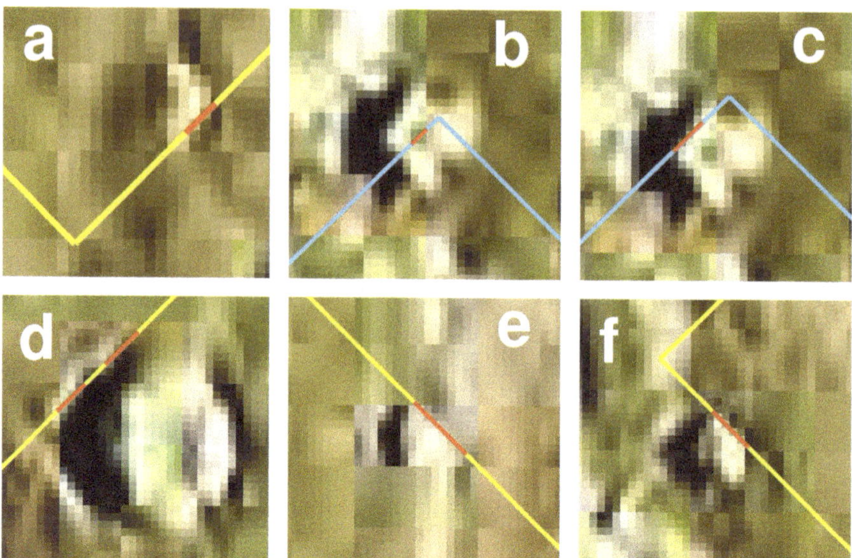

Fig. 2.14: *Set of 6 expanded squares in the second octave. Expanded squares have interval ratios and diagonal sizes (latitude degrees) of (a) 91:64 and 2.8438, (b) 193:128 and 3.0156, (c) 25:16 and 3.1250, (d) 7:4 and 3.5000, (e) 239:128 and 3.7344, and (f) 501:256 and 3.9141. All of the squares make alignments with craters having an arrow shape in their perimeters or interiors. The square in (b) aligns with the central region of the arrow shape within the crater. USGS Astrogeology.*

an interval ratio of 7:4 and a diagonal length of 3.5000 latitude degrees. The arrow shape of the crater in image e is very asymmetrical with a bevel occurring only on its north side. It points eastwards and aligns with a square having an interval ratio of 239:128 and a diagonal length of 3.7344 latitude degrees. In image f, a square with an interval ratio of 501:256 and a diagonal size of 3.9141 latitude degrees aligns to the northwest edge of a north-pointing arrow-shape in the perimeter of a crater which has 2 arrow shapes - one arrow pointing due north and the other pointing due west. This square is the final expanded square that I could find in the second octave to align with a linear segment of a landform.

In summary, all of the expanded squares in the second octave have interval ratios corresponding to harmonics which range from 7 to 501. Only the major 2nd interval of 9:8 is found in a standard chromatic scale. The interval of 7:4 is a variant of a minor 7th interval in music. The second octave also introduces arrow-shaped craters which play a big role in the higher octaves.

Third Octave Squares

As was the case with the second octave of expanded squares, the fundamental square for the third octave of expanded squares did not align in a significant way to any landforms. The first set of expanded squares that do make major alignments for this octave is shown in Fig. 2.15. The squares increase in size from Fig. 2.15a to i, and their interval ratios (with respect to the fundamental square for the third octave) and their diagonal sizes are given in the caption to Fig. 2.15. All of the alignments are with linear sections of crater perimeters or with linear features found within the crater's interior. All of the craters have an arrow shape either in their perimeters or interior to the crater. The southwest side of the square is aligned in images a and b, the northwest side in images c, g and h, the northeast side in images d and i, and the southeast side in images e and f.

This brings us to a very remarkable square (Fig. 2.16). It fits the northwest and northeast edges of 2 large arrow-shaped craters which seem to act as guides for the fitting of this square. The perfection of the fitting of this square provides excellent evidence that my choice of longitude for the centre of the series of squares is the one intended by the architects. This square has an interval ratio of 193:128 in the third octave

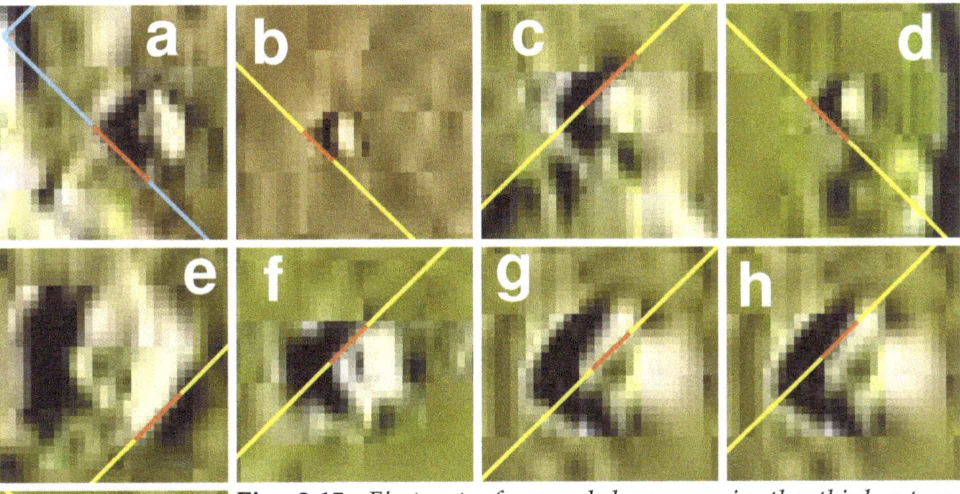

Fig. 2.15: *First set of expanded squares in the third octave. Squares have interval ratios and diagonal sizes (latitude degrees) of (a) 67:64 and 4.1875, (b) 287:256 and 4.4844, (c) 289:256 and 4.5156, (d) 81:64 and 5.0625, (e) 339:256 and 5.2969, (f) 23:16 and 5.7500, (g) 93:64 and 5.8125, (h) 377:256 and 5.8906, and (i) 95:64 and 5.9375. All alignments are with arrow-shaped craters. USGS Astrogeology.*

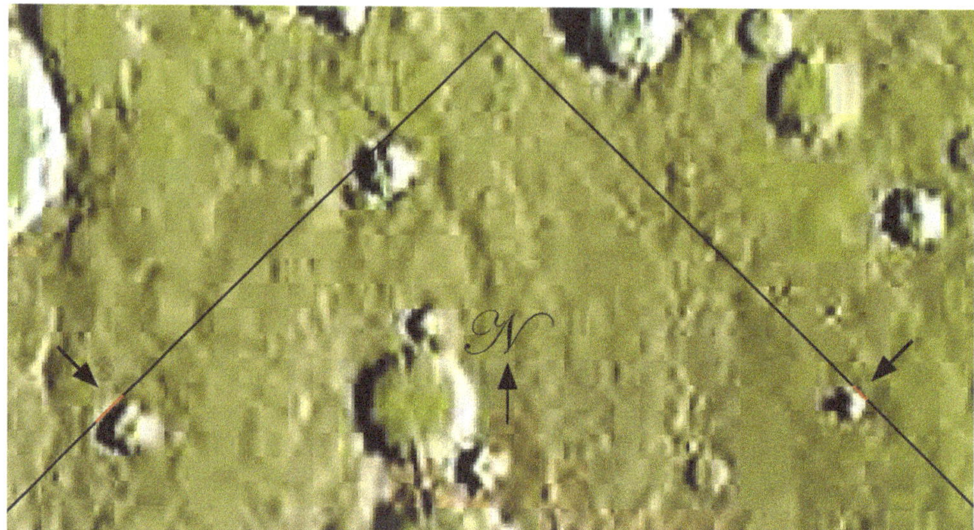

Fig. 2.16: *Expanded square in the third octave with an interval ratio of 193:128 and a diagonal size of 6.0313 latitude degrees. Note that it perfectly aligns with 2 large arrow-shaped craters (black arrows). USGS Astrogeology.*

which is only slightly larger than a perfect 5th interval ratio of 3:2.

The final 8 expanded squares which I could identify for the third octave are shown in Fig. 2.17. They increase in size from the square in Fig. 2.17a to the square in Fig. 2.17h. Their interval ratios in the third octave and their sizes in latitude degrees are given in the caption to the figure. Three of these squares (images a, e and h) align to linear perimeter sections of craters which have a clear arrow shape. In image b, the alignment is to the edge of a trench present in a large crater south of the Mystery Crater. In images c and g, the squares align to linear segments of crater perimeters. In image f, the square passes through the central peak of the same crater as in image d which has a covert arrow shape. The southwest side of the square is aligned in image b, the northwest side in images c, g and h, the northeast side in images d and f, and the southeast side in images a and e.

In summary, all of the squares of the third octave are harmonics of the third octave fundamental square. Most of the squares make alignments with arrow-shaped craters. Two of the squares were aligned to linear segments of crater perimeters of craters not having an arrow shape. One of the squares aligned to the edge of a trench inside a large crater, and another square aligned to the central peak of a crater. The two arrow-shaped craters in Fig. 2.16 seem to act as guides for the fitting of an expanded square, and this special square provides strong evidence for the accuracy of the longitude selected for the centre of the expanding series of squares from the Mystery Crater.

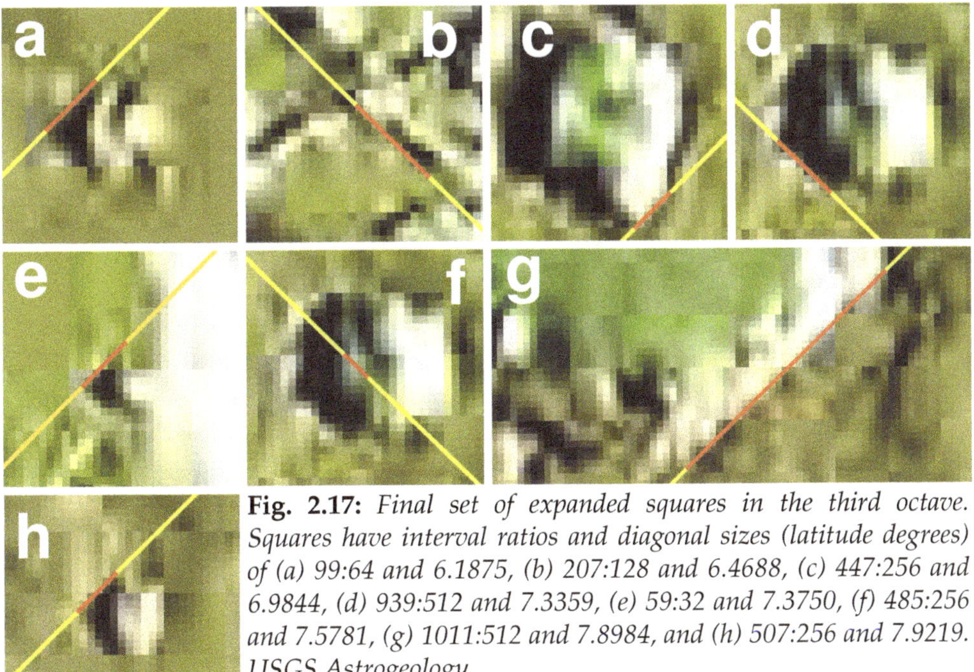

Fig. 2.17: *Final set of expanded squares in the third octave. Squares have interval ratios and diagonal sizes (latitude degrees) of (a) 99:64 and 6.1875, (b) 207:128 and 6.4688, (c) 447:256 and 6.9844, (d) 939:512 and 7.3359, (e) 59:32 and 7.3750, (f) 485:256 and 7.5781, (g) 1011:512 and 7.8984, and (h) 507:256 and 7.9219. USGS Astrogeology.*

Fourth Octave Squares

The sizes and interval ratios of the expanded squares for the first part of the fourth octave are listed in the caption to Fig. 2.18. The fundamental square for the fourth octave shown in Fig. 2.18a makes an alignment with the edge of a part of the Mangala Fossa which is a large fissure running east-west for hundreds of kilometers. The rest of the squares in Fig. 2.18 align with linear sections of crater perimeters. The craters in images b, d, e and f have covert arrow shapes, barely distinguishable in image b and blunted in image d. The southwest side of the square is aligned in image e, the northwest side in images c and g, the northeast side in images b, d and f, and the southeast side in image a.

A second set of expanded squares for the fourth octave is shown in Fig. 2.19, and their interval ratios and diagonal sizes are listed in the caption for the figure. The squares in images a and e make alignments with long linear sections of the perimeters of large craters, and the square in image f aligns with a linear segment of the perimeter of a small crater. The squares in images b and c make alignments with the central peaks of 2 craters. The square in image b is aligned to the southern edge of the peak whereas the square in image c passes through the centre of the crater's central peak. This is reminiscent of the role played by auxiliary craters in

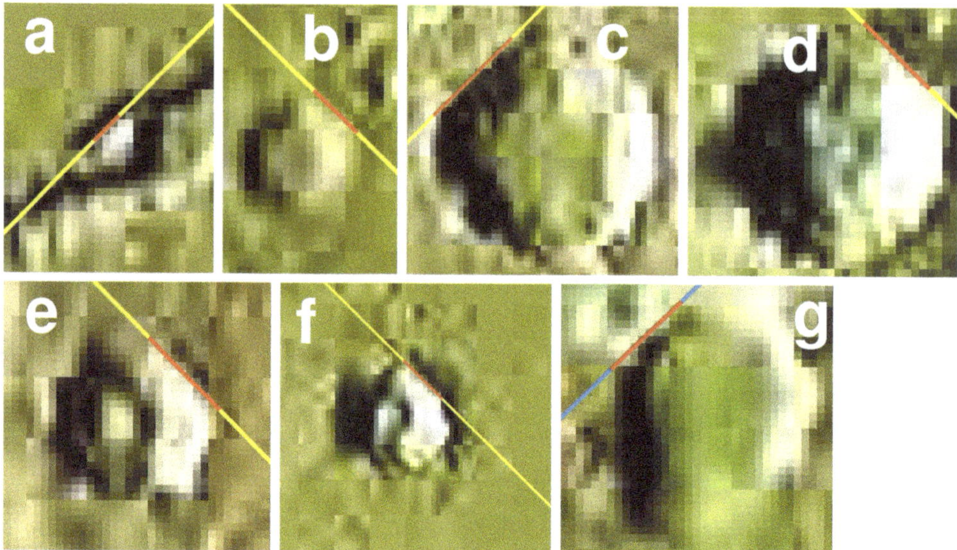

Fig. 2.18: *Expansions of the square in the fourth octave. Expanded squares have interval ratios and diagonal sizes (latitude degrees) of (a) 1:1 and 8.0000, (b) 35:32 and 8.7500, (c) 581:512 and 9.0781, (d) 147:128 and 9.1875, (e) 313:256 and 9.7813, (f) 21:16 and 10.5000, and (g) 675:512 and 10.5469. USGS Astrogeology.*

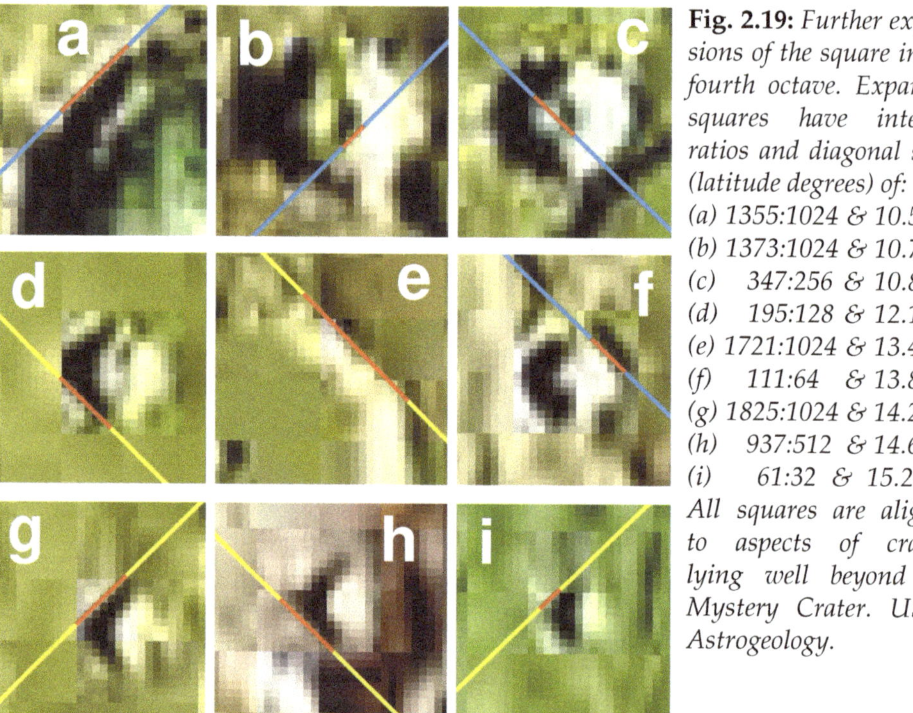

Fig. 2.19: *Further expansions of the square in the fourth octave. Expanded squares have interval ratios and diagonal sizes (latitude degrees) of:*
(a) 1355:1024 & 10.5859
(b) 1373:1024 & 10.7266
(c) 347:256 & 10.8438
(d) 195:128 & 12.1875
(e) 1721:1024 & 13.4453
(f) 111:64 & 13.8750
(g) 1825:1024 & 14.2578
(h) 937:512 & 14.6406
(i) 61:32 & 15.2500.
All squares are aligned to aspects of craters lying well beyond the Mystery Crater. USGS Astrogeology.

fitting polygonal shapes to craters as reported in *Intelligent Mars II*. The squares in images d, g, h and i align with the perimeters of arrow-shaped craters. The southwest side of the square is aligned in image e, the northwest side in images a, g and i, the northeast side in images c, d, f and h, and the southeast side in image b.

Fifth Octave Squares

As we come into the fifth octave of expanded squares, once again the fundamental square for the octave shows no good alignments to any linear edges of craters or other landforms. The sizes and interval ratios of the expanded squares that I could find for the first part of the fifth octave are listed in the caption to Fig. 2.20. All of these squares are aligned to linear sections of crater perimeters, some of which have the shape of an arrow. Some of the arrow shapes are quite covert such as the blunted

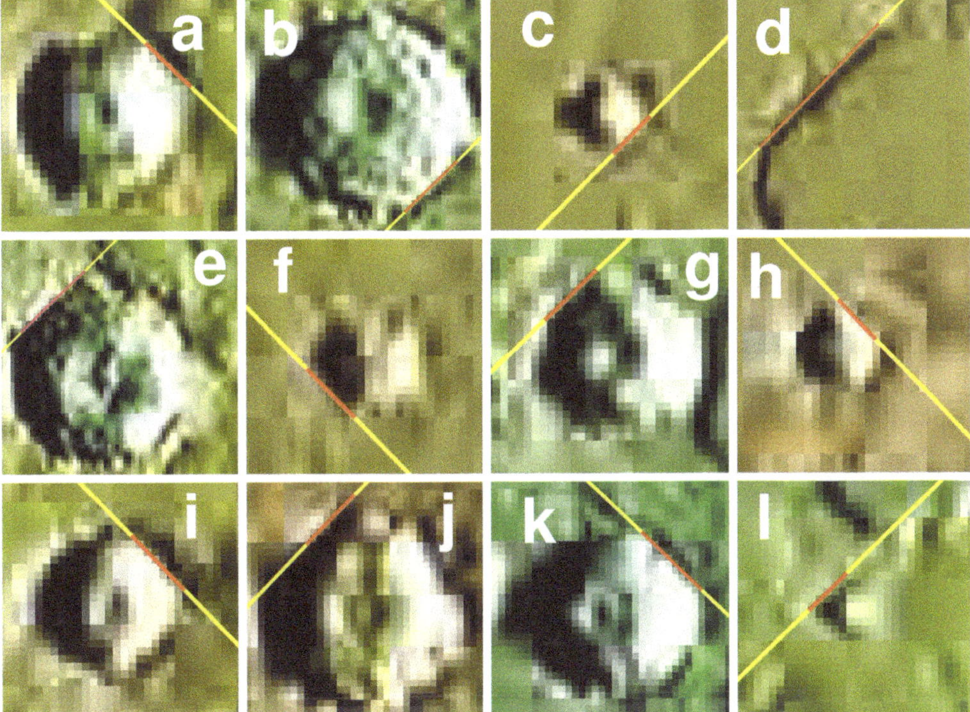

Fig. 2.20: *First set of expanded squares in the fifth octave. Squares have interval ratios and diagonal sizes (latitude degrees) of (a) 1155:1024 and 18.0469, (b) 1177:1024 and 18.3906, (c) 593:512 and 18.5313, (d) 1221:1024 and 19.0781, (e) 1365:1024 and 21.3281, (f) 171:128 and 21.3750, (g) 685:512 and 21.4063, (h) 1401:1024 and 21.8906, (i) 1495:1024 and 23.3594, (j) (1583:1024 and 24.7344, (k) 809:512 and 25.2813, and (l) 841:512 and 26.2813. USGS Astrogeology.*

arrow-shaped crater in image b and irregular arrow-shaped crater in image g. The southwest side of the square is aligned in images f, h and i, the northwest side in images b, e, g and l, the northeast side in images a and k, and the southeast side in images c, d and j.

There are several more expanded squares in the fifth octave which make alignments to linear perimeter sections of craters surrounding the Mystery Crater (Fig. 2.21). Their interval ratios and diagonal sizes are given in the caption. The square in image a is aligned to the northwest perimeter of an oddly shaped crater lying south of the Mystery Crater. The 30° S line of latitude is seen at the bottom of the image. The square in image b aligns to a straight line segment in the perimeter of a crater north of the Mystery Crater. The crater in image c is curious in that it has a channel emerging from it with a bearing angle of about -45°. It sort of resembles an apple on a stick, or a wine class without the bottom stand. A fifth octave square aligns to the linear section of its northeast perimeter. The squares in images d and e align to linear sections of the perimeters of very irregularly shaped craters. All the remaining squares align to arrow-shaped craters which are quite irregularly shaped. The square in image h aligns to the eastern side of an arrowhead belonging to an arrow-shaped crater which points due south. This is a rare orientation for an arrow-shaped crater since most of them seem to point east or west. The

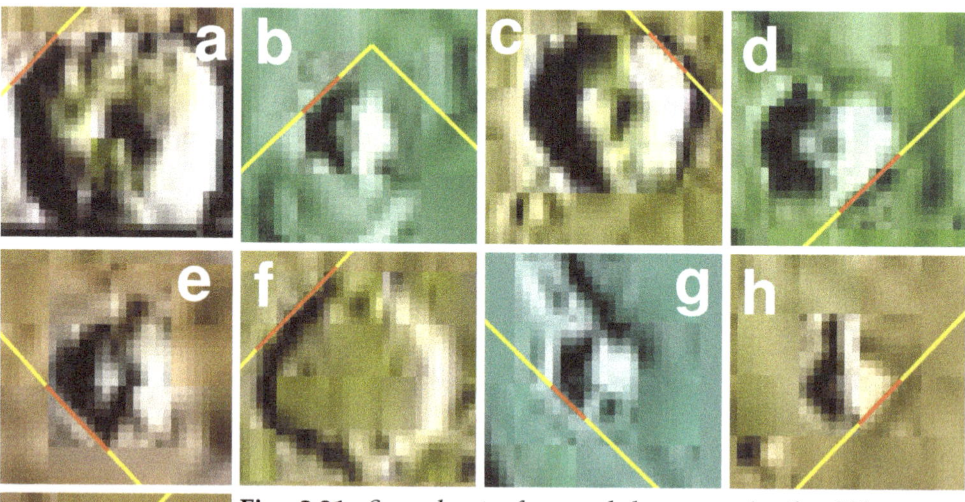

Fig. 2.21: *Second set of expanded squares in the fifth octave. Squares have interval ratios and diagonal sizes of (a) 437:256 and 27.3125, (b) 223:128 and 27.8750, (c) 457:256 and 28.5625, (d) 229:128 and 28.6250, (e) 3807:2048 and 29.7422, (f) 981:512 and 30.6563, (g) 63:32 and 31.5000, (h) 507:256 and 31.6875, and (i) 1017:512 and 31.7813. USGS Astrogeology.*

southwest side of the square is aligned in images c and e, the northwest side in images b and d, the northeast side in image g, and the southeast side in images a, f, h and i.

In summary, the expanded squares of the fifth octave all make alignments with linear sections of the perimeters of craters surrounding the Mystery Crater at a great distance. Most of the craters have an underlying arrow shape although some are cleverly disguised to hide this design.

Sixth Octave Squares

Alignments for expanded squares from the Mystery Crater were found to continue into the sixth octave. The first set of squares is shown in Fig. 2.22, and the interval ratios and diagonal sizes of the squares are listed in the caption to the figure. While arrow-shaped craters continued to play a major role, several of the alignments occurred with linear sections of the perimeters of craters of varying shapes. The square in image a is aligned to the west side of a north-pointing arrow shape inside a crater. The square in image c is aligned with the south side of an arrowhead pointing due west. This arrowhead is curious in that it belongs to an elevated structure rather than a crater, and the structure has the shape of a curved arrow. It is located north of the Mystery Crater. The square in image g is aligned to the linear interface between 2 craters lying side-by-side. The square in image h is aligned to the northwest linear side of a partly eroded crater. The southwest side of the square is aligned in images b, d, f

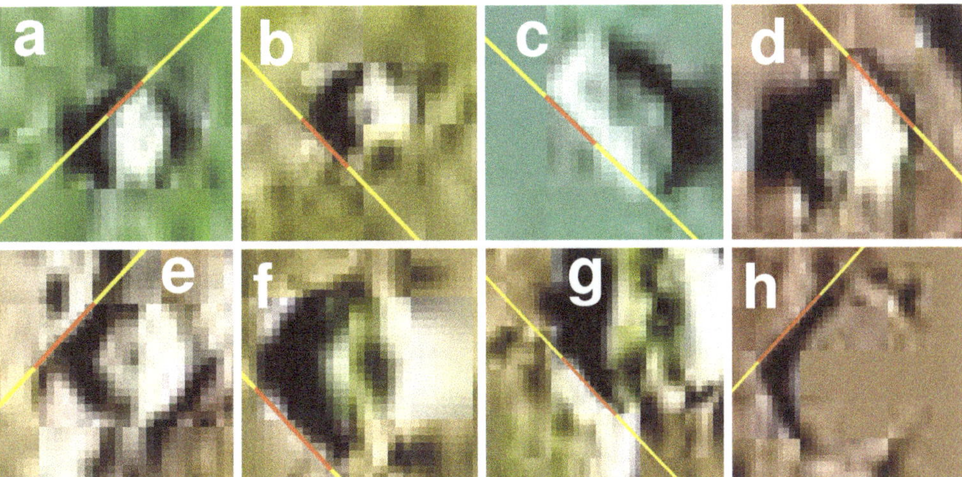

Fig. 2.22: *First set of expanded squares in the 6th octave. Interval ratios and diagonal sizes (latitude degrees) are (a) 1047:1024 and 32.7188, (b) 2103:2048 and 32.8594, (c) 2121:2048 and 33.1406, (d) 2363:2048 and 36.9219, (e) 611:512 and 38.1875, (f) 1231:1024 and 38.4688, (g) 1243:1024 and 38.8438, and (h) 175:128 and 43.7500. USGS Astrogeology.*

and g, the northwest side in image a, the northeast side in image c, and the southeast side in images e and h.

A second set of squares from the sixth octave is shown in Fig. 2.23 together with their interval ratios and diagonal sizes in the caption. Several squares align to craters which have a blunted arrow shape (images a, e, g and h). The square in image b is aligned to the southwest side of an arrow shape inside a crater. The crater in image g has the shape of a south-pointing blunted arrow. The square in this figure also aligns to a couple of outcroppings just outside the southeast perimeter of the Nicholson Crater (not shown). All of the remaining squares align to linear sections of crater perimeters (images c, d, f, i, j, k and l). In image i, the square aligns to a crater which has the shape of a flask. The square aligns to the top of the "flask". In image j, the side of an expanded square aligns to the northwest perimeter of the huge Nicholson Crater. The southwest side of the square is aligned in images a, b, c, e, f and g, the northwest side in images d, j, k

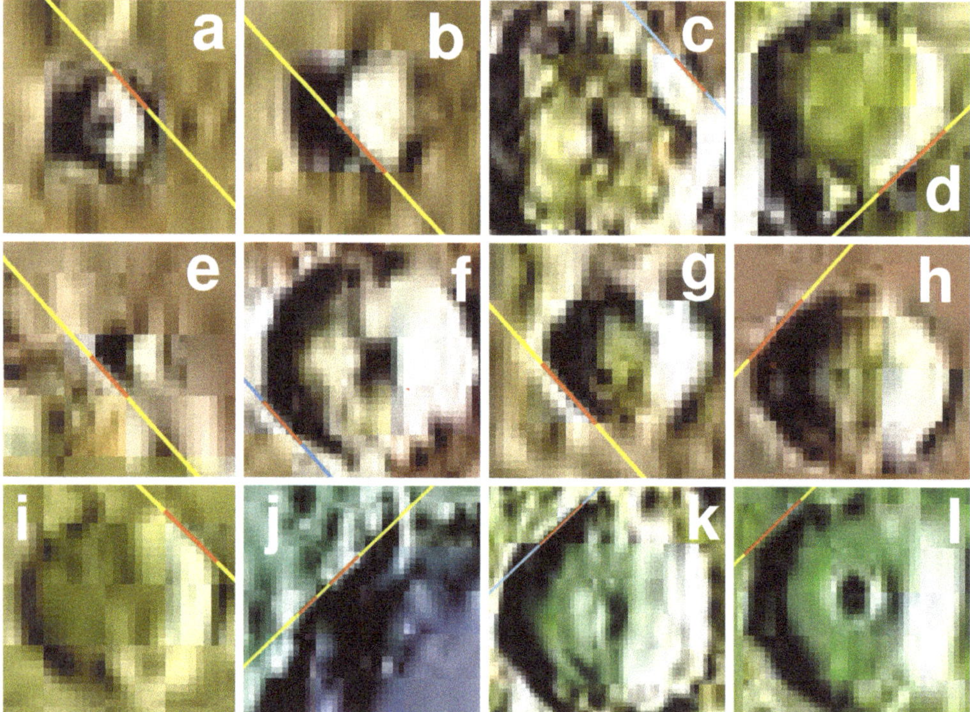

Fig. 2.23: *Second set of expanded squares in the 6th octave. Interval ratios and diagonal sizes (latitude degrees) are (a) 353:256 and 44.1250, (b) 1485:1024 and 46.4063, (c) 3095:2048 and 48.3594, (d) 789:512 and 49.3125, (e) 3245:2048 and 50.7031, (f) 215:128 and 53.7500, (g) 3501:2048 and 54.7031, (h) 1757:1024 and 54.9063, (i) 1879:1024 and 58.7188, (j) 1901:1024 and 59.4063, (k) 1943:1024 and 60.6563, and (l) 2039:1024 and 63.7188. USGS Astrogeology.*

and l, the northeast side in image i, and the southeast side in image h.

In summary, many of the expanded squares in the sixth octave make alignments with craters which have an arrow shape which is either obvious or else discernible when examined more closely. The remaining squares mostly align to linear sections of crater perimeters. One square aligns to a very clear arrow shape interior to the crater. Another aligns to an elevated landform which has the shape of a curved arrow.

Seventh Octave Squares

Alignments to linear features of the Martian landscape have also been found for a seventh octave of expanded squares. The first set of these is shown in Fig. 2.24 with their interval ratios and diagonal sizes listed in the caption. Most of these squares make alignments with linear sections of the perimeters of craters, some of which have a blunted arrow shape (images e and f). The square in image g aligns to a blunted arrow shape interior to the crater. It is the same square that aligns with the arrow-shaped crater in image f. The square in image b is significant because it aligns to a long linear portion of the southwest edge of Arsia Mons. The square in image d aligns with the bottom of the wall of a deep fracture on the west side of the Syria Planum which lies east of Arsia Mons. The square in image h aligns with the northeast linear section of the perimeter of the northern crater lying on Ulysses Tholus. The northwest side of the square is aligned in images e, f, and g, and the northeast side in images a to d, and h.

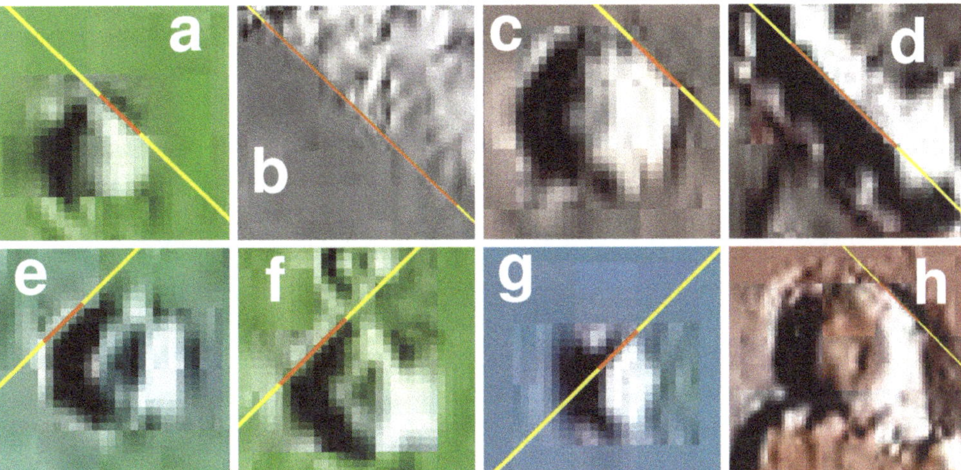

Fig. 2.24: *First set of expanded squares in the 7th octave with interval ratios and diagonal sizes (latitude degrees) of (a) 2085:2048 and 65.1563, (b) 551:512 and 68.8750, (c) 713:512 and 89.1250, (d) 1427:1024 and 89.1875, (e) 3001:2048 and 93.7813, (f) and (g) 1571:1024 and 98.1875, and (h) 3163:2048 and 98.8438. USGS Astrogeology.*

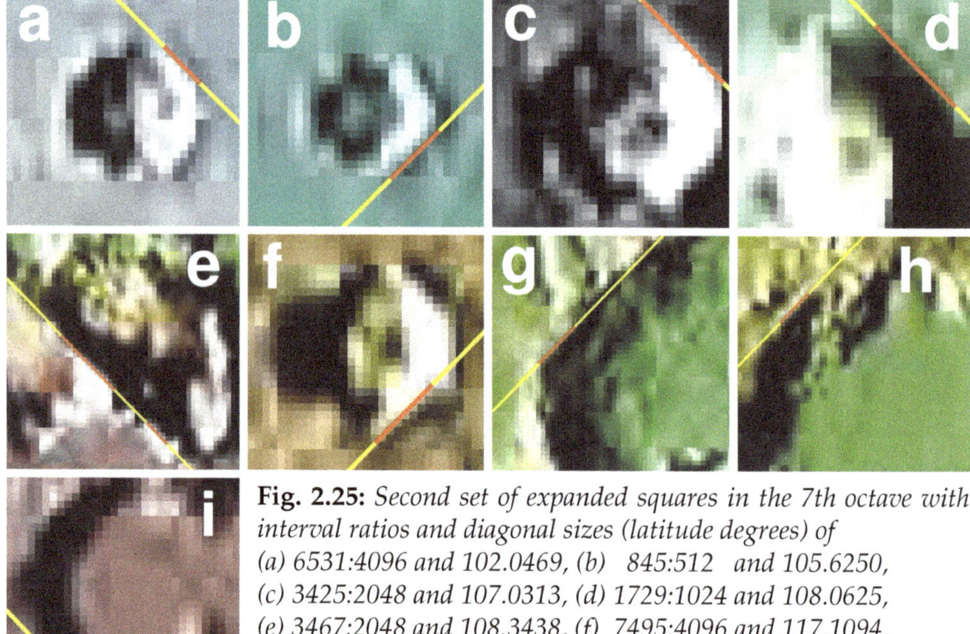

Fig. 2.25: *Second set of expanded squares in the 7th octave with interval ratios and diagonal sizes (latitude degrees) of (a) 6531:4096 and 102.0469, (b) 845:512 and 105.6250, (c) 3425:2048 and 107.0313, (d) 1729:1024 and 108.0625, (e) 3467:2048 and 108.3438, (f) 7495:4096 and 117.1094, (g) 963:512 and 120.3750, (h) 3891:2048 and 121.5938, and (i) 3913:2048 and 122.2813. USGS Astrogeology.*

The second set of squares lying within the seventh octave is shown in Fig. 2.25 with their measurements listed in the caption. Four of the squares align to arrow-shaped craters (images a, b, c and f). The crater in image a is the Pangboche Crater and the crater in image c is the Karzok Crater. Both of these craters lie on Olympus Mons. The square in image d aligns with a section of the northern perimeter of Olympus Mons. Close by is a section of the northern wall of the mountain which aligns to a larger expanded square (image e). The craters in images f, g and h lie to the west of the Mystery Crater. The square in image i aligns to the southwest side of a crater lying in the Noctis Labyrinthus region of Mars. The northwest side of the square is aligned in images b and f to h, and the northeast side in images a, c to e, and i.

In summary, the expanded squares in the seventh octave make alignments not only with craters but also with features of Olympus Mons and Arsia Mons, and with a landform east of the giant mountains.

Eighth Octave Squares

Alignments continue to occur with squares expanding into the eighth octave (Fig. 2.26). Their interval ratios and diagonal lengths are provided in the caption to the figure. The square in image a aligns with the

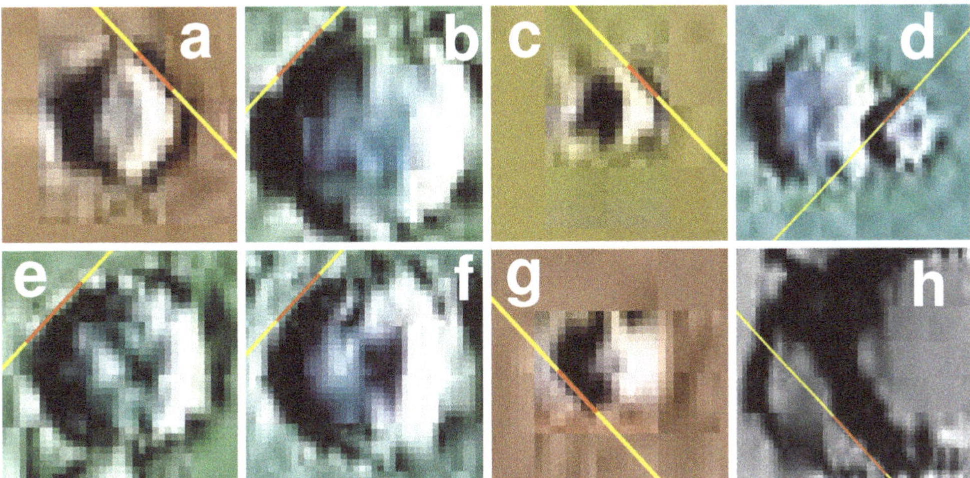

Fig. 2.26: *Expansions of the squares in the eighth octave. Expanded squares have interval ratios and diagonal sizes (latitude degrees) of (a) 4111:4096 and 128.4688, (b) 4217:4096 and 131.7813, (c) 33:32 and 132.0000, (d) 4247:4096 and 132.7188, (e) 4335:4096 and 135.4688, (f) 2207:2048 and 137.9375, (g) 4477:4096 and 139.9063, and (h) 4485:4096 and 140.1563. USGS Astrogeology.*

perimeter of a blunted arrow-shaped crater lying between Olympus Mons and Ascraeus Mons. The square in image b aligns with a long linear section of the perimeter of a large crater northwest of the Mystery Crater. The small crater in image c lies east of Olympus Mons. The interval ratio for the square aligned to this crater is 33:32. Note that 33 is a very important number in Freemasonry. In image d we see that the square aligns to an arrow shape inside a small crater. The craters in images e and f lie northwest of the Mystery Crater and both have a covert arrow shape. The square in image g aligns to an arrow-shaped crater lying to the northwest of Ascraeus Mons. In image h, the square is shown to align with a linear edge in the Ascraeus Mons Caldera. This is the last expanded square that I tried to find alignments for since the spherical nature of the planet was now starting to distort the fitted square by altering the bearing angle of the NW and NE sides of the square to about 42.3° instead of 45°. The northwest side of the square is aligned in images b and d to f, and the northeast side in images a, c, g and h.

The sizes of the squares in the eighth octave are enormous. The square in Fig. 2.26h has a diagonal size of 140.1563 latitude degrees which translates into a distance of more than 8300 km!

Summary and Conclusions

Although the Mystery Crater can be rightly classified as a "square" crater since it has an overall square shape and its northwest side is oriented at a clockwise 45 degree angle with regard to due north (the most common bearing angle for "square" craters), it does not share other properties of "square" craters which were reported in *Intelligent Mars II*. It was not possible to find a square which aligned to linear segments of at least 3 of the perimeter sides simultaneously while including the entire perimeter. After much experimentation, a square with a diagonal length of 1 latitude degree was selected to serve as a fundamental square. It was centred at 53°° E [DPPM] $2(e^2)$°° S. This square, and expanded squares in the first octave, were found to fit linear edges which have bearing angles of 45° in the clockwise or counterclockwise direction. The linear edges bounded structures internal or external to the crater or formed parts of the crater perimeter. All of the squares which fit either interior structures of the crater or the crater perimeter itself belonged to the first octave of squares. Another departure from the normal "square" craters was that many of the diagonals of the squares are not sized according to interval ratios corresponding to the standard chromatic scale in music. Rather, the diagonals of all of the squares, including those with chromatic scale interval ratios, are sized according to harmonics.

What really sets the Mystery Crater apart from the more normal "square" craters is the presence of craters and structures lying far beyond its boundaries which align to huge squares centred on the Mystery Crater. While these exterior craters are reminiscent of auxiliary craters which helped with the alignment of squares fitting standard "square" craters, they differ in that they align to squares which lie completely beyond the Mystery Crater perimeter. The shape of these alignment craters is also remarkable in that a large number of them have the underlying shape of an arrow. It would appear from the fitting of expanded squares that a major function of arrow-shaped craters is to align geometric shapes. In *Intelligent Mars II*, arrow-shaped craters or other structures with an arrow shape were found to mark important coordinates with their tips. The alignment function is, therefore, a hitherto unsuspected additional purpose of arrow-shaped craters.

Fig. 2.27 shows all of the expanded squares emanating from the Mystery Crater which could be aligned to at least one linear section of a crater perimeter or to other landforms. Those closest to the Mystery Crater tend to merge, especially for the first octave, due to the limited amount of resolution that could be achieved when such a large area of terrain is fitted to the page dimensions of this book. Only the interior of the Mystery Crater is visible within the fundamental square whose diagonal size is 1 latitude degree. What is truly remarkable is the relatively low degree of distortion of the squares over the 8 octave range.

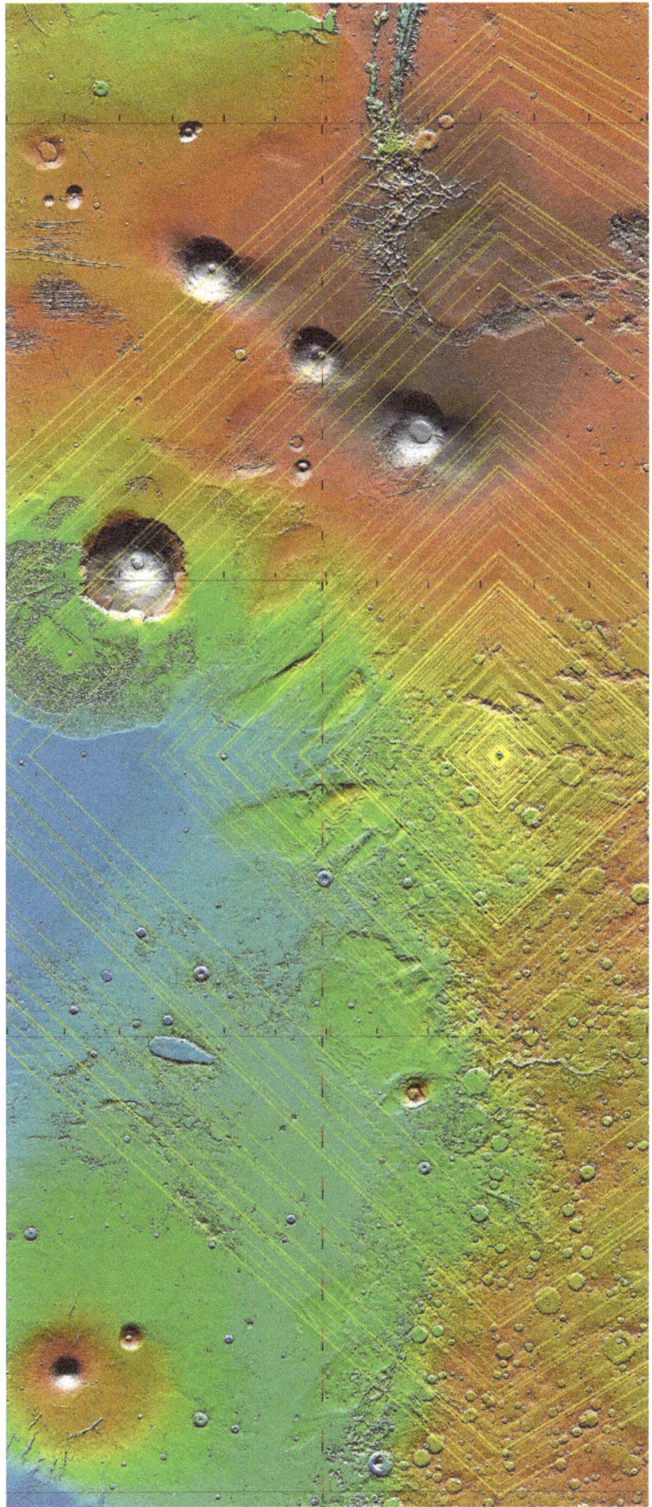

Fig. 2.27: *All expanded squares emanating from the Mystery Crater. Each square aligns to one or more linear components of craters or other landforms. The dark spot at the centre of all the squares is the area within the Mystery Crater which is bounded by a square whose diagonal is equal to one latitude degree. Note how the largest squares encompass Olympus Mons and the 3 Tharsis Montes. The largest square has a diagonal size of 140.1563 latitude degrees or 8307.72 km. USGS Astrogeology.*

This suggests that the location of the Mystery Crater was carefully chosen to minimize distortion. There are several possible models to choose from in order to represent a 2-dimensional square on a spherical surface. I chose to construct the squares by maintaining the integrity of the diagonals and using rhumb lines to construct the sides of the square from the vertices marked by the ends of the diagonals. A rhumb line is one in which the bearing angle is maintained over the entire course of a line between 2 points on a spherical planet. Rhumb lines are linear on a map which uses the Mercator projection as is the case for the MOLA (Mars Orbiter Laser Altimeter) maps of Mars between 30° S and 30° N. With this model, I discovered that the bearing angle of the northern sides of the squares stays within ±1° of 45° right up to an interval ratio of about 1.83 in the seventh octave. It then degrades to 43.3° by the beginning of the eighth octave and to 42.3° for the largest square that I found an alignment site for. This was the major reason why I decided to limit the squares to this final size. The bearing angle of the south sides stays within 1° of 45° until an interval ratio of about 1.34 in the fifth octave. It then degrades to 43° at an interval ratio of about 1.15 in the sixth octave and to 42° at an interval ratio of about 1.54 in the sixth octave. The last square that I found an alignment for one of its south sides was a square in the sixth octave with a south side bearing angle of 41.5° and an interval ratio of about 1.72 (see Fig. 2.23h). Despite the distortion in bearing angle, none of the alignments with south sides of the squares in Fig. 2.23 appeared to be misaligned. This suggests that I chose the same model for fitting expanded squares to the spherical surface of the planet as the one that was used by the architects.

The size and distance of aligned expanded squares from the Mystery Crater is truly mind-boggling. They cover about 8 octaves in magnitude. Their maximum diagonal size is at least 140 latitude degrees which translates into a distance of more than 8000 km. Hence, craters which align to these squares can be more than 4000 km away from the Mystery Crater. What is so important about the Mystery Crater that the Martian architects deemed it necessary to align so many squares over such huge distances to craters and other landforms?

It was at about this point that I started to think of Aum (also spelled Om). In Eastern religions, Aum is an extremely ancient concept that postulates a Divine Being creating the entire universe from a primordial sound. The intonation of the word "Aum" is thought to mimic this sound which is conceived to contain the entire spectrum of sound frequencies. The frequencies combine to create the matter and energy which form the stars, planets and galaxies. What better way to portray this than to have a fundamental square representing a fundamental frequency expand outwards to express a huge number of frequencies as harmonics? With this in mind, I decided to call the Mystery Crater the *Aum* Crater. As I continued my investigation into this crater, it became more and more obvious that this indeed was the right choice. This crater was created by the ancient architects to represent the concept of Aum.

The harmonic numbers of the squares go up logarithmically as the octave number increases (Fig. 2.28). This is to be expected since the area available for structures to align with also goes up logarithmically as the octave number increases. This allows for greater and greater resolution as the squares expand outwards, and makes way for very high numbered harmonics. For the entire dataset of expanded squares that aligned with structures, 120 unique harmonics are present. For numbers less than 100, 27 odd numbered harmonic numbers are represented. Hence, for this region of harmonic numbers, more than 50% of the possible harmonic numbers are represented since, when even numbered harmonics are encountered, they can be divided down by factors of 2 until an odd numbered integer is reached. It is possible that far more harmonic numbers are actually represented since there may be expanded square

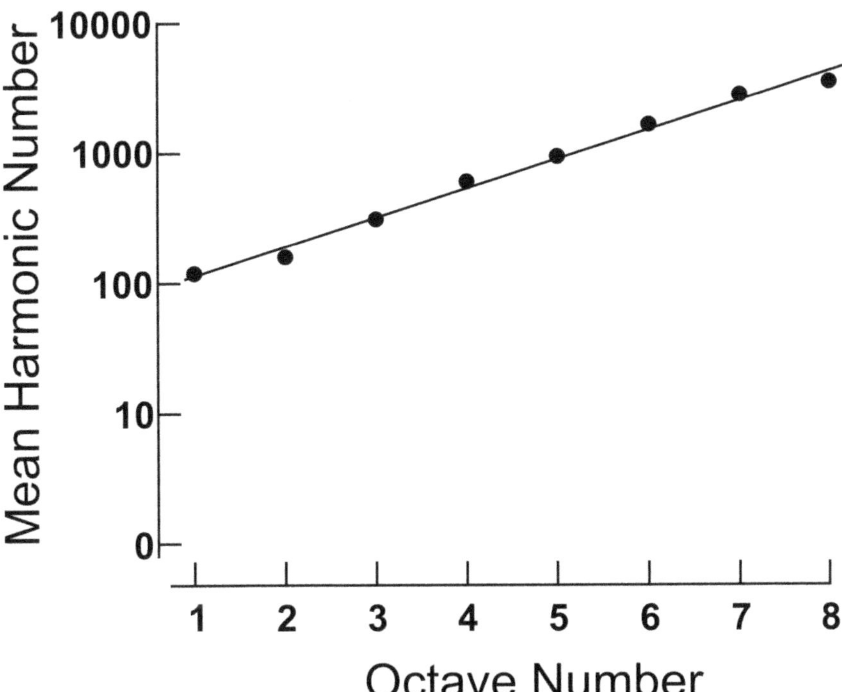

Fig. 2.28: *Plot of average harmonic number for the expanded squares within an octave vs. the octave number. The y-axis is on a logarithmic scale. The graph shows a linear relationship between the log value of the average harmonic number and the octave number.*

alignments to landforms or structures which are well below the resolution of the MOLA maps which I am using.

Before going on to the next chapter, there is a set of expanded squares with unique properties that set themselves apart from the squares which were sized according to harmonics. In this set, irrational numbers are used for diagonal sizing rather than pure integers. Thus the interval produced above the fundamental square is irrational! These squares are in close alignment with very important sites such as the centre of the pentagram fitting the Vitruvian Martian (see *Intelligent Mars I*). The simplest of these is a square whose northern vertex touches the equator (Fig. 2.29a). Its diagonal is exactly equal to $4.5e^2$ latitude degrees, which is twice the magnitude of the southern latitude [$2.25(e^2)°$ S = $2(e^2)°°$ S] of the centre of all the squares. Since the diagonal of the fundamental square is 1 latitude degree, the interval ratio of the expanded square is $4.5e^2$:1 in relationship to the fundamental of the first octave. Note that 4.5 is 1/8 the size of 36 which is the number of degrees in a pentagram star point. The significance of this square is tremendous. What it means is that the architects are showing that the planet itself has been created by an Aum sound which contains the irrational number e!

The northeast side of an expanded square with a diagonal size of $76\sqrt{2}$ latitude degrees (Fig. 2.29b) comes within 1.62 km of the pentagram centre which fits the body shape of the Vitruvian Martian. The same square also comes within 1.57 km of the centre of the square which fits the body shape of the Vitruvian Martian. Since the diagonal of the fundamental square is 1 latitude degree, the interval ratio of the expanded square is $76\sqrt{2}$:1 in relationship to the fundamental of the first octave. Thus it would appear that the Martian architects considered the Martian body to have been created by an Aum sound which contains the irrational number $\sqrt{2}$. This is especially meaningful for the square which fits the body shape of the Vitruvian Martian, since the length of the diagonal of a square is $\sqrt{2}$ times the side length. A couple more examples of important expanded squares whose sizes incorporate irrational numbers will be shown in Chapter 9.

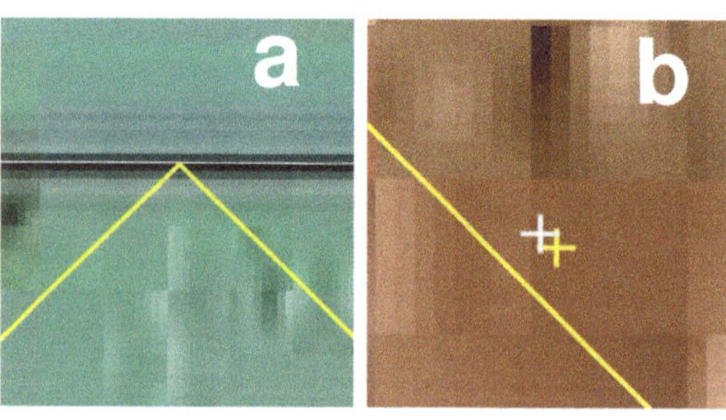

Fig. 2.29: *Image a: the northern vertex of an expanded square with an interval ratio of $4.5e^2$:1 sits on the equator. Image b: the northeast side of a square with an interval ratio of $76\sqrt{2}$:1 comes very close to the centres of the pentagram (lower cross) and square (upper cross) fitting the Vitruvian Martian. USGS Astrogeology.*

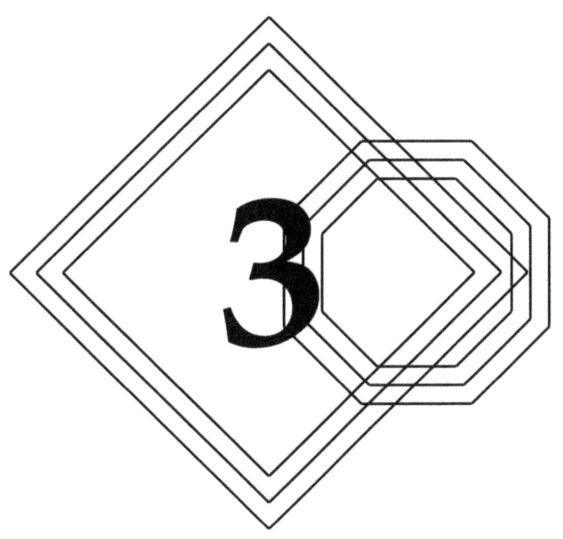

The 8-Sided Crater

After all of that work on the Aum Crater and its vast zone of influence, you would think that the analysis of this crater would be essentially complete. At last we can move on to discussing other things, right? Well.... what about that sizeable multi-sided crater lying within the east portion of the Aum Crater (Fig. 3.1)? Should it not be examined a little more closely to see if it has a larger story to tell? Is its presence in the Aum Crater more than just a fluke of a large impact creating a crater with linear sides? Just a quick look at its northeast side reveals the presence of a grey band lying within a bright coloured region. There also appears to be a light band of material in the dark northwest side of the crater wall. All of this suggests that, like the square, the multi-sided crater might be a template for the shape of a regular polygon which expands outwards to mark out harmonic intervals within and beyond the Aum Crater.

The first thing that had to be worked out for the multi-sided crater was the exact shape of the form it was meant to portray. At first I thought that the number of linear sides was 6. However, my attempts to fit a regular hexagon shape to the crater were not successful. Something was seriously wrong with the angle of the sides. It can be easily seen in Fig. 3.1 that the slope of the southwest side of a hexagon is too gentle to fit the southwest linear side of the crater. The interior angle of the southwest corner of the hexagon has to be larger in order to produce a good fit. Now the interior

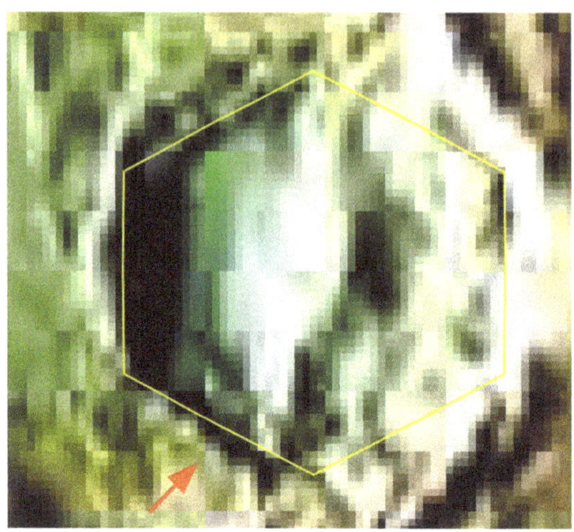

Fig. 3.1: *Multi-sided crater in the eastern half of the Aum Crater. Note how the slope of the southwest side of a hexagon is too gentle to fit the southwest side of the crater (red arrow). USGS Astrogeology.*

angles of a regular hexagon are 120°. A heptagon of 7 equal sides would produce angles of 128.57° but would be too irregular to fit a shape that appears to have an even number of sides. So the most logical shape to try next was the octagon, a polygon with 8 equal sides and 8 equal internal angles of 135°. Sure enough, that satisfied the slope of the southwest side of the crater. It turns out that the counterclockwise bearing angle of the slope is 45°. This angle fits the octagon because when two opposite sides of an octagon are aligned with latitude lines, the sides running at an angle to lines of latitude or longitude have a clockwise or counterclockwise bearing angle of 45°. It is interesting that this angle is exactly the same bearing angle as was found for the sides of the squares in the previous chapter. Now it remained to determine what might be the optimal location and size for a fundamental octagon shape to fit the crater.

As with the fundamental square used for the main part of the Aum Crater, I decided that the architects would have used meaningful coordinates for the centre of the octagon. The only meaningful longitude that lies in the central area of the interior crater is 208.1047° E because it translates into 60° E of the Dagger Peak Prime Meridian (DPPM). Looking closer at the 8-sided crater, it can be seen that there is a ridge which runs north-south in the centre of the crater. The line of longitude running through the midline of this ridge is none other than 60° E [DPPM]. This longitude is also exactly 3/8 of a degree of longitude from that selected for the fundamental square in the previous chapter. The denominator of this fraction would be an interesting way to refer to the 8 sides of an octagon. With all of this in mind, I decided on 60° E [DPPM] as the best choice of longitude for the centre of the fundamental octagon. Two meaningful latitudes are close to the central region of the internal crater.

These are $2(e^2)^{\circ\circ}$ S = 16.6254° S and $e\pi^\circ$ S [AMPL] = 16.6393° S. I chose the former since it was the same latitude that I used for the fundamental square in the previous chapter. It turned out that this was indeed the correct choice as will be seen at the end of the chapter where I present some very special octagons. The only thing remaining now for the choice of the fundamental octagon was its size.

There are 3 main parameters which could be used to size octagons: the side length, the width and the radius. I ended up choosing the radius since it gave the best results and it was analogous to the diameter of the square which is actually 2 radii of the square. I also reasoned that the size of the fundamental octagon and expanded octagons should be in units of latitude degrees like the series of expanded squares. It would be very appropriate then for the radius size of the fundamental octagon to be an even multiple or division of 1 latitude degree. The divisor or multiplier of 1 degree should be a power of the number 2 to allow for simple expansion of the octagon by octaves. A radius size of 1 latitude degree would be far too large for the fundamental octagon to fit within the boundaries of the interior crater. The same goes for 1/2 latitude degree. I therefore settled on a radius size of 1/4 degree for the fundamental octagon as it was small enough to easily fit inside the interior crater.

With all the parameters now selected, the next step was to see how the hypothesized fundamental octagon fit landmarks within the interior crater. The overlay of this octagon is shown in Fig. 3.2. The fit to structures within the crater is quite remarkable. Firstly it can be seen that the west side of the octagon aligns perfectly with the interface between black and green/grey-green areas which probably marks the region where the floor of the crater meets the crater wall. The southern half of the northwest side of the octagon also aligns with the interface between black and green areas, and perfectly follows the angle which the interface makes at the vertex where the west and northwest sides meet. The southern quarter of the northeast side aligns perfectly with the linear edge of a light grey area which probably marks a shallow depression. The east side of the octagon makes a perfect alignment with the edge of a dark grey area which probably marks the location of a deep depression. Notice how the angle between the northeast and east sides fits the angle of the grey regions perfectly. Finally, the southwest side of the octagon aligns with the interface between dark and lighter-coloured regions inside the crater. The goodness of fit of this octagon to crater structures suggests that it is well suited for the role of a fundamental octagon. This fundamental octagon can now be enlarged to test for the ability of expanded octagons to fit other landmarks within and beyond the crater in a manner similar to the expanded squares of the previous chapter.

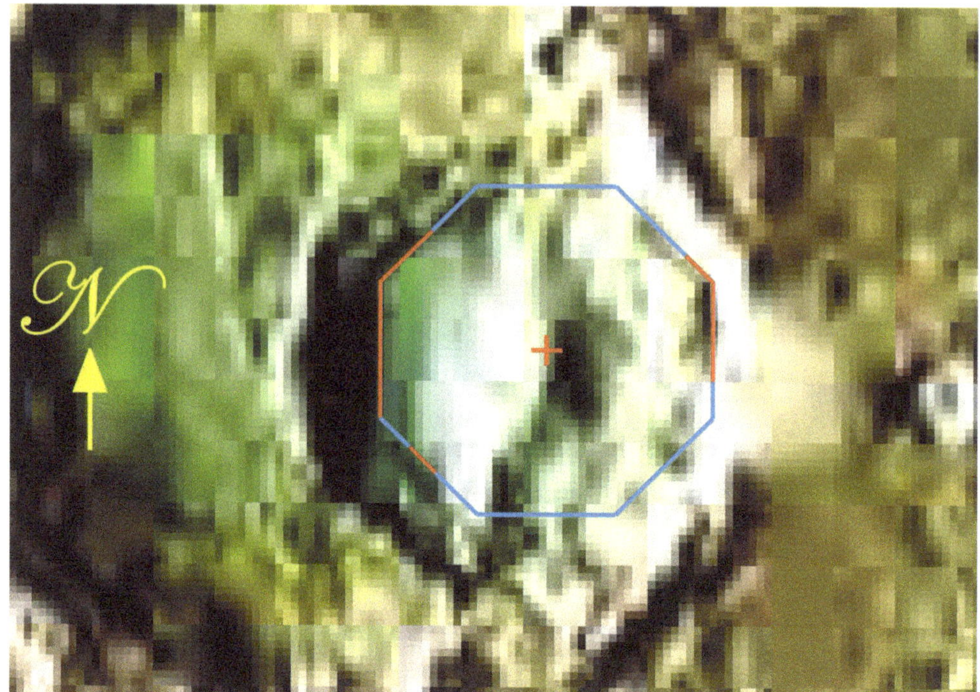

Fig. 3.2: *The fundamental octagon centred at the red cross (208.1047° E 16.6254° S). Red lines superimposed on the blue octagon outline demarcate regions of alignment with structures inside the multi-sided crater. Note how well the octagon follows the linear interface between black and green/grey-green areas on the west and northwest sides. The east side follows the linear edge of a dark grey area and the northeast side, the linear edge of a light grey area. See how well the angles are fit in these areas. USGS Astrogeology.*

Octagons in the First Octave

With the finding that interval values of expanded squares in the previous chapter were all harmonics of a fundamental square size analogous to the way music intervals are created, it was highly possible that expanded octagons would behave similarly. Harmonics are simply integer multiples of the fundamental frequency. Suppose we tune middle C on a piano to 256 cycles per second (Hz). The second harmonic would be 2 x 256 = 512 Hz and would be the C note one octave higher. It can be brought back into the first octave by dividing by 2 which would then return it to the fundamental frequency of 256 Hz. The third harmonic would be 3 x 256 = 768 Hz and would be the G key in the second octave above middle C. To bring this note into the first octave, we would have to divide it by 2, leaving us with a frequency of 384 Hz which would be the G note in the first octave above middle C. The fourth harmonic would be 4 x 256 = 1024 Hz which would be C in the third octave above middle C. It can be brought back into the first octave by dividing by 4, leaving us with the fundamental frequency of 256 Hz. The fifth harmonic would be 5 x 256 = 1280 Hz which would be the E note in the third octave above middle C. To bring it into the second octave, we would have to divide by 2, giving us a frequency of 640 Hz which would be the E note in the second octave. To bring it into the first octave we would divide the 1280 Hz frequency by 4 to get 320 Hz which would be the E note in the first octave. Thus harmonic interval ratios between frequencies are simply pure integers divided by a number large enough to bring them into the range of the desired octave. The divisor must always be a power of 2.

If we assume that the radius of an octagon is an analogue of sound frequency, then the fundamental octagon in Fig. 3.2 represents the fundamental frequency and all expanded octagons represent harmonic frequencies above this. To test whether expanded octagons are aligned to structures according to harmonic intervals, I examined octagons which were expanded by the harmonic interval ratios of 3:2, 5:4, 7:4, 9:8, etc. These are consecutive odd-numbered harmonics which are brought into the first octave by dividing by the power of 2 which would give the interval ratio a value in the range of 1 to <2. Even numbered harmonics become odd numbered harmonics when divided by sufficiently high powers of 2, so they would simply be duplicates of the odd numbered harmonics. Thus the 6th harmonic becomes 3 when divided by 2, the 80th harmonic becomes 5 when divided by 16, etc.

Fig. 3.3 shows the first set of expanded octagons for the first octave. The octagon interval ratios and radius sizes (in latitude degrees) are listed in the caption. The interval ratios of all of these octagons are harmonics ranging from 1 to 15. The harmonic values are brought into the first octave by dividing by the appropriate power of 2. Superimposed on each octagon are short red lines indicating alignments with crater features. As we proceed from the smallest to largest octagon, I will point out important alignments for each octagon. The smallest octagon is the fundamental octagon which was discussed above. It has an interval ratio of 1:1 so it is the first harmonic. Its alignments have already been discussed. The second octagon has an interval ratio of 9:8. Its northwest side aligns with the southeast edge of a band of lighter pixels. The third octagon has an interval ratio of 5:4. Its northwest side aligns with the northwest edge of the same band of lighter pixels which aligned with the northwest side of the previous octagon. The fourth octagon has an interval ratio of 11:8 and is coloured yellow for easy identification. Its northwest side aligns with the northwest edge of a band of dark grey pixels. Its southeast side aligns with a linear section of the Aum Crater perimeter. The fifth octagon has an interval ratio of 3:2. This is a very

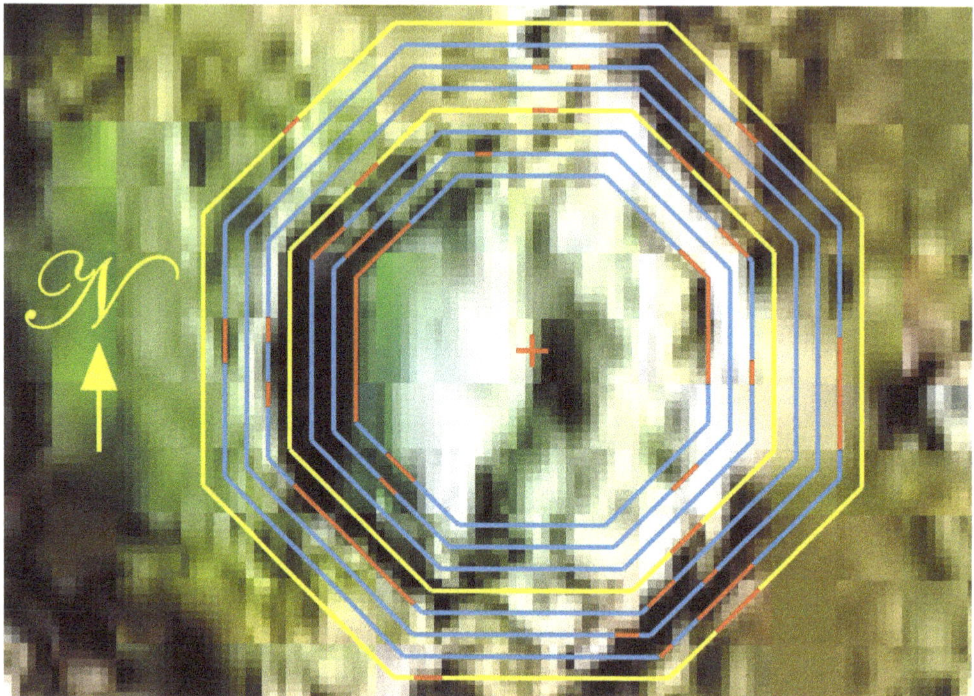

Fig. 3.3: *First set of octagons in the first octave. From smallest to largest, the interval ratios and radii sizes (latitude degrees) are (1) 1:1 and 0.2500 [fundamental octagon] (2) 9:8 and 0.2813 [9th harmonic] (3) 5:4 and 0.3125 [5th harmonic] (4) 11:8 and 0.3438 [11th harmonic] (5) 3:2 and 0.3750 [3rd harmonic] (6) 13:8 and 0.4063 [13th harmonic] (7) 7:4 and 0.4375 [7th harmonic] and (8) 15:8 and 0.4688 [15th harmonic]. Alignments are marked by short red lines superimposed on each octagon. USGS Astrogeology.*

important interval in music since it can be tuned by the human ear with a high degree of precision and was used to tune pianos in the era before electronic equipment was available. The southwest side of this octagon makes a very important alignment with a long linear portion of the perimeter of the interior crater. The sixth octagon has an interval ratio of 13:8. Its southeast side aligns with the southeast corner of a brown coloured area which seems to emanate from the Aum Crater's perimeter. The south side of the octagon aligns with a step in the southern perimeter of the Aum Crater. The 7th octagon has an interval ratio of 7:4. Its east side aligns to the long east side of a very light brown-coloured region. The southeast side of the octagon aligns in a couple of places to the outside edge of the southeast trench bordering the perimeter of the Aum Crater. The 8th and final octagon (yellow) has an interval ratio of 15:8. Its southeast side aligns with the southeast edge of a structure bordering the trench which is adjacent to the southeast side of the Aum Crater. The south side of the octagon aligns with a short linear section of the Aum Crater perimeter.

The multitude and quality of alignments to expanded octagons in Fig. 3.3 thus attest to the high probability that the octagon-shaped crater within the Aum Crater is intended to portray the primordial Aum sound by radiating octagons of increasing size over the Martian landscape in a manner similar to the squares in the previous chapter. The octagons so far are all simple harmonic multiples of the fundamental octagon. But before examining expanded octagons in higher octaves, there are still a few octagons which should be examined in the first octave. These are shown in Fig. 3.4 and consist of octagons with high numbered harmonic interval values. It would have been too crowded to show them in the previous figure. The smallest of these is an octagon with an interval ratio of 17:16. Its northwest side aligns with the interface between black and light-coloured bands which have a clockwise bearing angle of 45°. The northeast side of the octagon aligns with the interface between grey and white bands having an overall bearing angle of 45° in the counterclockwise direction. Going from small to large, the second octagon has an interval ratio of 19:16. Its northwest side aligns with northwest side of the light-coloured region which aligns with the previous octagon. The third octagon (yellow) has an interval ratio of 21:16. Its south side aligns with the linear interface between white and grey coloured areas. Its southwest side aligns with the interface between light grey areas and a black band whose bearing angle is 45° in the counterclockwise direction. The 4th octagon has an interval ratio of 23:16. It aligns in a major way with 2 long linear sections of the west perimeter of the interior crater. Its northwest side aligns with a short linear section of the interior crater perimeter. The 5th octagon has an interval size of 25:16. Its west side aligns with white bands which travel in a north-south direction. The 6th octagon (yellow) has an interval ratio of 27:16. Its west side aligns with the linear interface between light and dark coloured regions running in a north-south direction. Its east side aligns with the interface between light and dark brown regions travelling in a north-south direction. The 7th octagon has an interval ratio of 29:16. Its east side aligns with the east linear side of a dark brown region. The west side of the octagon aligns with the west linear edges of 2 grey areas. The 8th and final octagon has an interval ratio of 63:32 and a radius of 0.4922 latitude degrees. I chose this size rather than the 31st harmonic since it made a much better fit to the landforms. Its north side aligns to the interfaces between light and dark regions in 3 locations. Its east side makes a major alignment with the west side of the perimeter of a crater lying east of the Aum Crater.

Figs. 3.3 and 3.4 show that there are alignments for octagons having

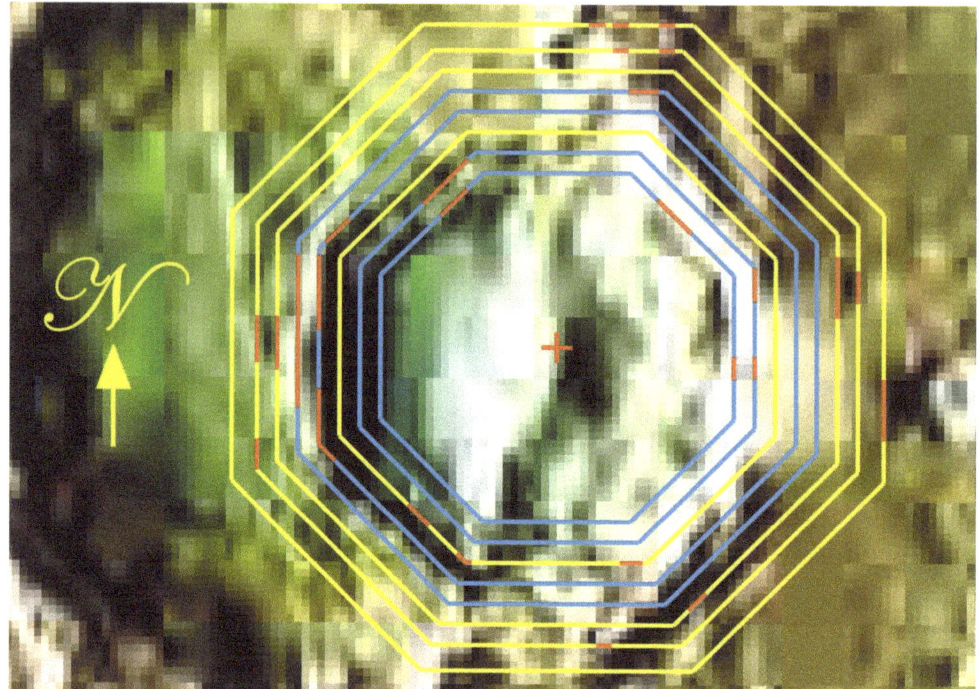

Fig. 3.4: *Octagons having high-numbered harmonic intervals in the first octave. From smallest to largest octagon, the interval ratios and radii sizes (latitude degrees) are (1) 17:16 and 0.2656 [17th harmonic] (2) 19:16 and 0.2969 [19th harmonic] (3) 21:16 and 0.3281 [21st harmonic] (4) 23:16 and 0.3594 [23rd harmonic] (5) 25:16 and 0.3906 [25th harmonic] (6) 27:16 and 0.4219 [27th harmonic] (7) 29:16 and 0.4531 [29th harmonic] and (8) 63:32 and 0.4922 [63rd harmonic]. Alignments are marked by short red lines superimposed on each octagon. USGS Astrogeology.*

interval ratios containing all the odd harmonic numbers between 1 and 29. The even numbered harmonics all revert to odd numbered harmonics when divided by the power of 2 required to bring them into the first octave. For instance, the 20th harmonic has to be divided by 16 in order to bring it into the first octave. The ratio 20:16 is the same as 5:4, so the 20th harmonic reverts to the 5th harmonic when brought into the first octave. The 31st harmonic did not appear to be aligned to any landform, but the slightly larger 63rd harmonic was very well aligned. The reason for this may be the presence of alignment sites below the resolution of the map for the 31st and other harmonic values. Whatever the case may be, the alignment of so many octagons which are sized to sequential harmonic intervals in the first octave offers huge support for the theory that the octagon-shaped crater was designed to represent the Aum sound by radiating out multiple octagons sized according to harmonic intervals which align to structures both within and outside the crater.

Second Octave Octagons

I decided to now express the interval ratios in octaves higher than the first octave by referring to the fundamental octagon from the first octave rather than to the fundamental of the current octave. This was done in order to make it easier to see all of the harmonic values that were found for the original octagon. The numerator now becomes the harmonic number with reference to the size of the fundamental octagon in the first octave. The denominator is 2 raised to the power necessary to bring the harmonic number (with reference to the fundamental of the first octave) into the range of the current octave. When an even numbered harmonic (with reference to the fundamental of the first octave) sits in the current octave with a denominator of 1, it cannot be further subdivided and simply remains in the current octave. Hence, even numbered harmonics are now valid as long as the denominator is 1. The first even harmonic to show up is the fundamental for the second octave since it is twice the size of the first octave fundamental and therefore has an interval ratio of 2:1.

Alignments were found for all expanded octagons in the second octave having interval ratios for all the odd harmonics from 3 to 31 and for the even harmonic of 2. These are shown in Fig. 3.5 for the east side of the octagons and in Fig. 3.6 for the west side of the octagons. In Fig. 3.5 it can be seen that most of the octagons align to the linear edges of structures inside the elongated crater which lies to the east of the Aum Crater. The 17th harmonic octagon (the 2nd octagon from the left) aligns to a length of bright pixels just north of the western end of the eastern crater. These pixels are mostly covered by the width of the octagon line. The 11th harmonic (the 8th octagon from the left) aligns to the linear eastern edge of a brown region north of the eastern crater, and also to the long linear western edge of a brown structure inside the crater.

The west sides of the octagons in the second octave are shown in Fig. 3.6. Many of the octagons are shown to align with linear edges of structures inside the Aum Crater as well as with its perimeter. Octagons which did not show any alignment in Fig. 3.5 do show at least one alignment in Fig. 3.6. The northeast side of the smallest octagon (2nd harmonic) shows an alignment with the northeast corner of a large structure lying beyond the Aum Crater. This was out of the picture in Fig. 3.5. The southwest side of the 3rd harmonic (9th concentric octagon) aligns to a long linear section of the Aum Crater perimeter. The southwest side of the 13th harmonic (11th concentric octagon, blue coloured) also aligns to a long linear section of the Aum Crater perimeter, and its west side aligns to a structure inside the Aum Crater. The west side of the 29th harmonic (3rd octagon from the left) aligns to a long

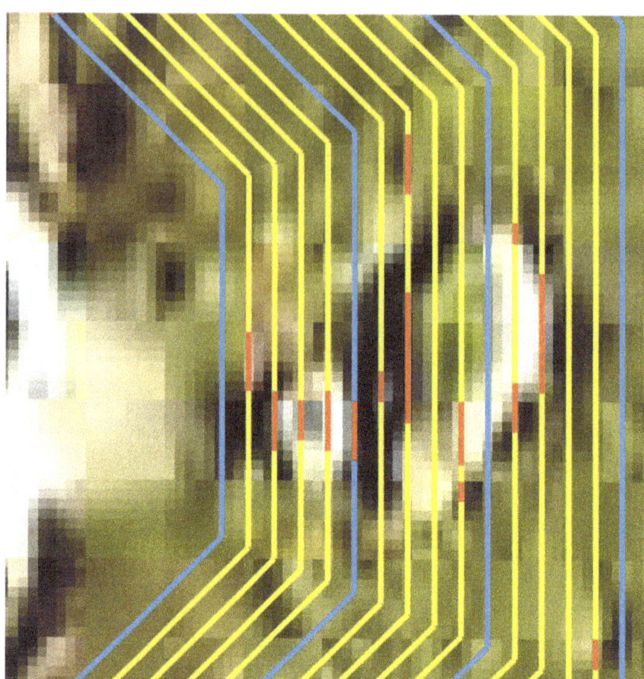

Fig. 3.5: *East side of octagons in the 2nd octave. Interval ratios and radii (latitude degrees) from smallest to largest are:*
(1) 2:1 = 2.0000, 0.5000
(2) 17:8 = 2.1250, 0.5313
(3) 9:4 = 2.2500, 0.5625
(4) 19:8 = 2.3750, 0.5938
(5) 5:2 = 2.5000, 0.6250
(6) 21:8 = 2.6250, 0.6563
(7) 11:4 = 2.7500, 0.6875
(8) 23:8 = 2.8750, 0.7188
(9) 3:1 = 3.0000, 0.7500
(10) 25:8 = 3.1250, 0.7813
(11) 13:4 = 3.2500, 0.8125
(12) 27:8 = 3.3750, 0.8438
(13) 7:2 = 3.5000, 0.8750
(14) 29:8 = 3.6250, 0.9063
(15) 15:4 = 3.7500, 0.9375
(16) 31:8 = 3.8750, 0.9688.
Short red lines mark alignments. USGS Astrogeology.

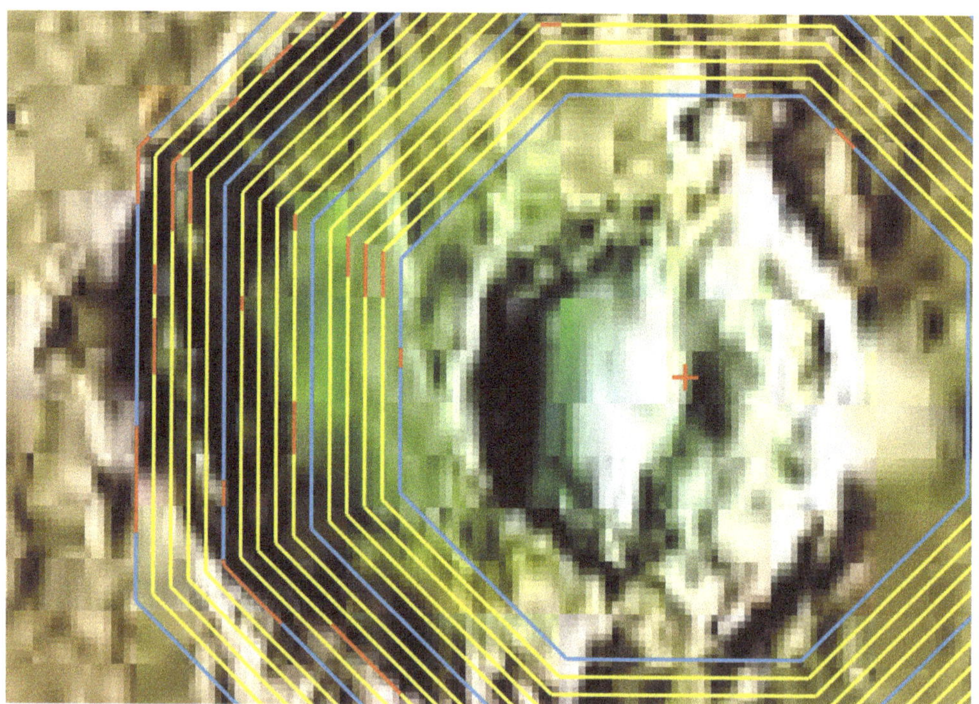

Fig. 3.6: *West side of octagons in the 2nd octave. Interval ratios and radii sizes are given in Fig. 3.5. Short red lines mark alignments. USGS Astrogeology.*

linear structure inside the Aum Crater near the northwest vertex, and its northwest side aligns to a couple of sections of the northwest perimeter of the crater. The west side of the 31st harmonic (first octagon on the left, blue coloured) aligns with a straight line section of the western perimeter of the Aum Crater and with the east side of a dark area south of this region. The longitude of the west side of this octagon is only 6.5 seconds of degree less than 60° E (Elysium Mons PM). The northwest vertex of the octagon aligns with the crater perimeter and with the west side of a small raised area on the lip of the crater perimeter.

Third Octave Octagons

A total of 32 expanded octagons were found to align to various landforms in the third octave. These produced harmonic interval ratios (with the fundamental octagon of the first octave) which ranged from 4 to 63 inclusively. The west side of the Aum Crater and the region west of here produced most of the alignments. Fig. 3.7 shows alignment examples for all of the expanded octagons except one. The interval ratio and radius size for each octagon are listed in the caption to Fig. 3.7. The smallest 6 octagons are aligned to steps in the west perimeter of the Aum Crater in the region where the perimeter protrudes outwards. Just to the north and slightly west of this is a tiny crater which produces alignments to the 7th and 8th octagons. To the west of this small crater there is a crater approximately 15 km in diameter. The 15th to 25th octagons align to steps in this crater's perimeter/apron and to the linear edges of formations within the crater. To the south and slightly west of this crater lies another small crater which aligns to the 27th and 28th octagons. All the remaining octagons align to the linear edges of landforms which travel in a north-south direction except for the 29th octagon which aligns to a step in the perimeter of a mid-sized crater south of the Aum Crater (not shown).

Eight octagons were aligned to sites with sides having bearing angles of 45° in the clockwise or counterclockwise direction. These are shown only for the 1st and 3rd octagons in Fig. 3.7. A total of 14 octagons aligned to landforms with their north sides, 11 octagons with their east sides and 11 octagons with their south sides. These are not shown.

Several sites have linear edges which could align to octagons with interval ratios that would fit in between those found for the octagons in Fig. 3.7. I did not pursue this further due to limitations of time and space.

Fourth Octave Octagons

Remarkably, alignments occur for the northeast sides of all possible

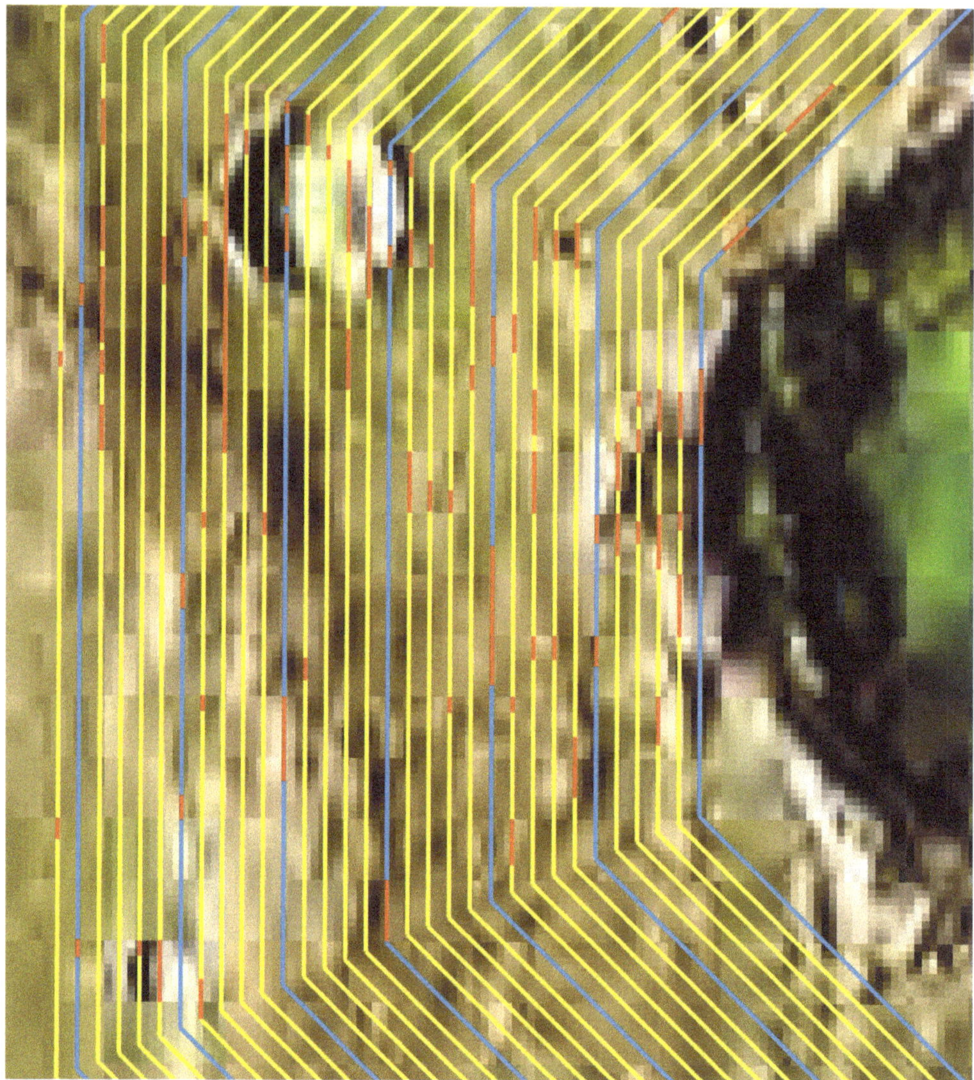

Fig. 3.7: *West side of octagons having harmonic intervals in the 3rd octave. Interval ratios (using the 1st octave fundamental) & radii (latitude degrees) from right to left are:*

(1) 4:1 = 4.0000, 1.0000 (2) 33:8 = 4.1250, 1.0315 (3) 17:4 = 4.2500, 1.0625
(4) 35:8 = 4.3750, 1.0938 (5) 9:2 = 4.5000, 1.1250 (6) 37:8 = 4.6250, 1.1563
(7) 19:4 = 4.7500, 1.1875 (8) 39:8 = 4.8750, 1.2188 (9) 5:1 = 5.0000, 1.2500
(10) 41:8 = 5.1250, 1.2813 (11) 21:4 = 5.2500, 1.3125 (12) 43:8 = 5.3750, 1.3438
(13) 11:2 = 5.5000, 1.3750 (14) 45:8 = 5.6250, 1.4063 (15) 23:4 = 5.7500, 1.4375
(16) 47:8 = 5.8750, 1.4688 (17) 6:1 = 6.0000, 1.5000 (18) 49:8 = 6.1250, 1.5313
(19) 25:4 = 6.2500, 1.5625 (20) 51:8 = 6.3750, 1.5938 (21) 13:2 = 6.5000, 1.6250
(22) 53:8 = 6.6250, 1.6563 (23) 27:4 = 6.7500, 1.6875 (24) 55:8 = 6.8750, 1.7188
(25) 7:1 = 7.0000, 1.7500 (26) 57:8 = 7.1250, 1.7813 (27) 29:4 = 7.2500, 1.8125
(28) 59:8 = 7.3750, 1.8438 (29) 15:2 = 7.5000, 1.8750 (30) 61:8 = 7.6250, 1.9063
(31) 31:4 = 7.7500, 1.9375 (32) 63:8 = 7.8750, 1.9688. USGS Astrogeology.

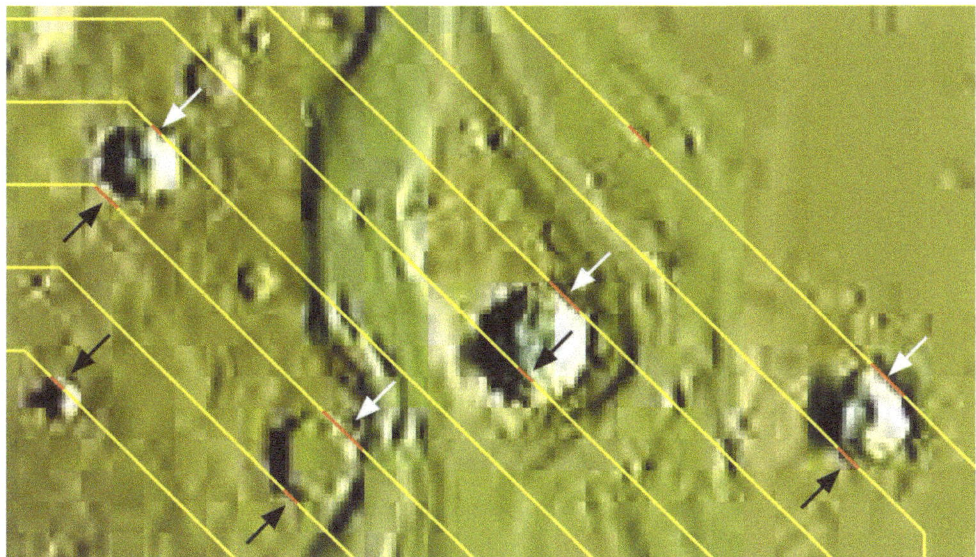

Fig. 3.8: *Northeast sides of octagons having harmonic intervals in the fourth octave. From smallest to largest, interval ratios and radii sizes (in latitude degrees) are (1) 8:1, 2.00 (2) 9:1, 2.25 (3) 10:1, 2.50 (4) 11:1, 2.75 (5) 12:1, 3.00 (6) 13:1, 3.25 (7) 14:1, 3.50 and (8) 15:1, 3.75. Arrows and short red lines mark alignments to linear sections of crater perimeters and linear edges of structures within craters. USGS Astrogeology.*

octagons in the fourth octave whose interval ratios have a denominator of 1 (Fig. 3.8). Since their denominators are 1, the harmonic intervals (with respect to the fundamental octagon in the first octave) reside in the current octave without having to be divided down from higher octaves. These octagons form a set of consecutive harmonic numbers in the fourth octave from 8 to 15. The smallest octagon (interval ratio = 8:1) aligns with the bottom edge of a crater wall of an arrow-shaped crater which points due east. The 3rd octagon from the left has an interval ratio of 10:1 and it aligns with the southwest edge of the crater apron of a crater north of the arrow-shaped crater, and also aligns with the northeastern perimeter edge for a crater lying to the east of the arrow-shaped crater. The 5th octagon from the left has an interval ratio of 12:1, and it aligns to the bottom edge of the crater wall of a larger crater further east and north. This edge is marked by the interface between a dark black/grey area and a light grey area inside the crater. The remaining octagons all make alignments to linear sections of crater perimeters.

Fig. 3.9 shows 2 arrow-shaped craters which align to octagons which have high numbered harmonic interval ratios. The smallest octagon in Fig. 3.9a actually lies within the 3rd octave rather than in the 4th octave. It aligns with the interface between light and dark areas near the southern

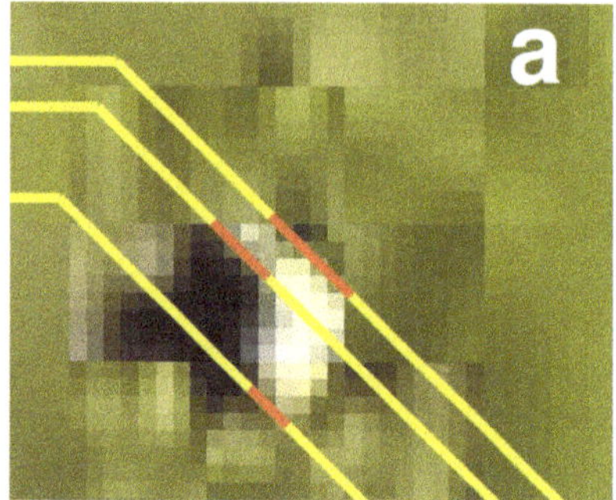

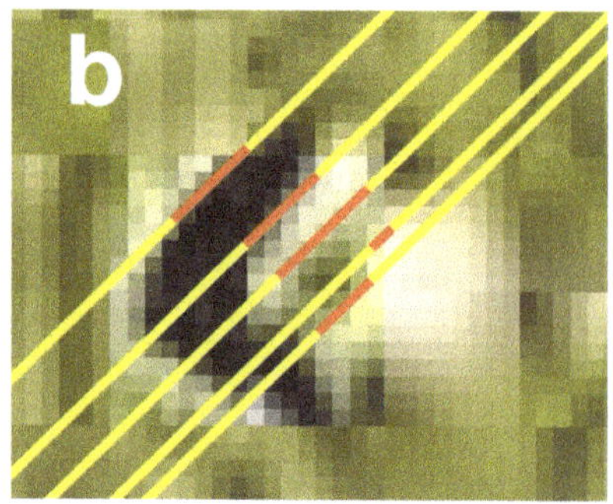

Fig. 3.9: *(a) Northeast sides of octagons from the 3rd (lowest octagon) and 4th (2 upper octagons) octaves. From left to right, interval ratios and radii sizes are 31:4 and 1.9375, 8:1 and 2.0000, and 65:8 and 2.0313. (b) Northwest sides of octagons having harmonic intervals in the 4th octave. From right to left, interval ratios and radii sizes are 157:16 and 2.4531, 79:8 and 2.4688, 10:1 and 2.5000, 81:8 and 2.5313, and 165:16 and 2.5781. Short red lines mark alignments. USGS Astrogeology.*

perimeter of the arrow-shaped crater. The 2nd octagon aligns with the interface between light and dark areas near the north side of the crater. The 3rd octagon aligns with the northern side of the arrowhead perimeter of the arrow-shaped crater. The smallest octagon (the rightmost) in Fig. 3.9b aligns with southwest linear edge of what appears to be a depression within the crater. The next octagon aligns with the edge of the same depression but at an area slightly to the northeast where it has a reduced width. The 3rd octagon aligns with the northwest linear edge of the depression. The 4th octagon aligns with the northwest edge of an arrow-shaped structure within the crater. The 5th octagon aligns with the northwest crater perimeter which forms the upper edge of the arrowhead of the arrow-shaped crater.

Alignments with another arrow-shaped crater are shown in Fig. 3.10a. The southeast side of the upper octagon aligns with the northern side of the arrowhead formed by the crater perimeter. The southeast side of the lower octagon aligns with the northwest side of an arrow shape interior to the crater. The northern sides of some of the largest octagons in the fourth octave align with steps in the perimeter, or with the edges of structures inside of a crater lying to the north of the Aum Crater (Fig. 3.10b). The 1st and 3rd octagons from the bottom align with the edges of what appears to be a depression in the centre of the crater.

The sizes and interval ratios of the octagons in Figs. 3.8 to 3.10 are provided in the captions to the figures. From these, it can be seen that in the fourth octave, the simple harmonics ranging from 8 to 15 are present as well as high numbered harmonics up to 165.

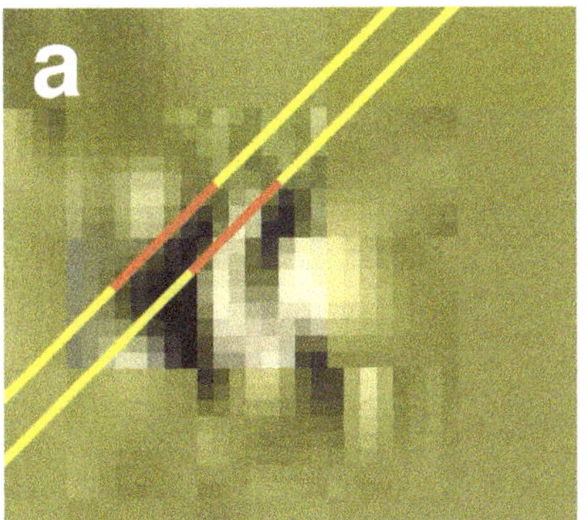

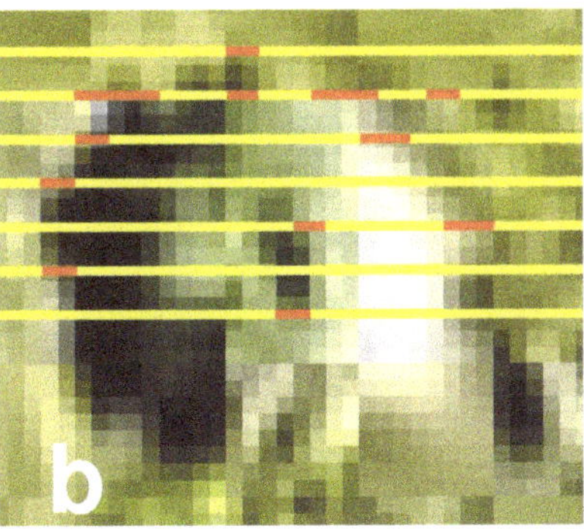

Fig. 3.10: *(a) Southeast sides of octagons having harmonic intervals in the 4th octave. From left to right, interval ratios and radii sizes are 67:8 and 2.0938, and 17:2 and 2.1250. (b) North sides of octagons from the 4th octave. From bottom to top, interval ratios and radii sizes are: (1) 15:1, 3.7500 (2) 121:8, 3.7813 (3) 61:4, 3.8125 (4) 123:8, 3.8438 (5) 31:2, 3.8750 (6) 125:8, 3.9063 and (7) 63:4, 3.9375. Short red lines mark alignments to linear elements. USGS Astrogeology.*

Fifth Octave Octagons

Like the fourth octave, alignments were found in the 5th octave for all the possible octagons having simple harmonic interval ratios with 1 in the denominator (Fig. 3.11). They form a set of consecutive harmonic numbers in the 5th octave from 16 to 31. The west side of the octagon for the 16th harmonic interval (the fundamental interval for the 5th octave) aligns with the perimeter of a substantial crater to the northwest of the Aum Crater (image a). Just to the northwest of this, the northwest side of the octagon for the 17th harmonic aligns with the perimeter of the giant Burton Crater (image b). The bottom edge of the Burton Crater wall aligns with the northwest side of the octagon for the 18th harmonic (image c). The south side of the octagon for the 19th harmonic (image d) aligns with the northern edge of what appears to a depression in the centre of a crater south of the Aum Crater. The north side of the octagon for the 20th harmonic aligns with a short linear section of the northern perimeter of a crater to the north of the Aum Crater (image e). In image f, the linear junction between 2 sizeable craters aligns to the 21st harmonic octagon just southeast of the Aum Crater. The west side of the 22nd harmonic octagon aligns with the linear west edges of 2 structures within a sizeable crater lying to the west of the Aum Crater (image g). The west side of the octagon for the 23rd harmonic (image h) aligns to the linear edges of structures within a crater just south of the crater in image g. Linear sections of the base of the northern wall of the giant Cobres Crater align with the northern side of the 24th harmonic octagon (image i). The east

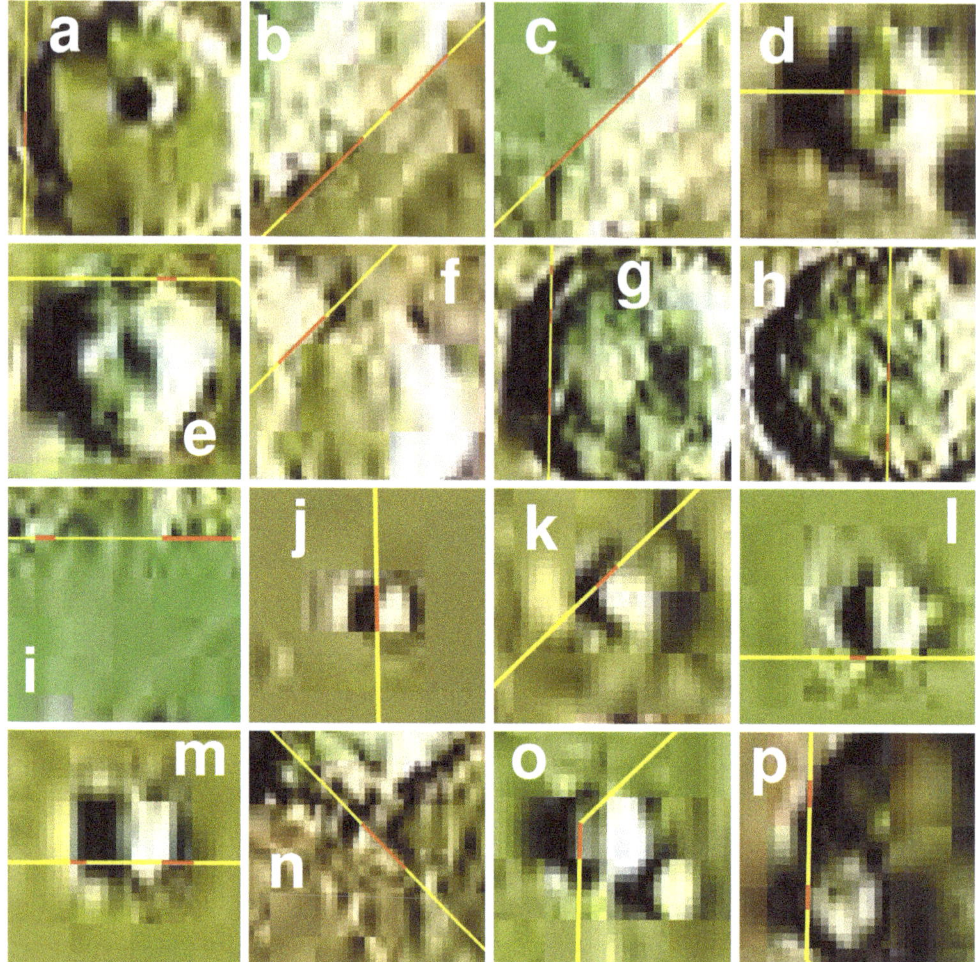

Fig. 3.11: *Octagons having harmonic intervals in the 5th octave. Interval ratios (using the 1st octave fundamental) and radii (latitude degrees) are:*
(a) 16:1, 4.00 (b) 17:1, 4.25 (c) 18:1, 4.50 (d) 19:1, 4.75 (e) 20:1, 5.00 (f) 21:1, 5.25 (g) 22:1, 5.50 (h) 23:1, 5.75 (i) 24:1, 6.00 (j) 25:1, 6.25 (k) 26:1, 6.50 (l) 27:1, 6.75 (m) 28:1, 7.00 (n) 29:1, 7.25 (o) 30:1, 7.50 (p) 31:1, 7.75. Red lines mark regions of alignment to octagons. USGS Astrogeology.

side of 25th harmonic octagon aligns with the linear interface between the white and dark halves of a small crater lying to the east of the Aum Crater (image j). The northwest side of the 26th harmonic octagon aligns with the linear northwest edge of a blunt arrow shape in the interior of a crater (image k). The north side of the 27th harmonic octagon aligns with a step in the perimeter of a crater lying to the north of the Aum Crater (image l). Steps in the perimeter of a small crater south of the Aum Crater align with the south side of the 28th harmonic octagon (image m). This crater looks artificial due the presence of multiple right angles in its perimeter. The southwest side of the 29th harmonic octagon aligns with the linear edge of a structure south of the perimeter of a medium-sized crater (image n). The west side of the 30th harmonic octagon aligns with the linear interface between dark and grey regions inside a small crater (image o). The western side of the 31st harmonic octagon aligns with the western side of a structure inside the western perimeter of the large Comas Sola Crater (image p). This side also aligns with a short linear section of the crater perimeter just to the north of the interior structure.

There are several expanded octagons having more complex harmonic interval ratios in the 5th octave (Fig. 3.12). The octagon in image a has an interval ratio of 281:16. Its northeast side aligns with the southern edge of the arrowhead of a crater which has the shape of a blunted arrow pointing due west. The southeast side of the octagon (interval ratio = 623:32) in image b aligns with a linear section of the crater perimeter shared by a small crater and a much larger crater which borders it to the south. In image c, the northeast side of an octagon with an interval ratio of 81:4 aligns with the linear northeast perimeter apron of a small crater. In image d, the northwest sides of 2 octagons are shown to align with the northern

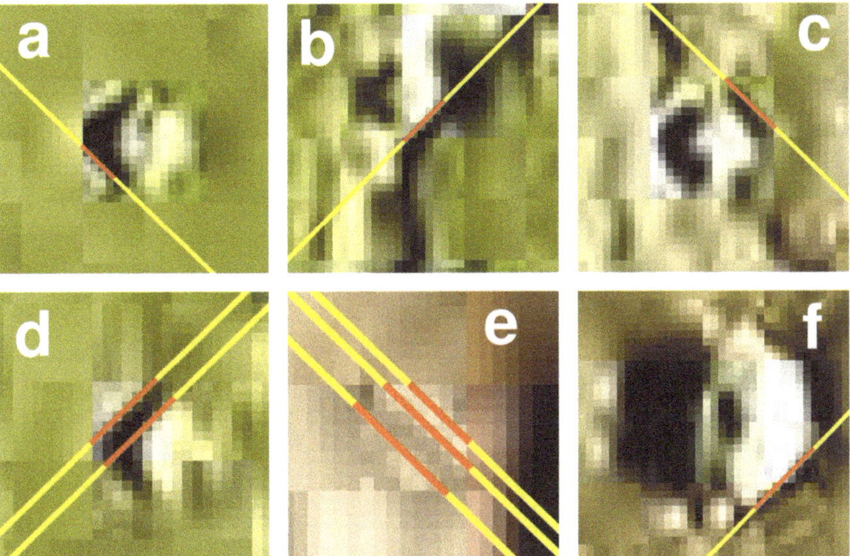

Fig. 3.12: *Octagons having more complex harmonic intervals in the 5th octave. Interval ratios (using the 1st octave fundamental) and radii (latitude degrees) are:*
(a) 281:16 = 17.5625, 4.3906 (b) 623:32 = 19.4688, 4.8672
(c) 81:4 = 20.2500, 5.0625 (d) [lower] 365:16 = 22.8125, 5.7031 and
[upper] 367:16 = 22.9375, 5.7344 (e) [lower] 365:16 = 22.8125, 5.7031,
[middle] 367:16 = 22.9375, 5.7344 and [upper] 23:1 = 23.0000, 5.7500
(f) 475:16 = 29.6875, 7.4219. Red lines mark alignments. USGS Astrogeology.

sides of arrow shapes, one forming an arrow interior to the crater, and the other, the perimeter of the crater. The northeast sides of the same octagons in image d are shown in image e (the 2 lowest octagons) where they bracket a grey-coloured band of terrain. Northeast of these is a third octagon which brackets a lighter band. This type of banding suggests that the region of the 5th octave may contain many more landforms or structures below map resolution which align to other octagons having high harmonic intervals. In image f, the southeast side of an octagon aligns with the linear southeast side of a crater.

Sixth Octave Octagons

There was no sign of a reduction in the existence of landforms which align to expanded octagons in the 6th octave. Linear landforms exist which align to all octagons having harmonic interval ratios with 1 in the denominator ranging from 32:1 to 63:1. Many of these are shown in Fig. 3.13. In image a, the southeast side of the fundamental octagon for the fifth octave aligns with a long linear stretch of the perimeter of a large crater. In image b, the south side of an octagon aligns with a linear step in the eastern perimeter of a crater. The southwest side of the crater in image c aligns with the northeast side of an octagon. A linear section of the northern perimeter of the crater in image d aligns with the north side of an octagon. In image e, the west side of the central peak of a crater aligns with the west side of an octagon. The north side of the octagon in image f aligns with the linear north side of a very small crater. The west side of the octagon in image g aligns with the base of the crater wall of the more northern crater and with the western edge of a channel which cuts through the northern perimeter of the more southern crater. A linear section of the northern perimeter of the crater in image h aligns with the northern side of an octagon. In image i, a linear section of the southeast perimeter of a crater aligns with the northwest side of an octagon. The west side of an octagon aligns with the linear east side of a small crater in image j. In image k, the southwest side of an octagon aligns with the linear outer edge of the lip of a large crater. The north and northeast sides of the octagon in image l aligns with several linear edges found in the Senus Vallis region of Mars. The west side of an octagon aligns with the long linear edge of the western perimeter of the crater shown in image m. In image n, the northeast side of an octagon aligns with the linear southwest side of a crater. The southwest side of the octagon in image o aligns with the arrowhead of an arrow-shaped crater pointing due east. In image p, the southwest side of an octagon aligns with the central peak and with a linear section of the northern perimeter of an irregularly

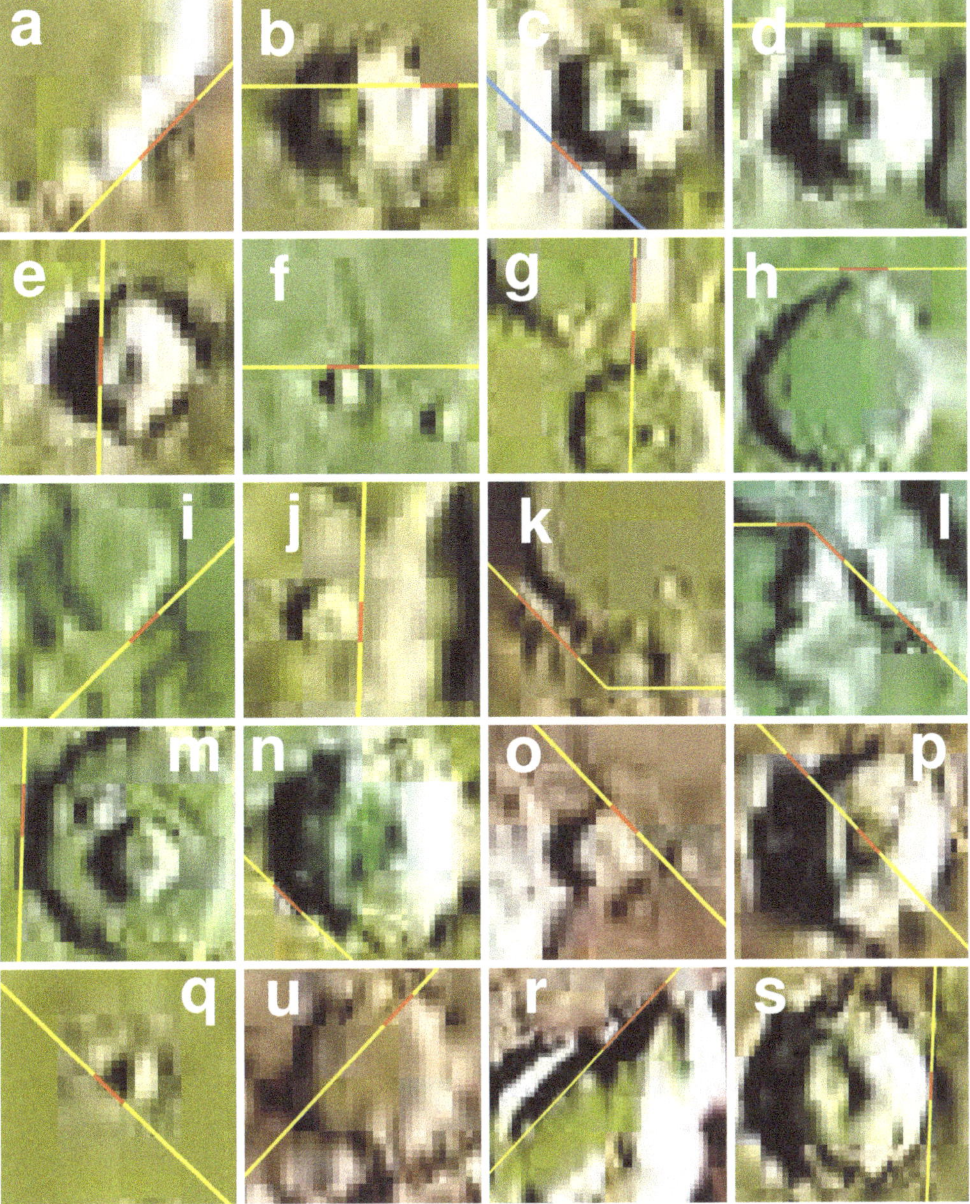

Fig. 3.13: *Octagons having simple harmonic intervals in the 6th octave. Interval ratios (using the 1st octave fundamental) and radii (latitude degrees) are:*
(a) 32:1, 8.00 (b) 33:1, 8.25 (c) 34:1, 8.50 (d) 37:1, 9.25 (e) 38:1, 9.50
(f) 41:1, 10.25 (g) 42:1, 10.50 (h) 44:1, 11.00 (i) 45:1, 11.25 (j) 47:1, 11.75
(k) 49:1, 12.25 (l) 50:1, 12.50 (m) 51:1, 12.75 (n) 53:1, 13.25 (o) 54:1, 13.50
(p) 56:1, 14.00 (q) 57:1, 14.25 (u) 59:1, 14.75 (r) 62:1, 15.50 (s) 63:1, 15.75.
Red lines mark regions of alignment to octagons. USGS Astrogeology.

shaped crater. The southwest side of an arrow-shaped crater in image q is shown to align with the northeast side of an octagon. The southeast side of the octagon shown in image u aligns with the southernmost linear edge of a channel which cuts through the northern perimeter of a crater. A wide section of the Sirenum Fossae is shown in image r. A long linear section of its northern perimeter aligns perfectly with the southeast side of an octagon. A linear section of the east perimeter of the large crater in image s aligns with the west side of an octagon.

Examination of octagons with more complex harmonic intervals reveals the presence of bands of different coloured areas of terrain which fit high numbered interval ratios (Fig. 3.14). In image a, the southwest sides of 4 octagons are shown to bound the edges of 3 linear bands. Harmonics with higher numbers than these are shown for the 5 octagons in image b which bound 4 linear bands with their northwest sides. Another band is shown in image d which is bounded by the southwest sides of 2 octagons. The presence of such bands offers much support to the hypothesis that landforms exist below map resolution which align to many more octagons having high harmonic numbers. Also shown in Fig.

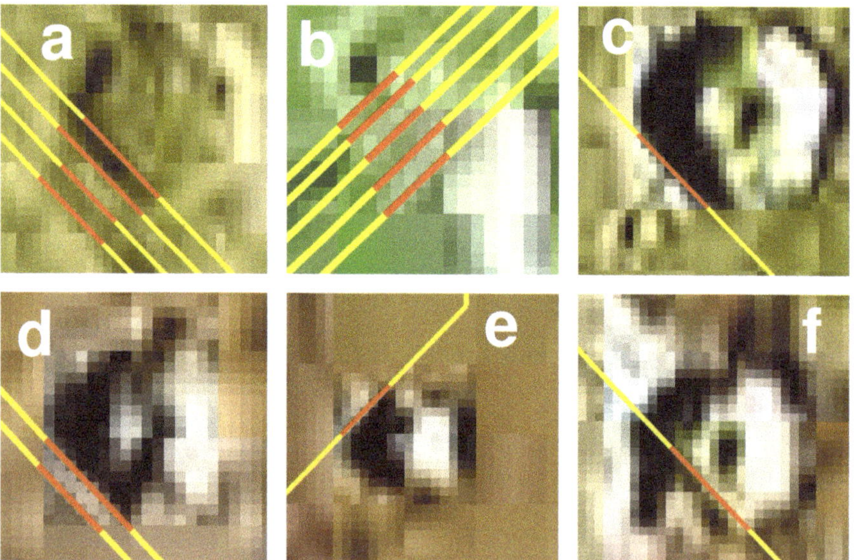

Fig. 3.14: *Octagons having more complex harmonic intervals in the 6th octave. Interval ratios (using the 1st octave fundamental) and radii (latitude degrees) are:*
(a) top to bottom - 287:8 & 8.9688, 36:1 & 9.0000, 289:8 & 9.0313, 145:4 & 9.0625
(b) bottom to top - 685:16 & 10.7031, 1373:32 & 10.7266, 43:1 & 10.7500, 1379:32 & 10.7734, 1381:32 & 10.7891
(c) 735:16 & 11.4844 (d) [upper] 747:16 & 11.6719 and [lower] 1497:32 & 11.6953
(e) 1527:32 & 11.9297 (f) 829:16 & 12.9531.
Red lines mark regions of alignment to octagons. USGS Astrogeology.

3.14 are octagons having high numbered interval ratios which fit long linear features of curiously shaped craters. In image c the southwest side of an octagon is shown to align with a linear section of the perimeter of a crater which has the shape of a wine glass complete with a stem than comes off at a 90° angle. The alignment is with the bottom of the wine glass shape. We have seen this crater before in Fig. 2.21c where the top of the glass shape is shown to align to the southwest edge of an expanded square. An arrow-shaped crater is shown in image e which points due west. The northern side of the arrowhead aligns with southeast side of an octagon. The southwest side of the octagon in Fig. 3.14f aligns with an arrow shape interior to the crater.

Seventh Octave Octagons

The large octagons in the 7th octave started to show notable distortion in their east and west sides as well as in their southern sides. The bearing angle of the east and west sides now exceeded 2 degrees so they deviated rather markedly from aligning with meridian lines. Also the bearing angles of the southeast and southwest sides were less than 43 degrees clockwise and counterclockwise respectively. However, the bearing angles of the northeast and northwest sides remained within 1 degree of 45° in octagons expanded up to the first 43% of the eighth octave. I therefore elected to trust only those alignments which occurred with the northeast, north and northwest sides of octagons expanded into the 7th and 8th octaves.

I checked for landform alignments to octagons having pure harmonic integer ratios with 1 in their denominators from 64:1 to 72:1 (Fig. 3.15). The octagon with an integer ratio of 64:1 is the fundamental octagon for the 7th octave (image a). Its northeast side aligns with a long linear section of the southwest perimeter of a large crater. The northwest side of an octagon with an integer ratio of 65:1 aligns with a linear anomaly in

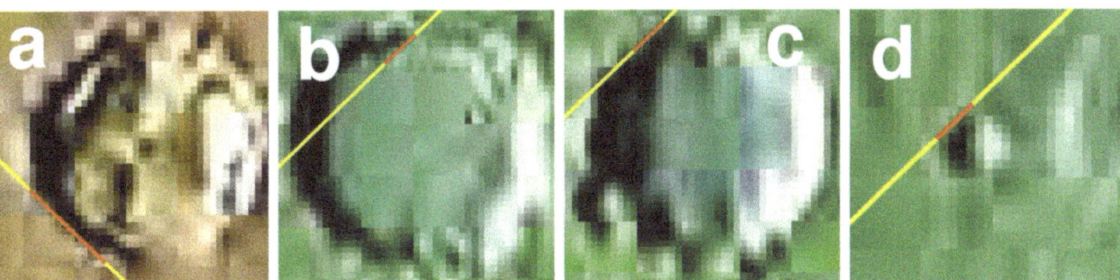

Fig. 3.15: *Octagons having harmonic intervals in the 7th octave. Interval ratios (using the 1st octave fundamental) and radii (latitude degrees) are:*
(a) 64:1, 16.00 (b) 65:1, 16.25 (c) 69:1, 17.25 (d) 72:1, 18.00.
Red lines mark regions of alignment to octagons. USGS Astrogeology.

the crater wall of a crater northwest of the Aum Crater (image b). The anomaly runs at a clockwise bearing angle of 45° along the inside of the wall from top to bottom. Just to the northwest of this crater is another crater of about the same size (image c). A linear section of its northern perimeter aligns with the northwest side of an octagon with an interval ratio of 69:1. The northwest side of an octagon with an interval ratio of 72:1 aligns with the linear northwest side of very small crater (image d).

I next observed that there were many other linear sections of crater perimeters that lay within the bounds of the 7th octave. These could be fit to octagons having high-numbered harmonics and are shown in Fig. 3.16 along with a large mesa whose northeast side is bounded by a long linear

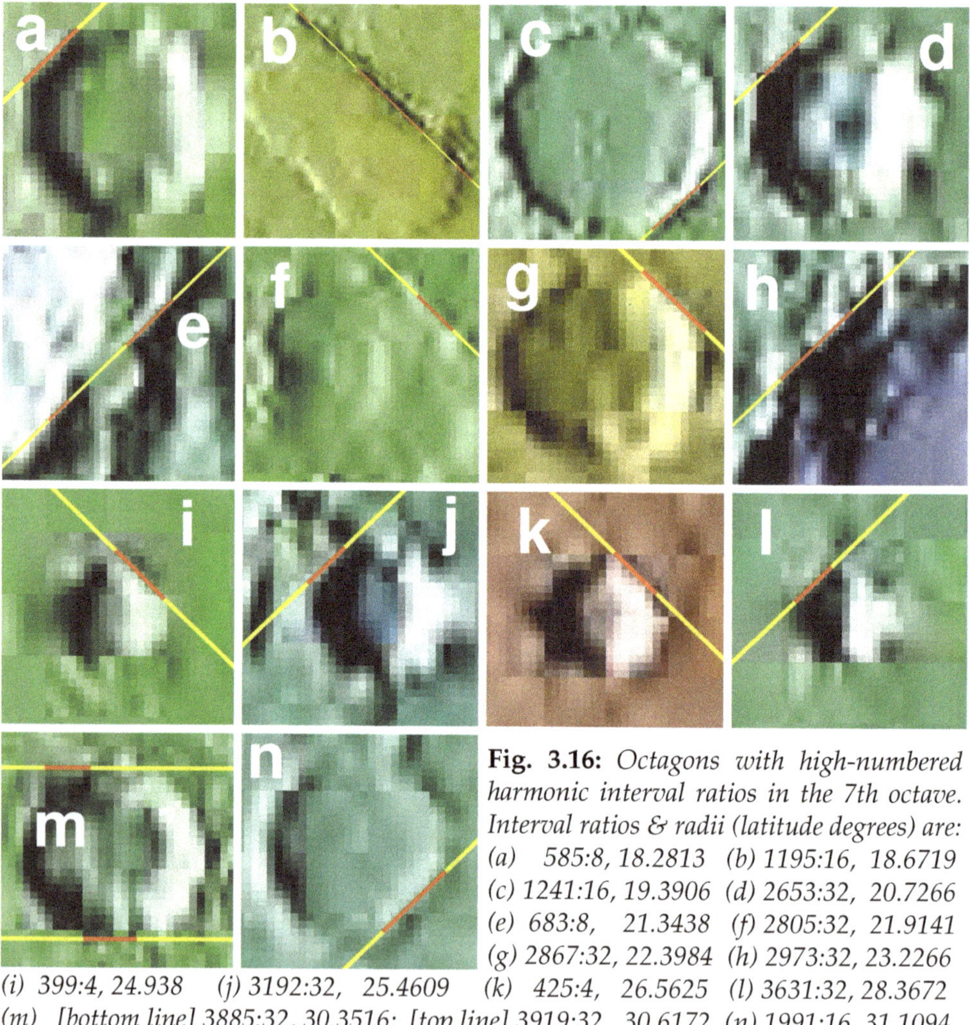

Fig. 3.16: *Octagons with high-numbered harmonic interval ratios in the 7th octave. Interval ratios & radii (latitude degrees) are: (a) 585:8, 18.2813 (b) 1195:16, 18.6719 (c) 1241:16, 19.3906 (d) 2653:32, 20.7266 (e) 683:8, 21.3438 (f) 2805:32, 21.9141 (g) 2867:32, 22.3984 (h) 2973:32, 23.2266 (i) 399:4, 24.938 (j) 3192:32, 25.4609 (k) 425:4, 26.5625 (l) 3631:32, 28.3672 (m) [bottom line] 3885:32, 30.3516; [top line] 3919:32, 30.6172 (n) 1991:16, 31.1094. Red lines mark alignments to octagons. USGS Astrogeology.*

edge (image b). The mesa aligns to an octagon having an interval ratio of 1195:16. Images e and h are of the southeast and northwest perimeters respectively of the huge Nicholson Crater which has been shown in *Intelligent Mars II* to fit 2 different sizes of squares. Both the north and south sides of the crater in image m have linear sections which align to octaves having harmonic interval ratios. The interval ratios and radii of all the octagons in Fig. 3.16 are provided in the caption to the figure, and they are distributed over the range of the 7th octave.

Although I could not find alignments for all the octagons in the 7th octave having a harmonic interval ratio with 1 in the denominator, this is partly due to my decision to no longer trust the course of the more southern sides of the octagon model that I was using to fit a 2 dimensional shape to a spherical surface. The architects may have used a different model to compensate for the distortion that I encountered. There also may be sites which align to these octagons which are below the resolution of the maps which I was using. Although the harmonic interval ratio of 72:1 was the last of this kind which I tested, the octagon with this ratio is one of the most important octagons associated with the Aum Crater. It has already been shown in Fig. 3:15d to fit the linear perimeter of a small crater, but its importance goes far beyond this alignment. It turns out that the north side (latitude = 0.0044° N) of this octagon aligns with the Martian equator almost perfectly, being only 0.26 km north of the equator (see Fig. 3.17). This octagon therefore provides overwhelming evidence that the Aum sound emanating in the form of octagons from the Aum Crater was meant to depict the creation of the entire planet itself! The dimensions of the octagon are further clues as to the intentions of the architects. The size is 72 times the size of the fundamental octagon fitting the Aum Crater itself. The number 72 is the size of the base angles of the star points of a pentagram. The radius of the octagon is 18 latitude degrees, where 18 is 1/2 the number of degrees of the star points in a pentagram. The reference to the pentagram is probably to show that the golden ratio φ is involved in the creation of the planet. The goodness of fit of this octagon to the equator is also very strong evidence that my choice of $2(e^2)^{\circ\circ}$ S for the latitude of the fundamental octagon was the latitude intended by the architects.

Eighth Octave Octagons

The fundamental octagon for the 8th octave shows a strong alignment with a long linear stretch of a large crater's perimeter (Fig. 3.18a), indicating that the architects continued to align landforms to expanded octagons from the Aum Crater with radii exceeding 1800 km. The octagon

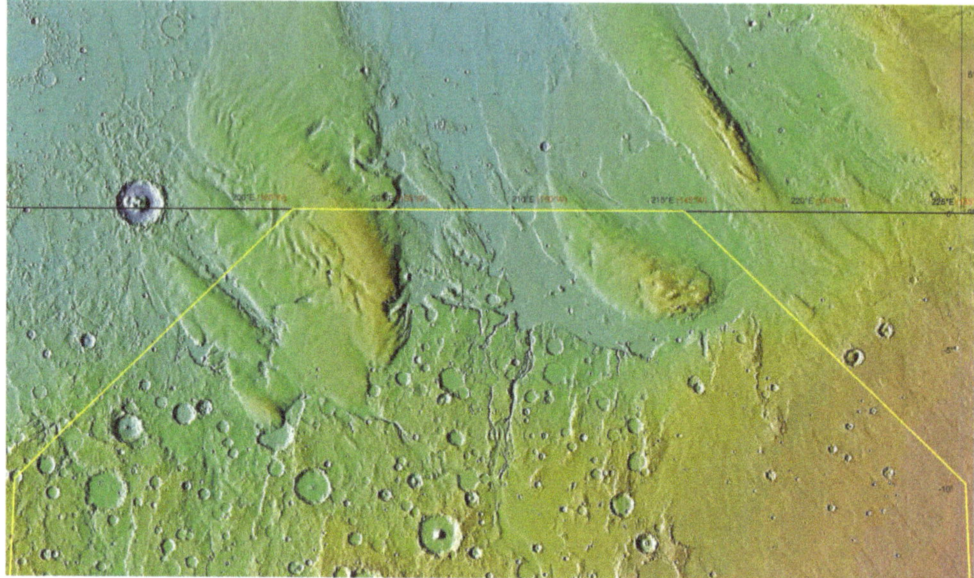

Fig. 3.17: *Octagon with an interval ratio of 72:1 and a radius size of 18 latitude degrees. The north side of the octagon passes through the equator and likely depicts the planet having been created from the Aum sound. USGS Astrogeology.*

in image b is shown to align with the southwest perimeter of the caldera of Biblis Tholus. In image c, an octagon is shown to align with the northeast perimeter of the southeast crater on Ulysses Tholus and another octagon is shown in image d to align with the northeast perimeter of the north crater on Ulysses Tholus. The next 4 figures align to structures on the top surface of Olympus Mons. The octagon in image e aligns to the linear section of the northeast perimeter of the Pangboche Crater. The octagon in image f aligns to the linear southwest perimeter of the main Olympus Mons Caldera. The octagon in image g aligns to the linear southwest perimeter of the sub-caldera lying in the northeast corner of the main Olympus Mons Caldera. The octagon in image h aligns with the linear northeast edge of the Karzok Crater. Thus the northeast edges of the Pangboche and Karzok craters located on Olympus Mons align to both octagons and squares (see Fig. 2.25, images a and c) emanating from the Aum Crater. The next figure shows an octagon which aligns with the northeast linear section of the perimeter of the Pavonis Mons Caldera (image i). The final 3 figures in Fig. 3.18 show alignments of octagons to features on Ascraeus Mons. The octagon in image j aligns with the southwest perimeter of the main caldera on Ascraeus Mons. The octagon in image k aligns with the southwest perimeter of the inner caldera of Ascraeus Mons. The 2 octagons in the final image of Fig. 3.18 bracket a linear band of material which appears to be elevated above the

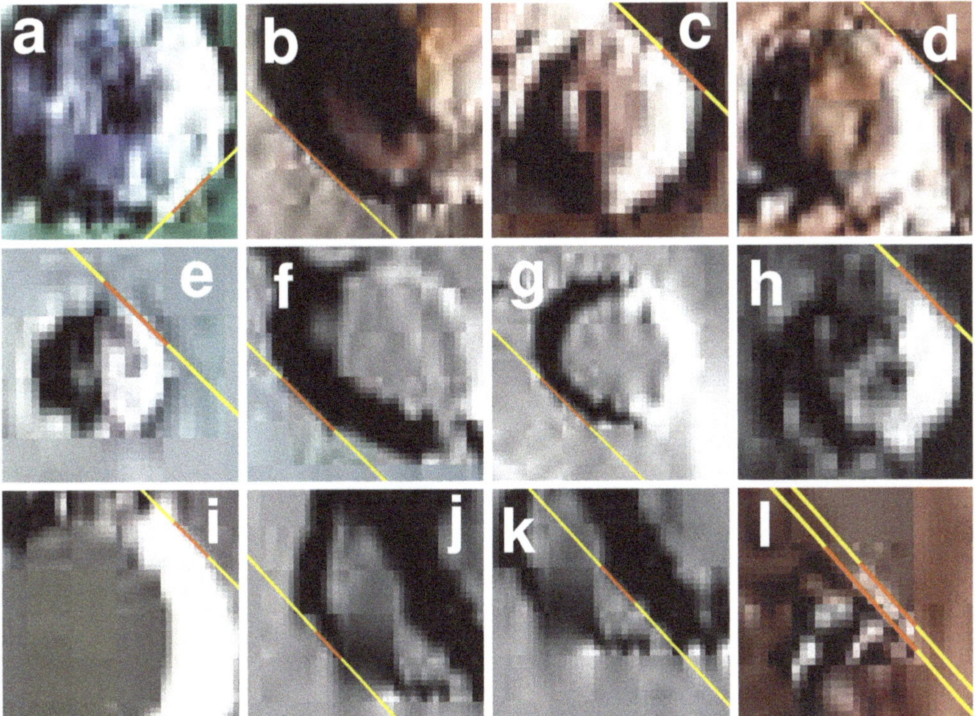

Fig. 3.18: *Octagons having harmonic intervals in the 8th octave. Interval ratios (using the 1st octave fundamental) and radii (latitude degrees) are:*
(a) 128:1, 32.0000 (b) 281:2, 35.1250 (c) 1221:8, 38.1563
(d) 2469:16, 38.5781 (e) 2553:16, 39.8906 (f) 1279:8, 39.9688
(g) 2633:16, 41.1406 (h) 2687:16, 41.9844 (i) 2751:16, 42.9844
(j) 7207:32, 56.2891 (k) 7225:32, 56.4453
(l) [lower line] 7723:32, 60.3359, [upper line] 3863:16, 60.3594.
Red lines mark alignments to octagons. USGS Astrogeology.

surrounding terrain. This band borders the northeast side of a trench close to the northeast edge of the Ascraeus Mons edifice.

Bands of different elevations which create alignment sites for octagons having high-numbered harmonics are shown in Fig. 3.19. Five different bands of light and dark areas on the southeast corner of the Ascraeus Mons Caldera are bounded by 6 different octagons.

Octagons with Irrational Intervals

Like the expanded squares, there is a set of expanded octagons which are not harmonics of the fundamental. Instead of being integer multiples of the fundamental, these octagons employ an irrational number to create the interval magnitude. All of the octagons of this set occur in the eighth

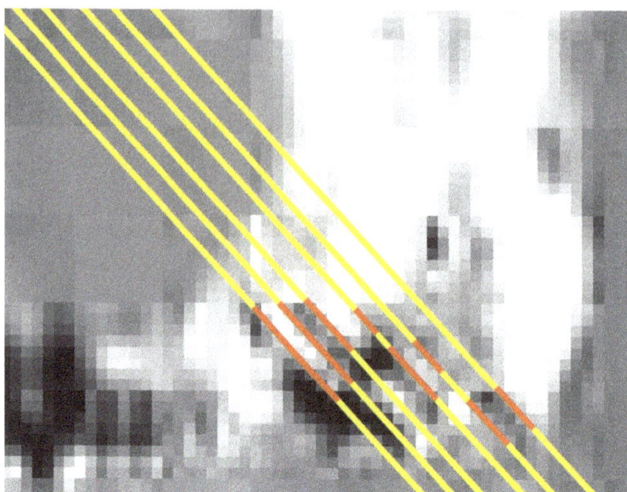

Fig. 3.19: *Octagons aligning with bands of different elevation on the southeast corner of the Ascraeus Mons Caldera. Interval ratios and radii (latitude degrees) for lowest to highest octagon are:*
(1) 7267:32, 56.77344
(2) 3635:16, 56.79688
(3) 7273:32, 56.82031
(4) 7277:32, 56.85156
(5) 455:2, 56.875
(6) 1821:8, 56.90625. Red lines mark alignments with bands. USGS Astrogeology.

octave. The first one that I will present passes exactly through the centre of the Pentagram Pyramid (Fig. 3.20)! This is probably the most important octagon of all, even more important than the octagon that aligns with the equator (Fig. 3.17). The interval ratio for this octagon has a value of

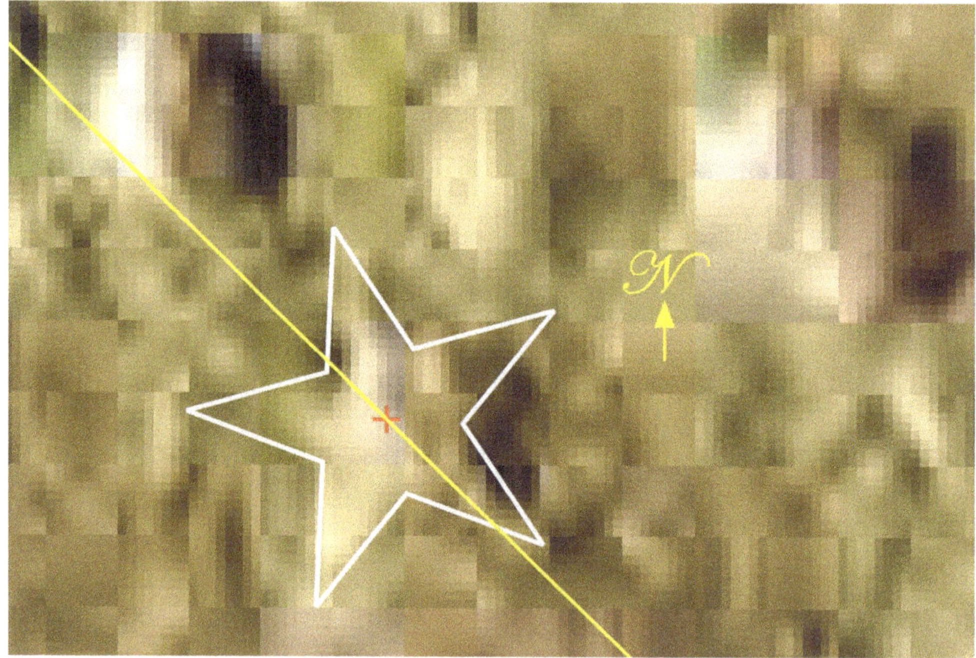

Fig. 3.20: *Octagon with an interval ratio of 1296φ:16 and a radius size of 18φ big latitude degrees. The northeast side of the octagon passes exactly through the centre of the Pentagram Pyramid. It very likely depicts the Martian body shape having been created from the irrational number φ. Red cross marks pentagram centre. USGS Astrogeology.*

1296φ:16 which is not a harmonic since the numerator is an irrational number rather than an integer. The square root of 1296 is 36 which is the number of degrees in a star point of a pentagram. Also 1296 is equal to 18 x 72 where 18 is 1/2 the number of degrees in a pentagram star point and 72 is the number of degrees in the base angles of a pentagram star point. The value of φ is encoded in the pentagram in many ways as was shown in *Intelligent Mars I*. The interval ratio of the octagon is also 81φ:1 which is notable because 81 is 3 x 27. The number 27 is 1/4 the number of degrees in the angles between the star points of a pentagram. The radius of the octagon is 20.25φ latitude degrees which can be translated into 18φ°°. The number 18 is equal to 1/2 of 36. Thus the octagon in Fig. 3.20 encodes the pentagram by referring to all 3 of its angles and by using the golden mean value of φ = 1.6180. The fact that its interval ratio is a function of φ has huge implications for this civilization's concept of Aum. Here the sound of Aum is depicted as creating the Pentagram Pyramid not from the integers of harmonics but from the irrational constant of φ. The Pentagram Pyramid itself has been shown to provide a key for the construction of the Vitruvian Martian in conjunction with Olympus Mons and the 3 Tharsis Montes (see *Intelligent Mars I*). It would therefore appear that another purpose of the alignment of the octagon in Figure 3.20 with the Pentagram Pyramid is to depict the creation of the Martian body shape from the Aum sound, particularly that aspect of the sound which embeds the value of the golden mean φ.

There are 4 other octagons of the eighth octave that use irrational numbers in their interval ratios. The northeast sides of these octagons all pass so close to the navel of the Vitruvian Martian that it is reasonable to assume that they were meant to portray the creation of the Vitruvian Martian from the irrational numbers contained in the octagon intervals (Fig. 3.21). The first of these (lowest) is an octagon which has the interval

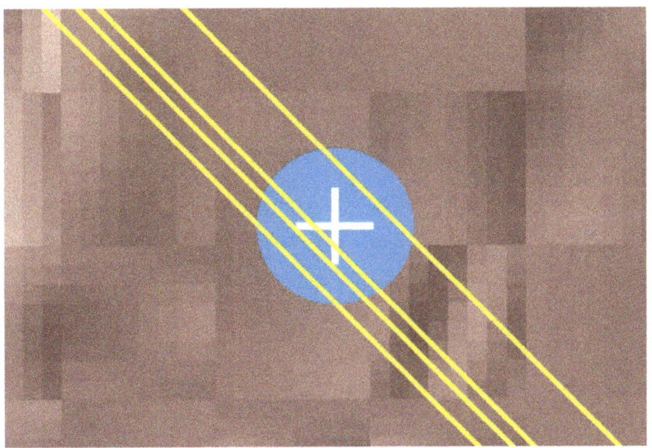

Fig. 3.21: *The northeast sides of 4 octagons pass very close to the centre of the Vitruvian Martian navel. The navel is depicted in blue with a white cross marking its centre. From lowest to highest, the interval ratios of the octagons are: 54π:1, 120√2:1, 98√3:1 and 1000e:16 = 62.5e:1. USGS Astrogeology.*

ratio of 54π:1 and a radius of 12π°° (i.e., big latitude degrees). The value of π is the ratio of the circumference of a circle to its diameter. The number 54 is 1/2 the value of the angle between the star points of a pentagram. The number 12 is a factor in all of the angles found in a pentagram. The Vitruvian Martian fits both a circle and a pentagram. This octagon comes within 1.85 km of the navel of the Vitruvian Martian. The second octagon has an interval ratio of $120\sqrt{2}$:1 and a radius of $30\sqrt{2}$ latitude degrees. The value of $\sqrt{2}$ is a factor in the diagonal of a square, and the Vitruvian Martian fits a square. This octagon comes within 1.03 km of the navel of the Vitruvian Martian. The third octagon has an interval ratio of $98\sqrt{3}$:1 and a radius of $24.5\sqrt{3}$ latitude degrees. The value of $\sqrt{3}$ is a factor in the height of an equilateral triangle, and the Vitruvian Martian fits an equilateral triangle. This octagon comes within 0.50 km of the navel of the Vitruvian Martian. Finally, the largest octagon (highest) has a remarkable interval ratio of 1000e:16 which can be reduced down to 62.5e:1. Its radius is 12.5e sacred degrees. This octagon comes within 1.56 km of the navel of the Vitruvian Martian.

Summary and Conclusions

The polygonal shape in the eastern half of the Aum Crater was best fit by a regular octagon which is an 8-sided geometric figure having 8 equal sides and 8 equal internal angles (135°). A fundamental octagon was determined by finding a radius size (a power of 2 of one latitude degree) just small enough to permit the octagon to fit within the crater, and by locating a centre having meaningful coordinates. The fundamental octagon meeting these requirements has a radius size of $2^{-2} = 0.25$ latitude degrees and coordinates of 60° E [DPPM] and 16.6254° = $2(e^2)$°° S, and has an orientation so that its north and south sides have a bearing angle of 90°. Other larger octagons with the same centre were then created to see if they aligned to landforms both within and beyond the Aum Crater perimeter.

Since the octagon is a 2 dimensional geometric shape, fitting it to a spherical surface distorts the shape of the octagon and the size of its internal angles. These problems become very significant with large octagon sizes and require choices to be made regarding their construction. The choices depend on what the architects would have considered to be the most important features of the octagon. I decided that the Martian architects would want to preserve the bearing angles of the sides as much as possible so that they could align the octagons to structures whose components had bearing angles which represented sacred geometry. The bearing angles for the sides of the octagons are 0° for the east and west sides, 90° for the north and south sides, and ±45° for the remaining 4 sides. The simplest model that I could devise which would reasonably maintain these angles was the construction of the octagon from 8 radius lines projecting to the 8 vertices of the octagon. Each radius line was constructed as a rhumb line (constant bearing angle). The radius line to the vertex between the N and NE sides of the octagon was given a clockwise bearing angle of 22.5° and the rest of the radii were set 45° apart from each other as they went counterclockwise around the octagon. The ends of the radius lines marked the vertices of the octagon which were then joined by rhumb lines to form the final octagon.

The chief strength of this model was the relative lack of distortion over a huge range of octagon sizes for the northern part of the octagons which included the NW, N and NE sides. The bearing angles of the NW and NE sides remained within 1° of 45° right up to the first 43 percent of the eighth octave. The only octagons exceeding the 1° deviation were those fitting the Ascraeus Mons Caldera (about 42°) and the northeast perimeter of Ascraeus Mons (about 41.3°). The bearing angles of the east and west sides of the octagon remained within 1.1° of 0° until the 6th octave and within 2.2° until the 7th octave. Similarly, the bearing angles of the SW and SE sides remained within 1° of 45° until the 6th octave and within 2.3° until the 7th octave. I therefore did not attempt to find alignments for the W, SW, S, SE and E sides for octagons in the 7th or 8th octaves as it was possible that the Martian architects had a more complex model to compensate for the greater distortion in these more southern sides.

For expanded octagons, I used the fundamental octagon from the first octave (i.e., with a radius of 0.25 latitude degrees) to measure interval ratios rather than the fundamental octagon for each octave. In this way, all harmonic values were with reference to the same fundamental octagon, which would be more analogous to the Aum sound generating harmonics from an extremely low frequency sound. Alignments were found for all octagons having simple harmonic interval ratios (those having 1 in the denominator) from 1:1, 2:1, ...64:1 and 65:1 which covered the range from the first octave to the beginning of the 7th octave, and also for octagons with harmonic interval ratios of 69:1, 72:1 and 128:1. In addition, many more alignments were found for octagons having more complex harmonic interval ratios (i.e., having an integer number greater than 1 in the denominator). Altogether, I was able to find good quality alignments for 197 octagons having harmonic intervals over the range of 8 octaves. The largest octagon (aligns with the northeast perimeter of Ascraeus Mons) has an interval ratio of 3863:16 and a radius size of 60.3594 latitude degrees which is 3578 km. The octagons were mainly aligned to linear stretches of crater perimeters and other landforms. These craters and other structures must therefore have been deliberately engineered to fit octagons with harmonic intervals in order to represent the concept of a primordial sound emanating from a single source to create the universe. The fact that the north side of an octagon having the very meaningful interval ratio of 72:1 aligns with the equator shows that the architects intended to depict the planet Mars as having been created by this primordial sound. Also the fact that an octagon with an interval ratio of $(36^2)\varphi$:16 passes through the centre of the Pentagram Pyramid shows that the architects intended to depict the Martian body shape as being a product of a component of the primordial sound containing the irrational number φ. The Pentagram Pyramid was shown in *Intelligent Mars I* to provide the measurements necessary for the construction of the virtual Vitruvian Martian with the building blocks of Olympus Mons and the 3 Tharsis Montes. The 4 octagons having interval ratios containing π, e, √3 or √2 which come very close to the navel of the Vitruvian Martian suggest that the Aum sound is also being depicted as creating the Martian body out of these irrational numbers. The many octagons having harmonic interval ratios which align to components of the giant mountains offer additional support for the concept that the Martian body was created from the sound of Aum. From all of this, it must be concluded that the placement and construction of the Aum Crater, a large number of other craters and landforms, the Pentagram Pyramid and the giant mountains must have been carefully planned as a single unit to depict Divine creation from the Aum sound. The scale of such a project is far beyond anything that our own technology could produce or even conceive.

Fig. 3.22 shows all of the enlarged octagons centred on the Aum Crater which were found to be aligned to landforms both within and mostly well beyond the crater. The low resolution of the map causes many of the octagons, especially the ones closest to the Aum Crater, to merge with one another. I used wide white lines for the fundamental octagons of the 6th, 7th and 8th octaves to orient the reader as to where distortion occurs. Octagons are relatively distortion free up to the 7th octave except for minor distortion in the 5 southern sides of the larger octagons in the 6th octave. The 3 northern sides of the octagons remain relatively distortion free except for those passing through Ascraeus Mons where the NE and NW sides deviate from a bearing angle of 45° by 3 to 4 degrees. From Fig. 3.22, it can be readily seen how the outward spread of the octagons from the Aum Crater passes through the equator and includes the pattern created by the 4 largest Martian mountains. The largest octagons also include the western portion of the Valles Marineris but I was not confident enough in my octagon model at that size to locate any alignment sites in the Valles Marineris region to the eastern octagon sides.

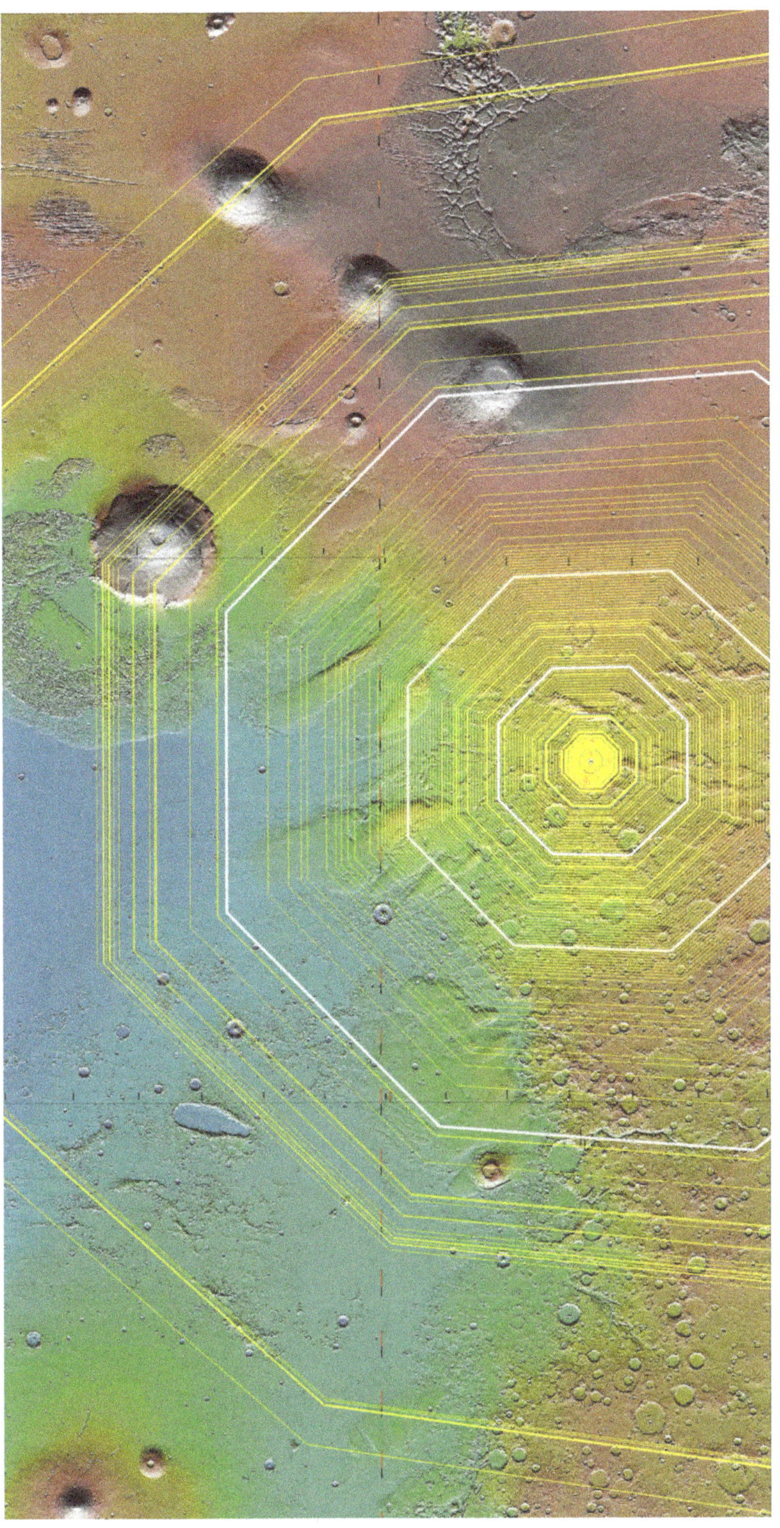

Fig. 3.22: *Entire dataset of octagons which were fit to Martian landforms. The white octagons are the fundamental octagons for the 6th, 7th and 8th octaves. All octagons below the 6th octave are relatively distortion free. A slight amount of distortion becomes noticeable for the 5 southern sides of the larger octagons in the 6th octave. The 3 northern sides of octagons are relatively free from distortion except for the octagons passing through Ascraeus Mons where the bearing angles of the NE and NW sides were between 41.3° to 42° instead of 45°. USGS Astrogeology.*

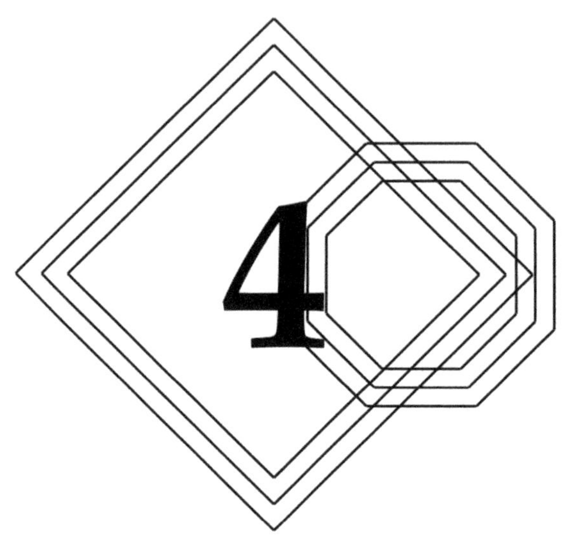

Solving the Aum Crater

Having finished fitting 8 octaves of squares and another 8 octaves of octagons to landforms surrounding the Aum crater, it is now appropriate to try to interpret all of the data as one unit so as to come to a better understanding of the purpose behind this enormous feat of architecture and engineering. So far we have evidence that the Aum Crater links together a vast region of the planet with 2 different geometric shapes emanating concentrically from 2 different regions within the crater. Both shapes expand in harmonic intervals from a fundamental size except for a subset of squares and octagons which form irrational intervals with the fundamental. I had already come to the conclusion that the series of expanding squares probably represents the concept of Aum. But now we also have a series of expanding octagons. Why would the architects have 2 geometric shapes representing the same thing? What could possibly be the purpose of such an arrangement?

It was at about this time that I happened to visit the home of a friend who was from India. While there, I noticed a strange symbol that had been mounted above the doorway to a room that his family used for meditation (Fig. 4.1). My curiosity prompted me to ask about it and I was told that it was a symbol that represented the word or sound for Aum (sometimes spelled Om). From my recollection of the yoga sessions that I took several years back, we chanted Aum as a representation of the Divine sound that brought the universe into existence. I always thought

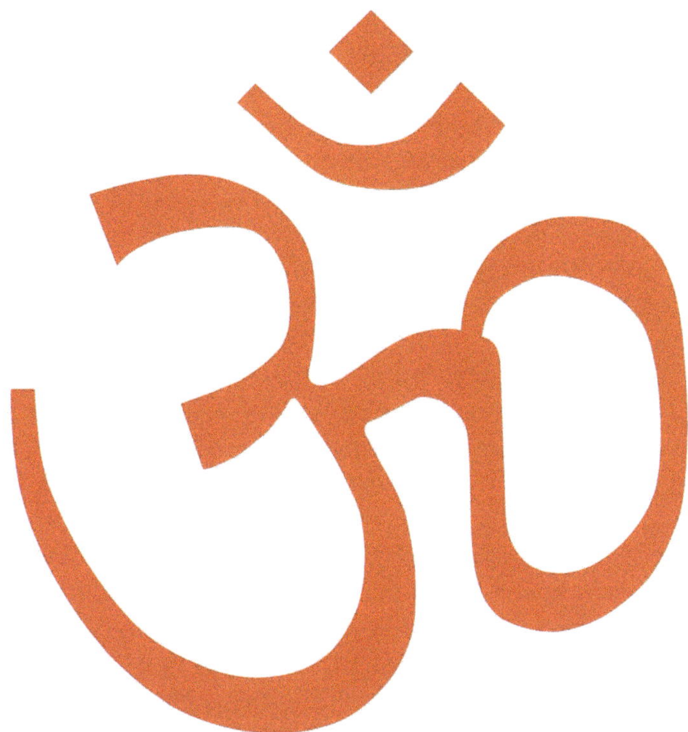

Fig. 4.1: *A common version of the Hindu symbol for Aum similar to the one in my friend's home. Many variations of the Aum symbol exist amongst the different religions and cultures of ancient and present times. Most of these have some representation of the "3" portion of the Aum symbol.*

of this original sound as an extremely low fundamental frequency sound from which an infinite number of harmonics arose to create all the vibrations associated with the subatomic particles and energy waves which in turn gave rise to all matter and forms of energy in the universe. This struck me as a very good parallel to what I was observing with the Aum Crater. Instead of sound frequencies, the sizes of geometric shapes are used to simulate harmonics. A harmonic is simulated when the size of a larger square or octagon is an integer multiple of the size of a fundamental square or octagon, just as a sound is a harmonic when it is an integer multiple of a fundamental frequency. In sound, the harmonic number is preserved when an integer multiple of a fundamental frequency is divided by a power of 2 to bring it into the current octave. The same concept applies to the size of a square or octagon when it is an integer multiple of the fundamental square or octagon. It simply has to be divided by a power of 2 to bring it into the current octave of sizes. All of the squares or octagons which showed alignments to straight line features of landforms, such as linear sections of crater perimeters, were found to have interval sizes which are integer multiples of a fundamental square or octagon divided by an integer power of 2. Thus harmonics are represented as a ratio of integers where the numerator is the number of the harmonic and the denominator is the divisor (a power of 2) required to bring the harmonic into the current octave. For example, the interval of 3:2 in the first octave is the third harmonic divided by 2 to bring it from the second octave into the first octave. The interval of 255:128 is the 255th harmonic brought into the current octave from an octave which is 7 octaves higher. The power of 2 in the divisor has to be large enough to bring the interval ratio, with respect to the fundamental of the current octave, to a value ≥ 1 and < 2. The power required is the number of octaves that the harmonic resides above the octave which contains the fundamental square or octagon.

The interval ratios that I used for the harmonics in the expanded squares were those determined for the fundamental square in each octave. The interval ratios that I used for the harmonics in the expanded octagons were those determined using the fundamental octagon for the first octave rather than the fundamental octagon in the current octave. In this way the harmonic numbers were always with reference to the same fundamental size of octagon rather than for a different fundamental size for each octave. I will call the system used for squares *System S*, and the system used for octagons *System O*. Both systems produce the same harmonic numbers when the harmonic numbers are large. However, when the harmonic number is smaller than or equal to its octave number in System S, the interval ratio from System S has to be multiplied by a power of 2 in order to bring it into System O. The power that is needed is equal to the octave number minus one. Take, for example, a square which has an interval ratio of 3:2 in the third octave using System S. Since the harmonic number 3 is equal to the octave number, the interval ratio has to be multiplied by 2 raised to the power of 3 – 1 = 2. Thus, 3:2 x 2^2 = 3:2 x 4 = 12:2. The interval ratio cancels down to 6:1 which gives the correct harmonic number for System O. Similarly the interval ratio of 5:4 in octave 6 for System S becomes 5:4 x 2^5 = 5:4 x 32 = 160:4. This interval ratio cancels down to 40:1 which gives the harmonic number of 40 for System O.

In doing a little research on the origins of the Aum concept, I found that it was extremely ancient, going right back to the origins of Hinduism and other Indian religions. What also piqued my interest was that it is composed of 3 sounds: A (aaa), U (ooo) and M (mmm), and is often intoned with these 3 sounds sung in sequence. The first sound A (aaa) is produced in the throat with the mouth wide open, and the second sound

U (ooo) is produced as the mouth moves towards closure which when reached, the third sound M (mmm) is produced. Hmmm... or should I say aaaooommm...? I seem to be missing a sound with my analysis of the Aum Crater. I only have 2 geometric shapes. Let's look again at the Aum Crater. We have a small octagon in a larger square. Suppose the square represents the open mouth. Then it would correspond to the A (aaa) sound. Since the octagon is smaller than the square, it might represent the mouth moving toward closure and would correspond to the O (ooo) sound. That leaves the M (mmm) sound without anything to represent it. It would seem that a third geometric shape should be present somewhere within the confines of the Aum Crater to correspond to the M (mmm) sound. The thought that the 3 sounds of Aum could be the solution to this enigmatic crater so intrigued me that I quickly went back to take a fresh look at this seemingly innocuous piece of Martian landscape.

Symbols Around the Aum Crater

While the arrays of squares and octagons provide a wealth of data about the Aum Crater, it is important to return to the vicinity of the crater itself to see if anything was missed. The hypothesis that this crater is associated with Aum has a lot of support now, so we can proceed more confidently in interpreting other evidence. During your examination of the various figures picturing the Aum Crater, did you notice anything peculiar about the surrounding landscape? What about the various structures, particularly those just beyond the northeast side of the crater? Are they simply hills of various shapes and sizes?

I started to notice that several of the shapes resembled symbols that we are familiar with here on Earth. The most amazing one that I found was a symbol for Aum! If you look at the structure pointed to by the white arrow in Fig. 4.2 you can make out the shape of a "3" rotated 90° clockwise so that its arms point exactly northwards. A southward extension is attached to the lower (west) part of the "3" shape. I have superimposed a solid white colour on top of this structure in Fig. 4.3 to display it more clearly. Compare it to the symbol in Fig. 4.1. Although the extension to the "3" portion is different, and the crescent and dot are missing, the structure in Fig. 4.3 bears a strong resemblance to the Hindu symbol for Aum, especially since the extension is attached to the lower part of the "3" portion. One can expect differences to arise over a span of billions of years. Besides there are many variations among cultures and religious groups in the symbol for Aum and some of them are rotated 90° in the clockwise direction just like the symbol seen in Figs. 4.2 and 4.3.

To the west and slightly north of the structure for the Aum symbol,

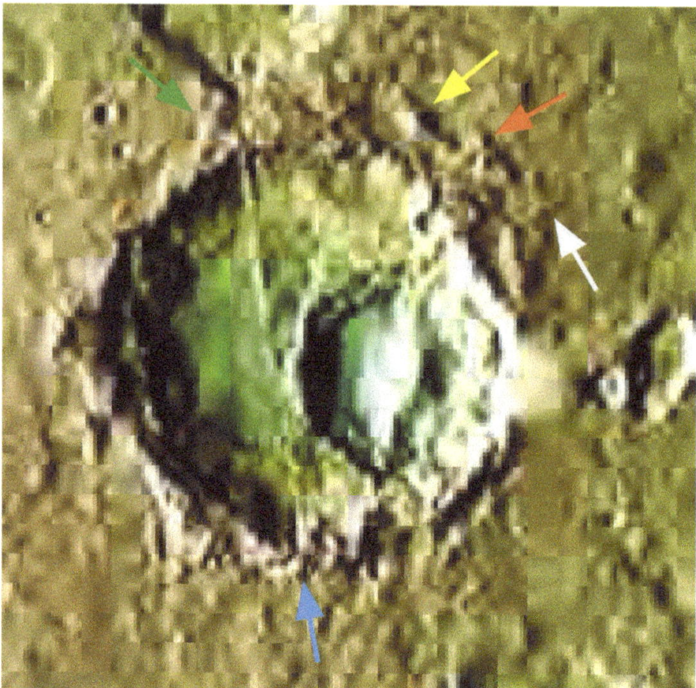

Fig. 4.2: *Symbols around the Aum Crater. The green arrow points to a capital alpha symbol, the yellow arrow points to a large pyramid, the red arrow points to a capital omega letter, the white arrow points to an Aum symbol, and the blue arrow points to a scorpion symbol. USGS Astrogeology.*

there is another object which appears to have the shape of the symbol for the capital letter omega of the Greek alphabet. This is pointed out by the red arrow in Fig. 4.2 and is superimposed by a solid red colour in Fig. 4.3. Just west of the northern vertex of the Aum Crater, there is another symbol which looks like a capital 'A' rotated counterclockwise by about 154°. It is pointed out by a green arrow in Fig. 4.2 and is superimposed with a solid green colour in Fig. 4.3. A similar symbol had been used around 1500 B.C. by the Phoenicians and Hebrews for aleph, the first letter of their alphabets, and it was rotated counterclockwise by about 130°. It looked similar to our capital 'A' except that the crossbar extended a bit beyond the 2 sides of the letter. The Greek capital letter 'A' for alpha was developed from the Phoenician alphabet. The Greek letters alpha and omega are heavily associated with Aum, being the first and last letters of the alphabet. Indeed, the lower case letter for omega (ω) is shaped like the Aum symbol but is rotated 90° in the clockwise direction from that shown in Fig. 4.1, just like that in Fig. 4.3. Aum encompasses everything from beginning to end, and since alpha represents the beginning and omega represents the end, they are an integral part of the meaning of Aum.

Just west of the omega symbol, there is a structure which looks very much like a pyramid. It is pointed out by a yellow arrow in Fig. 4.2 and is

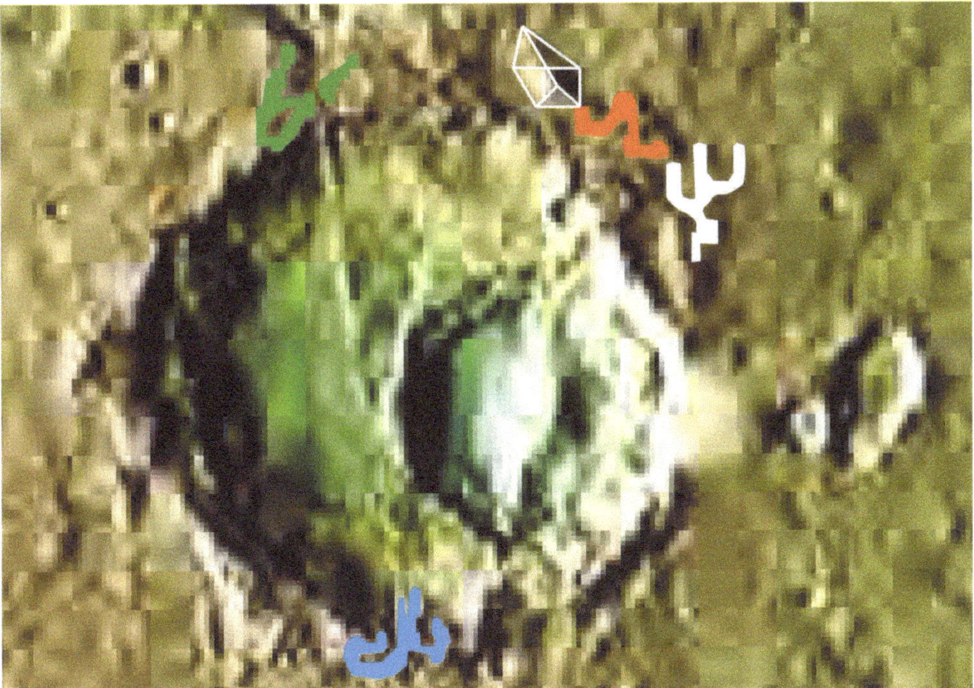

Fig. 4.3: *Symbols around the Aum Crater filled with solid colours to increase visibility. The alpha symbol is filled green, the omega symbol, red, the Aum symbol, white, and the scorpion symbol, blue. White lines demarcate the edges of a pyramid. USGS Astrogeology.*

outlined in Fig. 4.3 by white lines. The best fit to a geometric figure that I could come up with is to an irregular 5-sided pyramid. From this figure it would appear that the top of the pyramid forms a straight line rather than a peak. The east, south and southwest sides of the pyramid are very distinct but the northwest and northeast sides are less well delineated. The top line has a counterclockwise bearing angle of 30° and the east side has a bearing angle of 0°. The vertex created by the junction of the east and south sides forms a right angle whereas the vertex created by the junction of the south and southwest sides forms an angle of 120°. Each of the edges from the ends of the base of the south side to the peak forms an angle of 45° with the base. The angle where these edges meet at the peak is therefore 90°. The line along the top of the pyramid is parallel to the southwest side. The northeast side has a fairly distinct edge and it is best fit with a line having a counterclockwise bearing angle of 54°. The northwest side has the least well-defined edge but it seems to be best fit by a line having a clockwise bearing angle of 12°. This pyramid is huge by Earth standards. Its southern base is about 5.7 km long, the length of the southwest side is about 6.0 km and its total length is about 12 km.

At the southern vertex of the Aum Crater there is another important structure that seems to have the shape of a scorpion. This is pointed out by a blue arrow in Fig. 4.2 and is superimposed by a solid blue colour in Fig. 4.3. The presence of a scorpion structure immediately brings to mind Scorpio, the 8th astrological sign in the Zodiac. Even though a Martian Zodiac would not be expected to have the same constellations that we have now since these would have changed greatly in the course of 3 to 4 billion years, perhaps some of the symbols themselves may have persisted over time. It would seem appropriate to include the signs of the Zodiac around the Aum Crater since the Zodiac represents a complete cycle of time, all of the human personality traits, and all of the energies of the planets and constellations. All of these would be included in the concept of Aum. What is truly amazing is that in present day astrology, Mars is one of the planets that rules the sign of Scorpio together with Pluto. Mars also rules the sign of Aries, but I could not discern any symbol for Aries around the Aum Crater. There are definitely more symbols on the outskirts of the Aum crater, but they are either too indistinct or too esoteric for my limited knowledge to be interpretable. Hopefully others will be able to decipher at least some of them. In addition, there are some symbols inside the eastern part of the Aum Crater which I was unable to interpret.

The incredible presence of an Aum symbol and the alpha and omega symbols around the Aum Crater confirms its identity and purpose. I was right after all! This crater is intended to represent Aum in all its aspects. At this point, it might be useful to examine more in depth what is known about the Aum concept before continuing on with our analysis of the Aum Crater.

An Aum Review

A very thorough description of Aum comes from the ancient Hindu texts of the Upanishads. In these writings, Aum can be considered to be God in the form of sound. It might also be conceived of as the first manifestation of the Divine. In Sanskrit, it is called Anahata Nada which means "unstruck sound". This type of sound is considered to be simply vibratory and is not produced by things striking each other. The entire universe arises from the myriad of frequencies contained in this sound. All dimensions of time (past, present and future) and all actualities and potentialities are contained in Aum. When we create the Aum sound we are trying to be in sympathy with the combined sound of the vibrations of all the atoms in the universe. In doing so we are attempting to awaken our connection with the Source of all existence, the absolute Divine which is also our true selves as the One.

There are many different ways to produce an Aum sound. The resulting sound can be either close to a monotone throughout, or it can start off at one frequency and move through a gradual rise in frequency to end with a final tone higher than the first. In the latter case the interval is usually close to a perfect 4th. Adding further to this complexity, in the latter case the Aum sound is generally broken up into 3 components or syllables, each representing its own state of consciousness:

1. the 'A' sound is the starting sound considered to be the primal sound of the universe. It is also the first letter of the Sanskrit alphabet 'aa' and represents the first state of consciousness (jagruti) which is the waking state focused on physical reality perceived through the 5 senses. This sound resonates in the throat. It sounds like 'AH' which is a mantra in itself and is used in many of the names for God such as Allah, Buddha, Yahweh, Krishna etc. and also in sacred words such as Amen and Alleluia. It is most often associated with the heart chakra, with love and with compassion.
2. the 'U' sound is the Sanskrit letter 'au' and it is formed with the lips moving towards closure. It signifies the dream state of consciousness (swapna) which is composed of thoughts, emotions, and images behind closed eyes. It sounds like the 'o' in "who" and vibrates at the back of the mouth. This sound starts at the back of the palate and rolls forward to the front of the mouth being shaped by the gradually closing lips.
3. the 'M' sound is the Sanskrit letter 'ma'. It is sounded with gently closed lips and signifies the state of deep sleep (sushupti) which is a dreamless state. It is a thoughtless state of consciousness where one is not conscious of anything in particular, but is simply conscious. The 'M' sound resonates at the front of the mouth and vibrates the entire skull.

There is also a 4th state of consciousness in the chanting of Aum and that is the state known in Sanskrit as turiya. Like the 3rd state of consciousness it is devoid of all thoughts, but now the individual has risen to a transcendental state in which the chanter is united to the Absolute in profound Oneness. This state occurs in the silent period after the M sound of Aum. It is the "unstruck sound" part of Aum in which we become aware of the divine vibration of the cosmos. It is signified by the dot in the Hindu symbol for Aum (Fig. 4.1) and is separated from the other 3 states by a crescent which represents illusion. The other 3 states are represented by the 3 curves in the Aum symbol. In the Hindu version of the Aum symbol, the lower curve of the "3" represents the waking state, the upper curve of the "3" represents the dream state, and the curve behind the "3" represents the state of deep sleep.

Reassessing the Aum Crater

Now let's see how well the Aum Crater matches with this more complete description of our terrestrial concept of Aum. Earlier in this chapter I conjectured that the square might represent the open mouth and would correspond to the A (aaa) sound, and that the octagon might represent the mouth moving toward closure and would correspond to the O (ooo) sound. But I had nothing to represent the M (mmm) sound. We need to go back to examine the Aum Crater more carefully to see if I missed something.

I was always mystified by the presence of a thick white streak passing through the middle of the 8-sided crater. Being white, it probably represents elevated terrain, or the rising side of a ridge which then slopes downward at the dark region to the east. The white/dark streak widens considerably when it approaches the centre of the crater. After staring at this on and off for several days, an answer as to its purpose finally dawned on me. Could this be a representation of closed lips? The white side could represent the lower jaw as well as the lower lip since it is wider than the dark side. As you move to the northern wall of the crater and beyond, the white part of the streak takes on the shape of a braid or perhaps a chain. This braid then radiates outwards towards the pyramid, the omega symbol and the Aum symbol. It's as if it was intended to form a bridge between the symbols and the white/dark streak.

If my hypothesis about the white/dark streak representing closed lips is correct, then what about the Aum Crater and its internal octagonal crater? A light was starting to turn on in my head. The Aum Crater must be a representation of the shape of the lips of the mouth as the Aum sound passes through the various components of its intonation! In this scenario, the top of the mouth is at the eastern part of the Aum Crater and the bottom is at the western part. The perimeter of the Aum Crater is the shape of the lips at the beginning of the AH of the Aum sound. As the sound shifts to U, the lips take on an octagonal shape represented by the 8-sided crater. Finally, they close for the M sound and this is represented by the thick white/dark band in the centre of the 8-sided crater. The different centres of the Aum Crater for the square vs. the 8-sided crater even takes into account the shifting of the centre of the oral opening upwards as the lower jaw comes to partial closure for the final parts of the Aum intonation. The series of squares emanating from the Aum Crater would represent the sound coming out of the open mouth and would carry the AH sound outwards towards the surrounding terrain, the equator, the Pentagram Pyramid and the giant mountains. The series of octagons emanating from the octagonal crater would carry the U (ooo) sound outwards also to the surrounding terrain, the equator, the Pentagram Pyramid and the giant mountains. The M sound would be retained by the white/dark streak representing closed lips and would resonate inside the Aum Crater and inside the planet itself. The white western side of the streak would represent the lower jaw and lip, and the dark eastern side would represent the upper lip. Note that the centre of the fundamental octagon occurs at the junction of the white and dark sides (Fig. 3.2).

I immediately went to a mirror to examine the shape that my lips took when I intoned the Aum sound. What about the wide open mouth at the start of the A sound? My lips took the shape of an imperfect square with the upper and lower corners of the 'square" being flattened, just like the east and west corners of the Aum Crater! Then as the sound proceeded to "ooo" my lips took the shape of an octagon with 2 sides missing at the right and left corners of my mouth. This made me think that the Martian anatomy might be different, or perhaps the architects took some artistic license. Whatever the case, the similarity to the Aum Crater was substantial enough to give high credibility to my interpretation of this crater as representing a mouth chanting Aum.

The braid is still kind of a mystery though. I think it has to do with the silent period following the Aum chant. This is when the chanter is in the 4th stage of consciousness, connected with a perfectly silent mind to the Absolute as represented by the Aum symbol and the alpha and omega symbols. The presence of a 5-sided pyramid may be associated with the use of irrational numbers in the interval ratios of key squares and octagons. The bearing angle of the NE side is +54° which is 1/2 the size of the angle between the star points of a pentagram. The pentagram represents φ, π and $\sqrt{5}$. The south side of the pyramid is 1/2 of a square, and its base is the diagonal of the square. This would represent $\sqrt{2}$ since the length of the diagonal is equal to the side length multiplied by $\sqrt{2}$. The bearing angle of the pyramid top and west side is +30° which would represent $\sqrt{3}$ since $1/(\tan(30)) = \sqrt{3}$.

In conclusion, it appears that the Aum Crater is a depiction of the Martian mouth intoning Aum. It took several years for me to finally find plausible centres for the expanding series of squares and octagons and to work out the concept of the crater representing Aum. The Aum Crater is undoubtedly a key component in the Martian architecture which was created to depict Martian spirituality. We will now examine other Martian architectural feats, some of which are of the same order of magnitude as the Aum Crater.

Arrow Craters

The presence of arrow-shaped craters on the surface of Mars offers us unequivocal proof that intelligent life exists or has existed on another planet in the solar system. Despite this, the existence of these craters has never been acknowledged either by academia, by NASA or even by independent investigators. It is as if humanity has a filter installed in its collective vision that removes these craters from its sight or causes them to be seen as normal, circular craters caused by impacts in the past. Arrow-shaped craters are clearly present on the MOLA maps which have now been taken down from the USGS Internet site. These MOLA maps are still available on the Wikipedia site as quadrangular maps of Mars but this may not last. I have not seen the arrow-shaped craters on any of the pictures and maps of Mars which are currently available to the public on NASA websites. For instance, the THEMIS maps now provided (https://planetarynames.wr.usgs.gov/Page/MARS/target) show only rounded out versions of the arrow-shaped craters.

I have presented numerous pictures of arrow-shaped craters aligning to squares and octagons emanating from the Aum Crater in Chapters 2 and 3. Did you consider them to be artificially created arrow-shaped craters or did you think they were just a fluke of nature? Or did your mind simply interpret them as being round? Fig. 5.1 shows several variants of arrow craters in the region occupied by squares and octagons emanating from the Aum Crater. All of these craters have an arrow shape in their outer

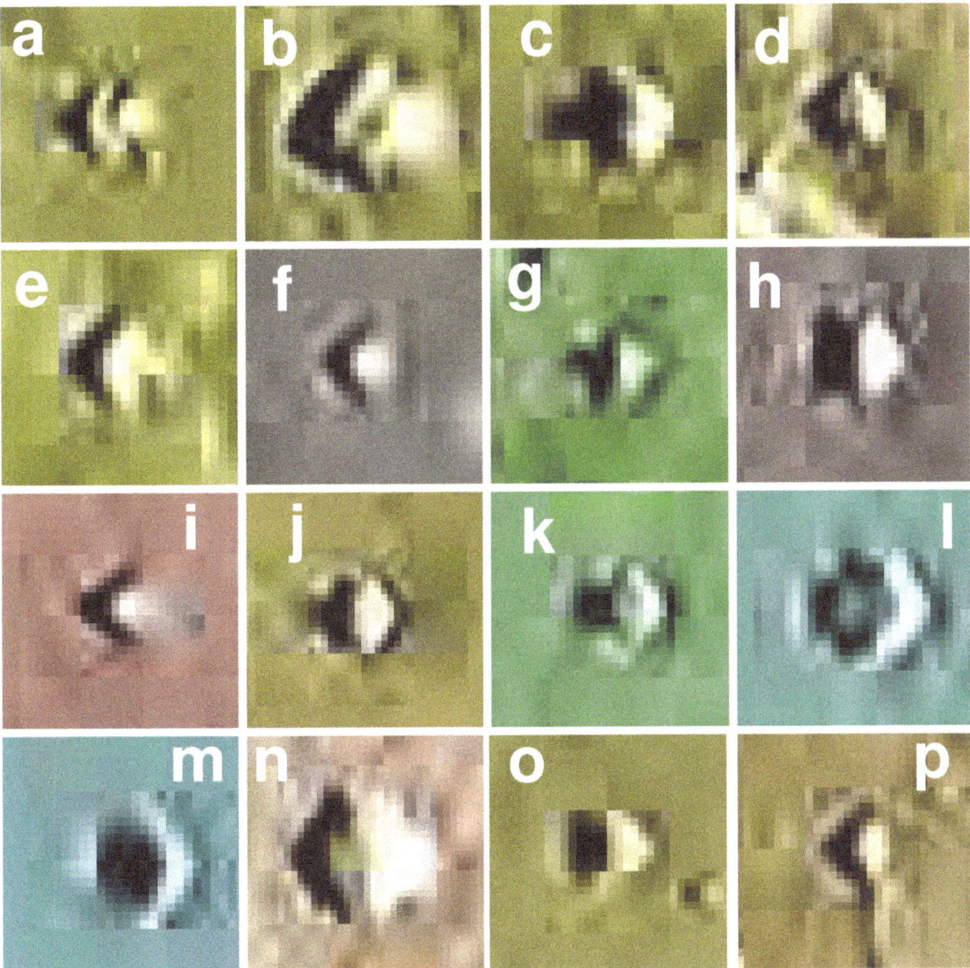

Fig. 5.1: *Arrow-shaped craters in the region containing squares and octagons expanded outwards from the Aum Crater. Notable coordinates of arrow tips are:*

(a) 62° E [EMPM]
(b) $13\pi^{\circ}$ W [PMPM], $10 \ln(2)^{\circ}$ S [AMPL]
(c) e/π sacred rad E [CEPM], $1/(e\sqrt{2})$ rad S
(d) $52^{\circ\circ}$ E [EMPM], $5\sqrt{3}^{\circ}$ S [AMPL]
(e) 1 rad E [EMPM], $1/(10\sqrt{2})$ big rad S [AMPL]
(f) $9.5^{\circ\circ}$ E [PCPM], $2\varphi^{\circ\circ\circ}$S [AMPL]
(g) $1/(e\varphi)$ sacred rad E [DPPM]
(h) $6\sqrt{3}^{\circ}$ E [PCPM], $10\sqrt{2}^{\circ\circ}$ S
(i) 16° W [PMPM], $1^{\circ\circ}$ S [AMPL]
(j) $\sqrt{2}\sqrt{5}^{\circ}$ S
(k) $13\Phi^{\circ}$ W [EMPM]
(l) $7\pi^{\circ}$ N [AMPL]
(m) $13(\varphi^2)^{\circ\circ}$ E [EMPM]
(n) $(\varphi^2/2)^{\circ\circ}$ W [EMPM]
(o) $13^{\circ\circ}$ E [EMPM]
(p) $2\pi^{\circ\circ}$ E [CEPM], 7° E [EMPM], $4\sqrt{3}^{\circ}$ S [AMPL]. *USGS Astrogeology.*

perimeters. The craters in images a to e are within 300 km of the Aum Crater, whereas those in images f to h are about 2700 km away. Perhaps your mind did not permit you to interpret these craters as having true arrow shapes since we have been taught that naturally formed craters are all perfectly round. Your mindset may have forced you to conclude that any deviation from roundness was only illusory and was simply the result of a round shape being modified by natural forces such as uneven density and shape of the local terrain in the strike area, or to eons of erosion. There are a few things to note about the craters in Fig. 5.1. They all possess shafts albeit relatively short for images d, i and p. They all point due east or due west which is a common finding for the vast majority of arrow-shaped craters. Very rarely an arrow crater points due north or due south. This fact alone proves the artificiality of these craters. If we look at the entire set of arrow-shaped craters on Mars, the probability of more than 90% of them pointing due east or west under random conditions is far less than 1 in a trillion trillion. Another interesting feature is that the bearing angles of the northern edges of the arrowheads in Fig. 5.1 are all 45° clockwise or counterclockwise. This is also true for their southern edges except for the arrowheads in images b, c and i. The southern edges of the arrowheads in images b and i have a bearing angle of 60° counterclockwise and the southern edge of the arrowhead in image c has a bearing angle of 30° clockwise. The angle of 45° is a common finding for the vast majority of arrowheads in the vicinity of the Aum Crater, making them ideal alignment sites for squares and octagons emanating from the Aum Crater.

From past experience with arrow shapes, it was found that the tips of the arrows were located at meaningful coordinates. This suggested that the arrow shapes were not used to point to objects in the distance, but to delineate important locations with the tips of their arrows. In *Intelligent Mars II* it was found that the tip of the triangle located on the edge of the Sharonov Crater perimeter pointed out a prime meridian (the Sharonov Triangle PM). Arrow shapes associated with the Denning and Savich Craters pointed out the important latitudes of 18° S and 27° S respectively. For these reasons, I measured the coordinates of the arrow tips in Fig. 5.1 and found that all of them marked a very meaningful longitude or latitude or both (see caption to Fig. 5.1). Note that the upper case Greek letter Φ is used to represent the inverse of the golden ratio φ (i.e., $\Phi = 1/\varphi = 0.6180$).

Arrows Inside Craters

Besides the existence of craters whose perimeters are shaped like an arrow, there are a number of craters that have an arrow shape in the crater's interior (Fig. 5.2). This phenomenon has been encountered before with an east-pointing arrow shape located in the caldera of Issedon Tholus (*Intelligent Mars II*). The tip of that arrow marked the remarkable

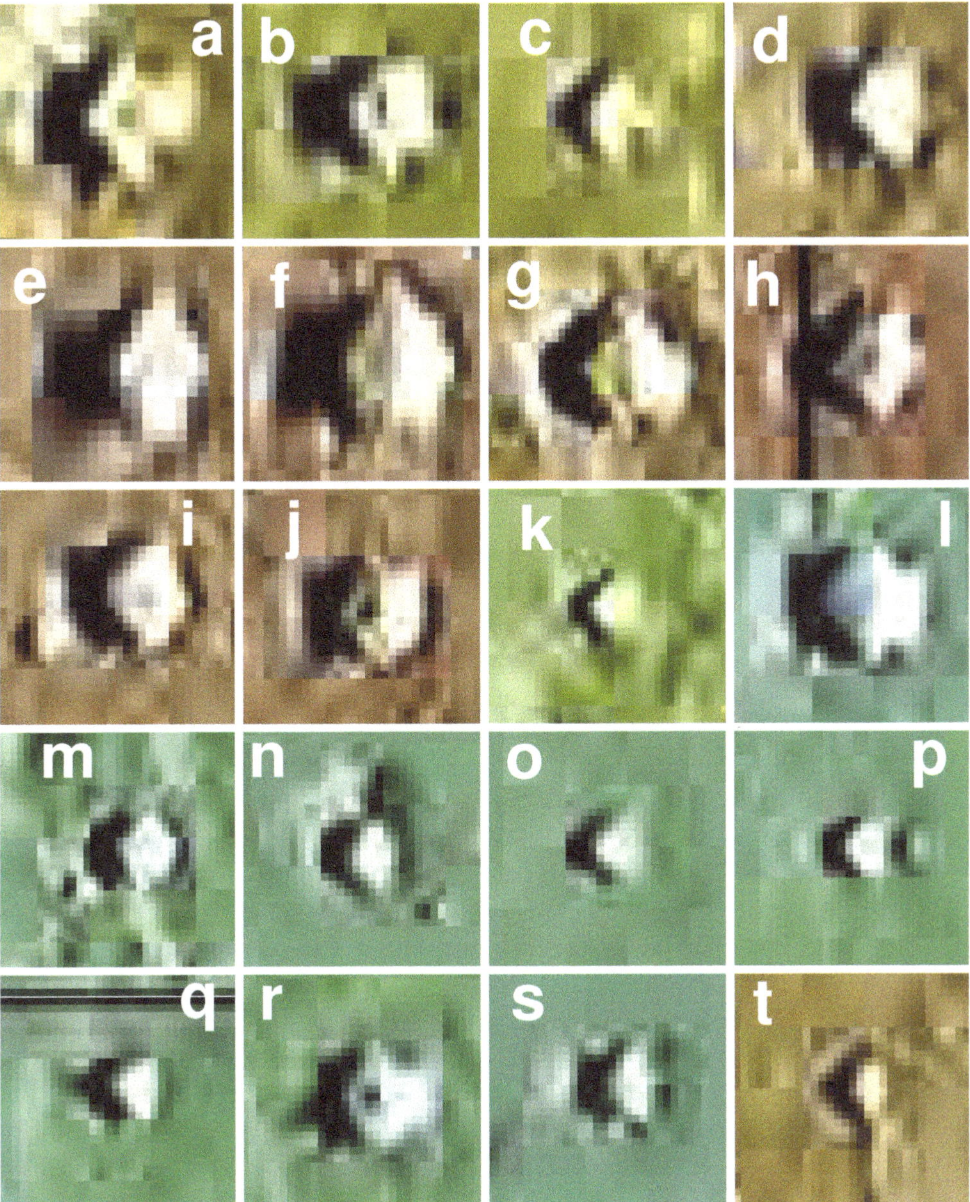

Fig. 5.2: *Arrow shapes in crater interiors in the region containing squares and octagons expanded outwards from the Aum Crater. Notable coordinates of arrow tips are:*

(a) $54\sqrt{3}^{\circ}$ *W [SToPM]* *(b)* 9.5° *S [AMPL]*

(c) $\sqrt{2}/\varphi$ *big rad E [DMPM],* $1/(10\sqrt{2})$ *big rad S [AMPL]*

(d) $23.5^{\circ\circ}$ *S* *(e)* $10\varphi^{\circ\circ\circ}$ *S [AMPL]*

(f) $62\varphi^{\circ}$ *W [STrPM]* *(g)* $54\Phi^{\circ\circ}$ *E [CEPM],* $7e^{\circ}$ *S [AMPL]*

(h) $17\Phi^{\circ\circ\circ}$ *W [DMPM]* *(i)* $\sqrt{2}^{\circ\circ}$ *W [DMPM]*

(j) $(\varphi^2)^{\circ\circ\circ}$ *W [DMPM],* $6^{\circ\circ}$ *S [AMPL]* *(k)* $(\pi/\sqrt{5})^{\circ\circ\circ}$ *S [AMPL]*

(l) $25\Phi^{\circ\circ\circ}$ *E [CEPM],* $(\varphi^2)^{\circ\circ\circ}$ *N [AMPL]* *(m)* $(e^2)^{\circ}$ *W [CEPM],* $10\Phi^{\circ\circ\circ}$ *N [AMPL]*

.....Cont'd page 88

.....Cont'd from page 87

(n) $(\varphi^2/5)°$ S *(o) $1^{\circ\circ\circ}$ E [CEPM], $(\sqrt{2}/4)^{\circ\circ\circ}$ N*
(p) $2^{\circ\circ}$ E [EMPM], $(e/4)^{\circ\circ\circ}$ S *(q) $\sqrt{2}°$ E [DPPM]*
(r) $3\sqrt{3}^{\circ\circ\circ}$ W [DMPM], $14.5^{\circ\circ\circ}$ N [AMPL] *(s) $100\Phi^{\circ\circ\circ}$ W [PMPM]*
(t) $2e^{\circ\circ}$ E [DPPM], $4\sqrt{3}°$ S [AMPL]. *USGS Astrogeology.*

coordinates of 36° W [SToPM] 36° N. These arrows are always whitish in colour on the MOLA maps which could signal either an elevation or a rising slope. An astounding feature of the arrows in Fig. 5.2 is that they all point due west except for the arrows in images e and f which point due north. Of course, the probability of this happening by randomness is so small that we can only conclude that they have been artificially constructed. Except for the north-pointing arrows, all of the arrow edges present linear lengths with a bearing angle of 45° in the clockwise or counterclockwise direction so they can be used as alignment sites for either the squares or octagons emanating from the Aum Crater. The east and west sides of the north-pointing arrows in Fig. 5.2 have bearing angles of plus or minus 30°. The tips of the arrows all mark very meaningful longitudes and/or latitudes which are listed in the caption to the figure.

Craters Shaped Like Blunted Arrows

Besides the arrow-shaped craters and craters with arrows in their interiors, there is another major class of craters possessing an arrow shape. These are craters which have the shape of a blunted arrow. Examples of such craters are shown in Fig. 5.3. These craters are either within or in close proximity to the region of the planet which contains squares and octagons expanded outwards from the Aum Crater, and all point either to the east or to the west. All of the edges of the blunted arrowhead shapes have bearing angles of 45° in the clockwise or counterclockwise direction except for the northern edges of the arrowheads in images e and i which have a clockwise bearing angle of 60°. The angle of 45° makes them ideal for alignment to squares and octagons expanded outwards from the Aum Crater. The edges of the blunted arrowheads can be extrapolated to meet at the virtual tips of their arrows (see Fig. 5.4), allowing tip coordinates to be calculated. It was found that all of these virtual tips mark either a meaningful longitude or latitude or both. These are listed in the caption to Fig. 5.3.

Non-standard arrow craters

There is yet another class of craters containing an arrow shape and that is

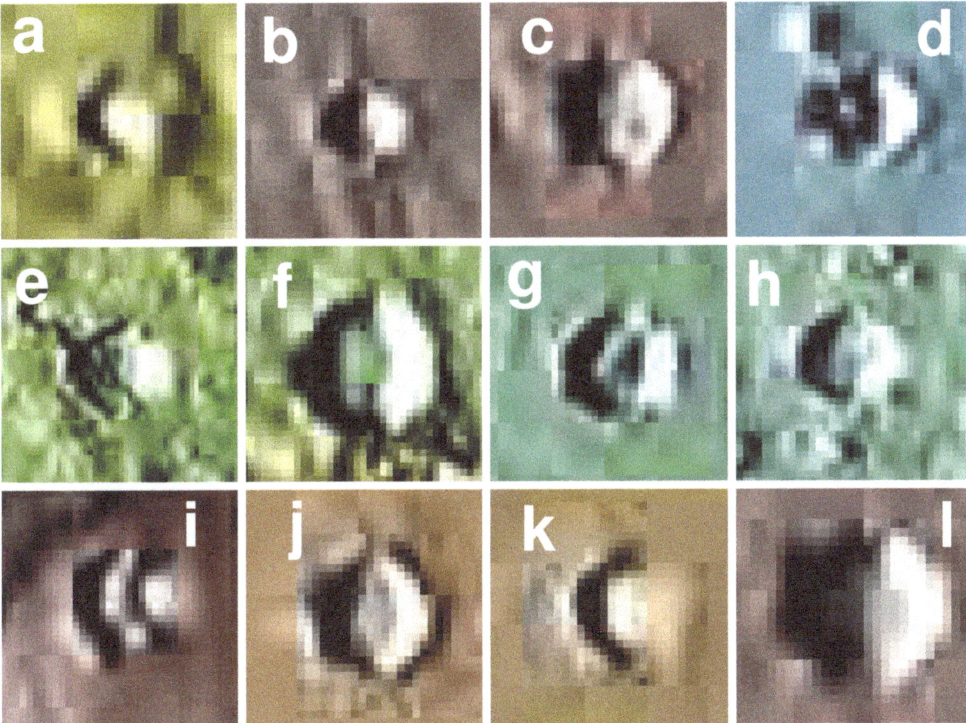

Fig. 5.3: *Craters whose perimeters form a blunted arrow shape located in or near the region containing squares and octagons expanded outwards from the Aum Crater. Notable coordinates of the arrow tips produced by extrapolating the arrowhead sides are:*

(a) $50^{\circ\circ}$ E [CEPM]
(b) $67\varphi^{\circ}$ E [DMPM], $11^{\circ\circ}$ S [AMPL]
(c) $55\Phi^{\circ\circ}$ W [STrPM], $3\pi^{\circ}$ S [AMPL]
(d) $18\sqrt{3}^{\circ\circ\circ}$ E [EMPM]
(e) $153^{\circ\circ}$ W [STrPM], $(\sqrt{2}/2)^{\circ}$ S [AMPL]
(f) $(\varphi^2/3)^{\circ}$ W [CEPM], $(e/2)^{\circ\circ}$ S [AMPL]
(g) $7e^{\circ\circ\circ}$ E [CEPM]
(h) $51\sqrt{2}^{\circ}$ W [PMPM]
(i) $(2\pi^2)^{\circ\circ}$ E [PCPM]
(j) $2\varphi^{\circ}$ W [PCPM], $9\sqrt{3}^{\circ}$ N
(k) $1/\pi$ rad E [PCPM]
(l) $(\sqrt{2}/2)^{\circ}$ W [PCPM], $25\Phi^{\circ}$ S.
USGS Astrogeology.

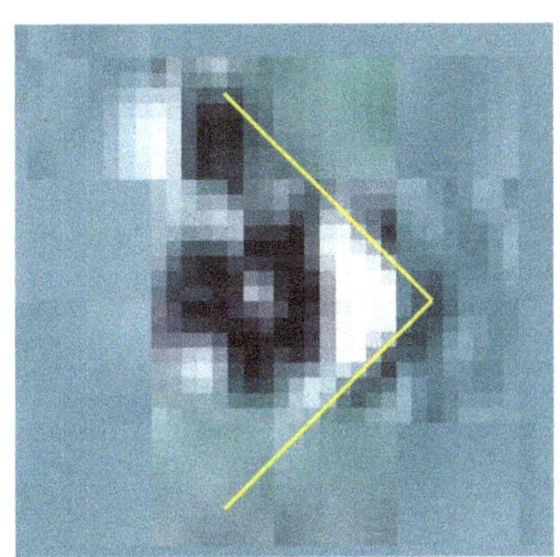

Fig. 5.4: *The arrowhead edges can be extrapolated to mark out the virtual tip of a blunted arrow-shaped crater. The coordinates of the virtual tip can then be calculated. This arrow crater is the same as in image d in the previous figure. USGS Astrogeology.*

a class of crater having an arrow shape that is non-standard or asymmetric. Sometimes the shape is quite covert, requiring a careful examination to point it out. Examples of such craters are presented in Fig. 5.5. All of these craters are within or close to the region containing squares and octagons expanded outwards from the Aum Crater. In image a, the arrow tip is extremely elongated. For the northern side, a 45° bearing angle is present near the tip whereas further to the west there is a linear section that has a bearing angle of 60°. The crater in image b is basically a covert south-pointing arrow with truncated sides, leaving an opening where the tip should be. The arrowhead in the crater in image c is pointing east. The longitude for the 2 lines which fit its tip is given in the caption to

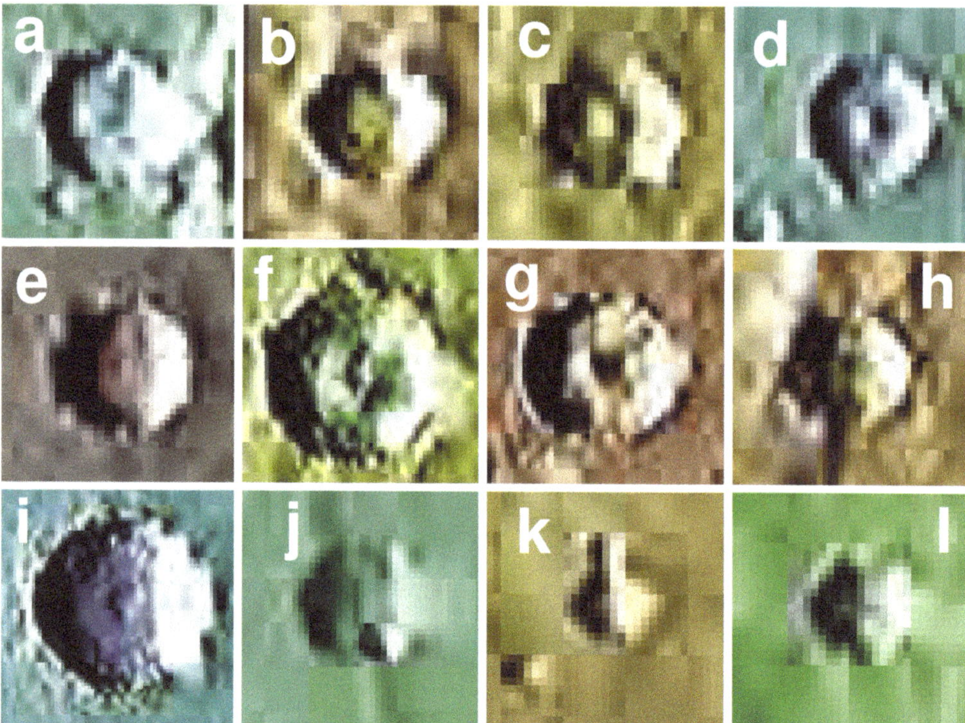

Fig. 5.5: *Craters having non-standard arrow shapes, some of which are covert. Notable coordinates of the arrow tips or tips obtained by extrapolating linear edges are:*
(a) $13(\pi^2)^\circ$ W [SToPM] *(b)* $17\sqrt{3}^{\circ\circ}$ E [CEPM]
(c) $12(\pi)^\circ$ W [PMPM] *(d)* $108^{\circ\circ}$ W [SToPM]
(e) $50\sqrt{5}^\circ$ E [CEPM], $8\varphi^\circ$ S *(f)* $31(\varphi^2)^{\circ\circ\circ}$ W [SToPM], $e/(10\pi)$ *big rad* S [AMPL]
(g) 108° W [PCPM], $33\Phi^{\circ\circ}$ S *(h)* $33(\varphi^2)^\circ$ W [SToPM], $11\sqrt{5}^\circ$ S
(i) $22\varphi^{\circ\circ}$ W [PCPM], $13\Phi^{\circ\circ\circ}$ N *(j)* $25.5^{\circ\circ}$ E [CEPM], $(2/3)(\varphi^2)^{\circ\circ\circ}$ S
(k) West pointing arrowhead: $16\sqrt{5}^\circ$ W [PCPM], 29° S
South pointing arrowhead: $33e^\circ$ W {SToPM], $13\sqrt{5}^\circ$ S
East pointing arrowhead: $28.5^{\circ\circ\circ}$ W [PCPM], 29° S
(l) $27\sqrt{2}^{\circ\circ\circ}$ W [PCPM], $(\varphi/e)^\circ$ S [AMPL]. *USGS Astrogeology.*

Fig. 5.5. The south side of the main arrowhead has another very small arrowhead projecting outwards which points southeast. Its coordinates are the meaningful values of $13(\varphi^2)^{\circ\circ\circ}$ W [PCPM] $6\sqrt{2}^{\circ\circ\circ}$ S [AMPL]. The crater in image d looks round until you realize that its western edges form a blunt arrowhead pointing west. When the edges are extrapolated to a tip, the longitude of the tip is the highly significant value of 108°° W [SToPM]. Here the numerical value of 108 is being honoured despite the use of big degrees. The number 108 in regular degrees is the value of the internal angles of a pentagon.

In the second row of Fig. 5.5, the crater in image e also looks round but forms a covert arrow with an extremely blunted arrowhead which points due east. There is a very asymmetrical arrow shape in image f which points due north. The east side of the arrowhead is much shorter than the west side, and the tip of the arrowhead is well west of the north-south midline of the crater. The crater in image g is a good example of a covert arrow crater. It simulates a round crater but its perimeter has linear sections on the northeast and southeast sides. Both of these linear sections have bearing angles of 45°and form a tip if extrapolated eastwards. The virtual tip of the arrow has the remarkable coordinates of 108° W [PCPM], $33\Phi^{\circ\circ}$ S. As mentioned above, 108° is the value of the internal angles of a pentagon. The number 33 is highly important in Freemasonry. The crater in image h has an asymmetrical covert arrowhead pointing due east. The arrowhead is very blunted and the linear part of the arrowhead's southern side is shorter than the linear part of its northern side. These linear sections of the north and south sides have bearing angles of 45° which can be extrapolated to a tip pointing due east. Once again we encounter the number 33, this time in the longitude measurement from the Sharonov Tower [$33(\varphi^2)^{\circ}$ W].

The third row presents some more highly unusual arrow craters. In image i we see an arrow crater with what at first appears to be a rounded tip. There is, however, a short linear region in the northern side of the arrowhead with a clockwise bearing angle of 45°, and a much longer linear region in the southern side of the arrowhead with a counterclockwise bearing angle of 45°. These linear portions can be extrapolated to form a west-pointing arrow tip with coordinate values of $22\varphi^{\circ\circ}$ W [PCPM] $13\Phi^{\circ\circ\circ}$ N. The arrow crater in image j looks like a rocket or missile travelling due west. The arrow crater in image k is most unusual in that it presents 3 arrow tips, each of which has meaningful coordinates (see caption to Fig. 5.5). The arrow tips point east, south and west. The arrow crater in image l points due west. The shaft of the arrow narrows and then expands. All of the actual and virtual arrow tips of the craters in Fig. 5.5 mark a meaningful longitude or latitude or both. These are listed in the caption to Fig. 5.5.

Arrow craters in other regions

Craters with arrow shapes are not restricted to the region associated with squares and octagons expanding outwards from the Aum Crater. Many of these craters can be found in one form or another in most regions of the planet. For the purposes of this book, I restricted myself to the area between the latitudes of 30° N and 30° S as the Mercator MOLA maps made coordinate calculations easier. Fig. 5.6 presents a sampling of these craters. All of these craters point due east or due west except for the crater in image a which points due south. The arrowhead in this crater has a blunt tip, and the sides have bearing angles of 30° with respect to due north. Other craters whose arrowhead sides have bearing angles of 30° are those in images f and j. The arrowhead in image l is located just west of the Denning Crater. Its sides are most unusual since they have bearing angles of 54° with respect to due north. This angle size is 1/2 the size of the internal angles of a pentagon or of the angles between the star points in a pentagram. The latitude of the tip of this arrowhead is very remarkable in that it is equal to atan(1/(φ√5)) = 15.4504°° S. Although it is in big degrees, the numerical value of this latitude is the same as that for the angle by which the pentagram pyramid is rotated counterclockwise from having a star point aimed due north. This value is also the °W longitude of the pentagram pyramid from the Pavonis Mons Prime Meridian. It should be noted that atan(1/(φ√5)) is virtually indistinguishable from 25Φ = 15.4508, so both sacred formulae were probably intended to be represented.

All of the remaining arrow shapes in Fig. 5.6 have bearing angles of 45° which is the bearing angle most often found for the arrow shapes surrounding the Aum Crater. Each of the arrow shapes in Fig. 5.6 are found in the perimeters of the craters except for Fig. 5.6e which has its clearest arrow shape inside the crater. Like all of the other craters with arrow shapes, the tips of the arrowheads in Fig. 5.6 have either a meaningful longitude or a meaningful latitude, or both. These coordinate values are listed in the caption to the figure.

Summary and Conclusions

An extremely important category of craters exists on Mars which has many subtypes. They can collectively be called arrow craters since they all possess 2 linear regions which, when combined, will form an arrowhead shape or provide a template for 2 lines to extrapolate to form a

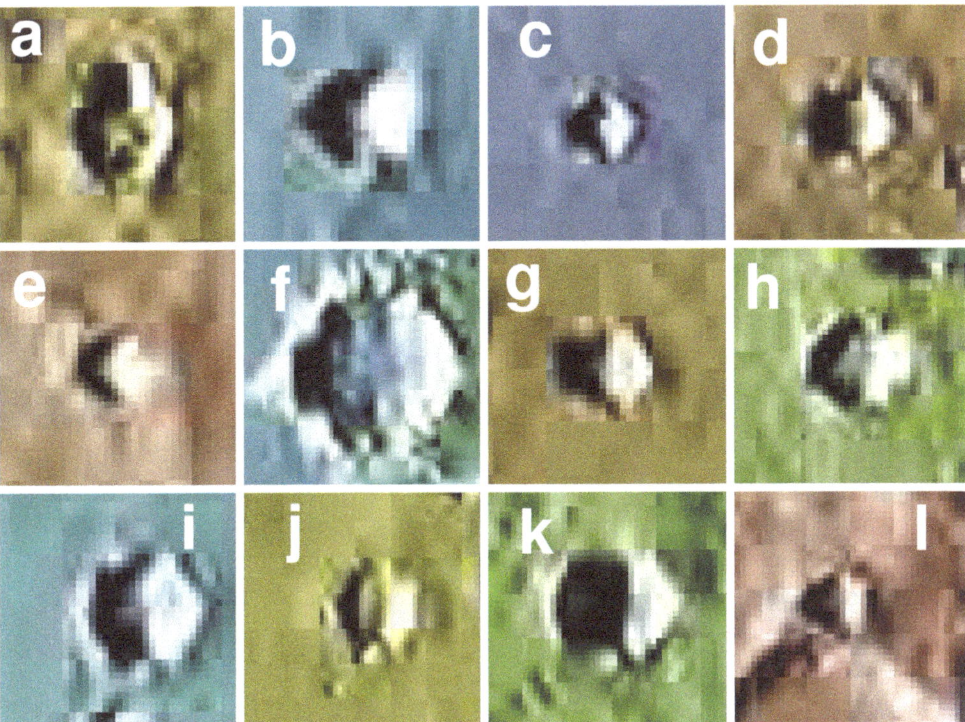

Fig. 5.6: *Craters with arrow shapes found in regions not surrounding the Aum Crater. Notable coordinates of the arrow tips are:*
(a) $34\pi^{\circ\circ}$ *W [DMPM],* $(\varphi/2)^{\circ\circ\circ}$ *N*
(b) $49\pi^{\circ}$ *W [PCPM]*
(c) $14e^{\circ}$ *W [CEPM],* $11e^{\circ\circ\circ}$ *N [AMPL]*
(d) $11.5^{\circ\circ\circ}$ *S*
(e) $4\sqrt{2}^{\circ\circ}$ *S*
(f) $2\varphi^{\circ\circ}$ *E [SToPM],* $1/(\sqrt{2}\sqrt{3})$ *big rad N*
(g) $5\Phi^{\circ\circ}$ *E [STrPM]*
(h) $39\sqrt{2}^{\circ\circ}$ *E [PCPM],* $6e^{\circ}$ *N [AMPL]*
(i) $15(\pi^{2})^{\circ\circ}$ *E [EMPM]*
(j) $6e^{\circ\circ}$ *E [SToPM]*
(k) $21\sqrt{3}^{\circ\circ\circ}$ *E [STrPM],* $1/\pi$ *big rad S*
(l) $14(e^{2})^{\circ\circ}$ *W [DPPM],* $atan(1/(\varphi\sqrt{5}))^{\circ\circ}$ *S*
USGS Astrogeology.

virtual arrowhead as in the case of blunt arrowheads. These craters very often contain a region which acts as the shaft of the arrowhead, but sometimes the shaft is absent or very distorted. Sometimes the arrow crater will look almost perfectly round except for the 2 linear regions forming the arrowhead (or blunt arrowhead). Most often the arrowhead is formed by the perimeter of the crater but there are also several instances in which the arrowhead is completely buried within the interior of the crater.

The tips of the arrowheads always occur at a meaningful longitude or latitude or both. This reveals that a major purpose of the arrowheads is to pay homage to one or more numerical values. There is no evidence that the arrow craters were used to create a coordinate grid, since most of the coordinate values are sacred formulae rather than integers or integers plus 1/2. Sometimes the value is in radians rather than degrees. Sometimes the value of big or sacred degrees is a meaningful integer number which is used for symbolic purposes only. For instance, the meaningful integer value of 108 big degrees for the arrow tip in Fig. 5.5d does not resolve to the proper magnitude for the internal angle of a pentagon. Instead it resolves to 121.5°. Hence, the use of 108 in this instance simply pays homage to the number 108.

Another role that the arrow craters play is to provide a linear edge for alignment to a virtual geometric figure such as an octagon or square. The high prevalence of arrow shapes in the region surrounding the Aum Crater having linear edges with clockwise or counterclockwise bearing angles of 45° makes it possible to fit multiple sizes of squares and octagons expanding outwards from the Aum Crater. It was found that some of the arrow shapes have linear edges with bearing angles of 30° or 60°, but in the region surrounding the Aum Crater, it was often only one of the sides of the arrow shape with such bearing angles, and that side was not in a position to align to a square or octagon radiating outwards from the Aum Crater. In areas which are not in the vicinity of the Aum Crater, there are a few arrow shapes with sides having bearing angles of 30° and other angle sizes, but even here, a bearing angle of 45° was a common finding for arrow shapes. The arrow shapes in these areas were possibly intended to fit geometric figures but I have not explored this possibility yet.

A third purpose of the arrow shapes may have been to pay homage to sacred geometry in the bearing angles of their sides. Thus 45° would pay homage to the square (45° is the angle created by the diagonal of a square), 30° and 60° would pay homage to the equilateral triangle (60° is value of the internal angles of an equilateral triangle and 30° is 1/2 of this value). The bearing angle of 54° for the arrow crater near the Denning Crater (see Fig. 5.6"l") would pay homage to the pentagram (54° is 1/2 the size of the angles of the internal pentagon or the angles between the star points of a pentagram).

The arrow craters offer immediate proof of the existence of intelligent life on Mars in the distant past which may persist to the present day. These craters could not have arisen by natural forces alone since impacts result in craters which are round. Even if abnormalities in the terrain at the point of impact might have resulted in linear edges on occasion, for this to happen repeatedly for over 100 craters is inconceivable. Besides, the direction of the arrowhead is overwhelmingly either due east or due west with a few pointing due north or due south. This could only have resulted from intelligent engineering.

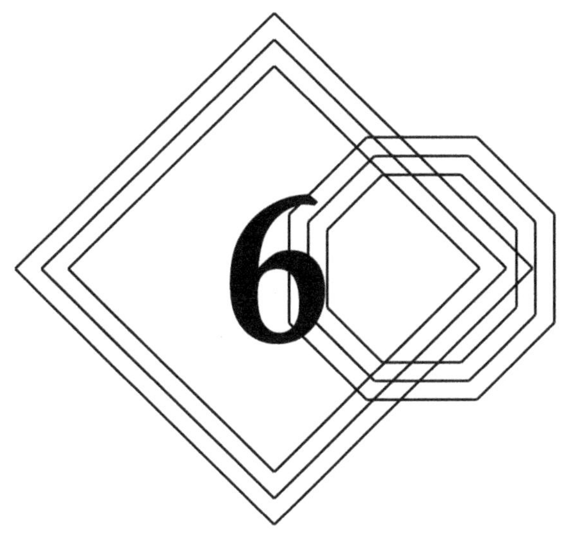

Grooves, Ridges and 7.5

This chapter will deal with the observation of very long linear grooves and ridges on the surface of Mars found at several different locations around the planet. The initial reaction is to regard the grooves and ridges as natural phenomena but when you examine them in more detail, it becomes very difficult to account for them by any kind of natural process. Some of the grooves are extremely long, one of which travels intermittently for over 2500 km. They usually go for some distance, stop, and then start over again several kilometers away. Different grooves can be anywhere from 1 to 3 km in width. They cannot be fractures since they are too linear and too uniform in width. They are highly unlikely to be caused by ancient glaciers since glacial grooves found on Earth only travel for about 120 meters at most and are no more than about 10 meters wide. The Martian grooves also maintain bearing angle independent of latitude over great distances so they follow rhumb lines rather than great circles. It is very difficult to imagine a natural phenomenon that could give rise to extremely lengthy rhumb lines.

Grooves on Pavonis Mons

The first groove that caught my attention was a long groove on the upper surface of Pavonis Mons (yellow arrows, Fig. 6.1). This groove starts at the south end of Pavonis Mons and travels up the mountain in a northwest direction until it reaches the outer edge of the peak of the mountain. It starts up again on the southern wall of the caldera, disappears on its bottom surface and re-emerges on the northern wall of the caldera. Then it travels intermittently across the northern section of the mountain, reaches the outer edge of the mountain and then continues into some rough terrain further north. At first glance, it appears as though this entire set of intermittent grooves marks out the path of a single straight line which approximately bisects the mountain into 2 halves. Indeed, a perfectly straight line can be superimposed on the groove, demonstrating its linearity. I did not draw the line on the figure because I did not want it to hide the groove which is only about 1.5 km wide for its northern half and about 3.0 km wide for its southern half. However, the north and south grooves do not align since the line passes through the centre of the groove for the southern half of the mountain and runs along the western edge of the groove for the northern half of the mountain. Closer examination reveals that there is a ridge line following the western side of the northern section of the groove which is absent for the southern section of the groove. Together, the groove and ridge line of the northern side create about the same width as the groove on the mountain's southern side. The ridge line is shown more clearly in Fig. 6.2 where 2 white arrows are used to mark its northern and southern ends. The amazing thing about this ridge line is that its length measures out to be 118.46 km which is almost exactly equal to 2 latitude degrees or 118.55 km. Hence, the ridge line on Pavonis Mons could be used by overhead spacecraft for calibration purposes. A short section of a ridge line is also present on the southern edge of the caldera which is in line with the ridge line on the northern side of the mountain.

The total length of the northern groove line from the rough terrain beyond the northern edge of Pavonis Mons to the groove mark touching the southern edge of the caldera is about 320 km. This length is long enough to allow a fairly accurate estimate of its bearing angle. I measured the bearing angle of this particular groove line to be 7.5° in the counter-clockwise direction. At first this did not seem too remarkable until I discovered that the bearing angle of the vast majority of the other grooves was also 7.5° either in the clockwise or counterclockwise direction.

If we focus on an area north of the main caldera just beyond the northern edge of what appears to be the remnants of an older larger caldera, we see that the midline of the groove passes directly through the survey centre (upper yellow cross in Fig. 6.3) of Pavonis Mons which is 100° E of the Crater Edge Prime Meridian. Going south, the groove line then passes through the northern perimeter of the Pavonis Mons Caldera at a latitude of 0.9101° N = $(\varphi/2)^{\circ\circ}$ N. The groove line continues through a groove in the southern wall of the Pavonis Mons Caldera and terminates

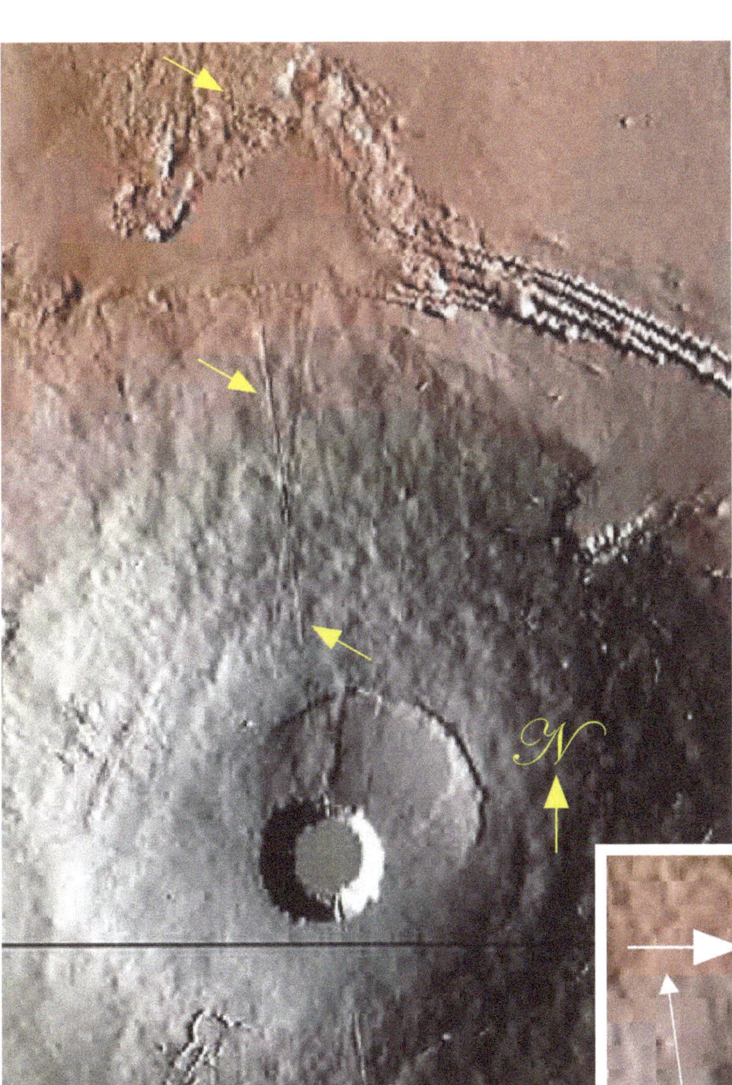

Fig. 6.1: *Long linear groove (see yellow arrows) across the surface of Pavonis Mons. It extends beyond the mountain to the north. The bearing angle of the groove is 7.5° in the counterclockwise direction. The width of the groove is about 1.5 km for the northern half of the mountain and about 3.0 km for the southern half. The horizontal black line is the equator. USGS Astrogeology.*

2°

Fig. 6.2: *A ridge line on the northern half of Pavonis Mons lies parallel to the groove line (yellow arrows). It shows up as a lighter line which runs along the western border of the groove line. The northern and southern ends of the ridge line are marked by horizontal white arrows. The ridge line is 2 latitude degrees in length.*

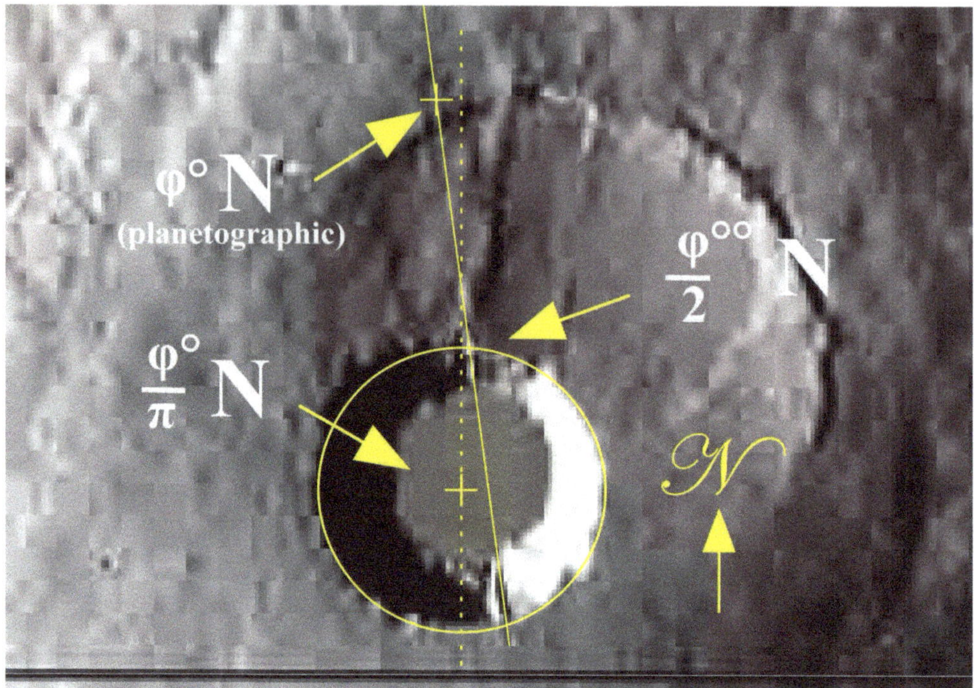

Fig. 6.3: *Pavonis Mons northern groove line marked with a solid yellow line. It passes directly through the survey centre for Pavonis Mons (upper yellow cross) and intersects the northern perimeter of the Pavonis Mons Caldera at a latitude of (φ/2)°° N. The lower yellow cross marks the centre of the Pavonis Mons Caldera. The dashed yellow line lies on the Pavonis Caldera Prime Meridian. USGS Astrogeology.*

at the end of a short groove just beyond the caldera southern perimeter.

Hence, the groove on the northern half of Pavonis Mons seems to act as a marker of the survey centre of the mountain and also marks the latitude of (φ/2)°° N in conjunction with the northern edge of the Pavonis Mons Caldera. The latitude value of (φ/2)°° N offers strong support for the use of big degrees and also offers a great deal of support to my theory in *Intelligent Mars II* that Pavonis Mons is associated with the golden ratio φ. In *Intelligent Mars I*, it was pointed out that the survey centre of Pavonis Mons is at φ° N (planetographic coordinates). Planetographic coordinates take into account the fact that the planet is an oblate spheroid rather than a perfect sphere. The other 2 coordinate values in Fig. 6.3 are in planetocentric coordinates which treat the planet as a perfect sphere of radius R, the equatorial radius of Mars. The dashed vertical line in Fig. 6.3 is 100° E of the Elysium Mons Prime Meridian and passes through the centre of the Pavonis Mons Caldera (lower yellow cross) which is at a latitude of (φ/π)° N. It also passes through the northern perimeter of the

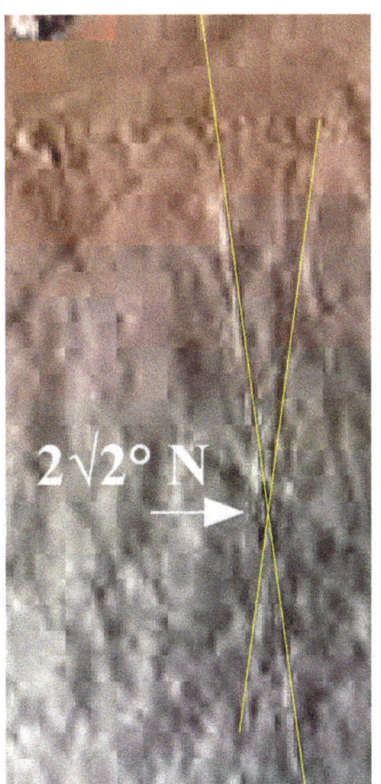

Fig. 6.4: *A second groove line on the northern half of Pavonis Mons intersects the first at 2√2° N. It has a clockwise bearing angle of 7.5°. USGS Astrogeology.*

Pavonis Mons Caldera at a latitude of (φ/2)°° N. The circle in Fig. 6.3 which fits the perimeter of the caldera has a diameter value (46.84 km) which is very close to the length of π/4 latitude degrees (46.55 km). Thus both φ and π are honoured by Pavonis Mons, but the emphasis appears to be on φ.

Approximately halfway up the northern section of the mountain, the long groove is intersected by a second groove which has a bearing angle of 7.5° in the clockwise direction (Fig. 6.4). This groove extends southward from the northern edge of the mountain to about 2/3 of the way to the caldera region. The latitude of the point of intersection is 2.8284° N which is exactly equal to 2√2° N. The longitude of the intersection point is 1/(e√5) = 0.1645° W (Pavonis Mons PM).

Finally, the fit of the groove on the southern half of the mountain to a straight line is shown in Fig. 6.5 (upper image). When this straight line is extended southwards for more than 1300 km, it aligns perfectly to 4 short grooves (Fig. 6.5 lower image). To the south of these 4 grooves is another set of 4 grooves (lowest yellow line) which does not align to the first set. The straight line which fits them is about 1 km to the west of the line extended from Pavonis Mons. Note that the lower image is more than 2 degrees east of the upper image so the dashed line does not actually connect at the junction between the 2 images, but simply indicates that the line is common to both images.

Other Grooves

So far we have seen 2 groove lines and 1 ridge line on Pavonis Mons with a bearing angle of 7.5° in the counterclockwise direction, and 1 groove line with a bearing angle of 7.5° in the clockwise direction. One purpose of the northern grooves seems to be to mark out latitudes and longitudes that are related to sacred geometry. One of them passes through the survey centre for Pavonis Mons which marks the Pavonis Mons Prime Meridian. There are also short groove lines with a bearing angle of 7.5° in the counterclockwise direction much further south of Pavonis Mons. The next

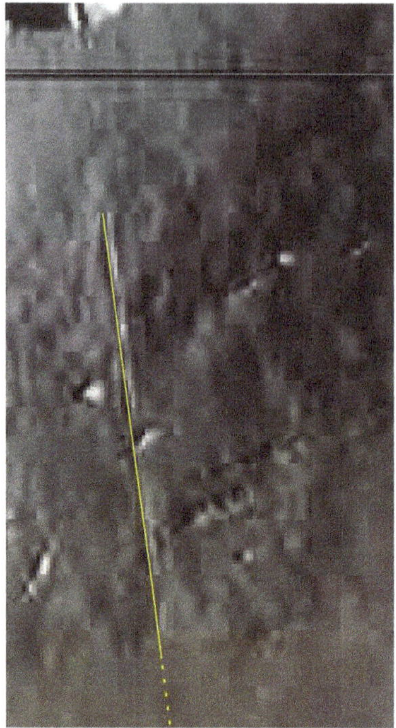

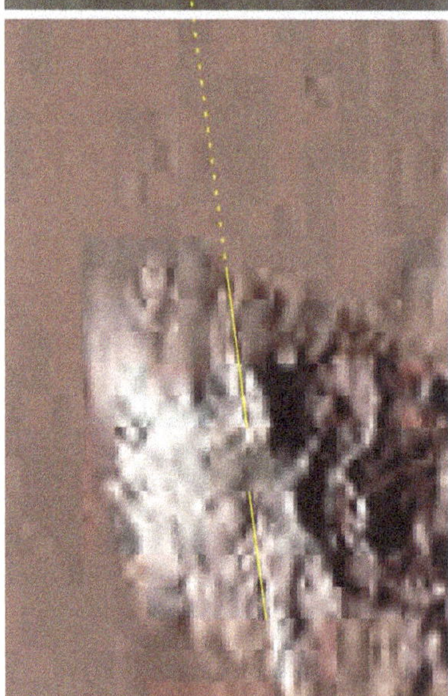

Fig. 6.5: *A line fitting the groove on the southern half of Pavonis Mons (upper figure) is aligned to 4 short grooves far south of the mountain (lower figure). Another 4 grooves are not aligned (lowest yellow line). USGS Astrogeology.*

question to ask is whether or not linear groove lines occur elsewhere or are these isolated phenomena? To answer this question, I decided to systematically scan the planet between 30° S and 30° N starting at 0° E and working eastward until I covered 360° of longitude. The first interesting landmarks that caught my attention were a series of short grooves and ridges between 51° - 52° E and 25° - 29° S (Fig. 6.6). Many of these occurred in and near crater walls. I marked the grooves with yellow arrows in Fig. 6.6 and the ridges with red arrows. I found that I could link all of the grooves with a single straight line having a bearing angle of 7.5° in the counterclockwise direction. A similar line could be drawn which connected all the short ridges. This line did not coincide with the line connecting the grooves, but lay about 0.5 km to the west of it. When these 2 lines were extrapolated to cover the entire distance between 30° S and 30° N it was found that the lines aligned with the perimeters of several craters. There were 9 craters which aligned with the line connecting the ridges and these are shown in the left column of Fig. 6.6. There were also 7 other craters whose perimeters aligned to the line joining the grooves (right column in Fig. 6.6). Was the alignment of so many craters a chance occurrence or had I discovered another function of the grooves and ridges, namely to act as guidelines for the location of craters under construction? Since the edge of a crater is a very much smaller target than the entire crater itself, this result seems unlikely to have occurred randomly.

The next extremely interesting groove line that I came across was composed of a series of short grooves and ridges

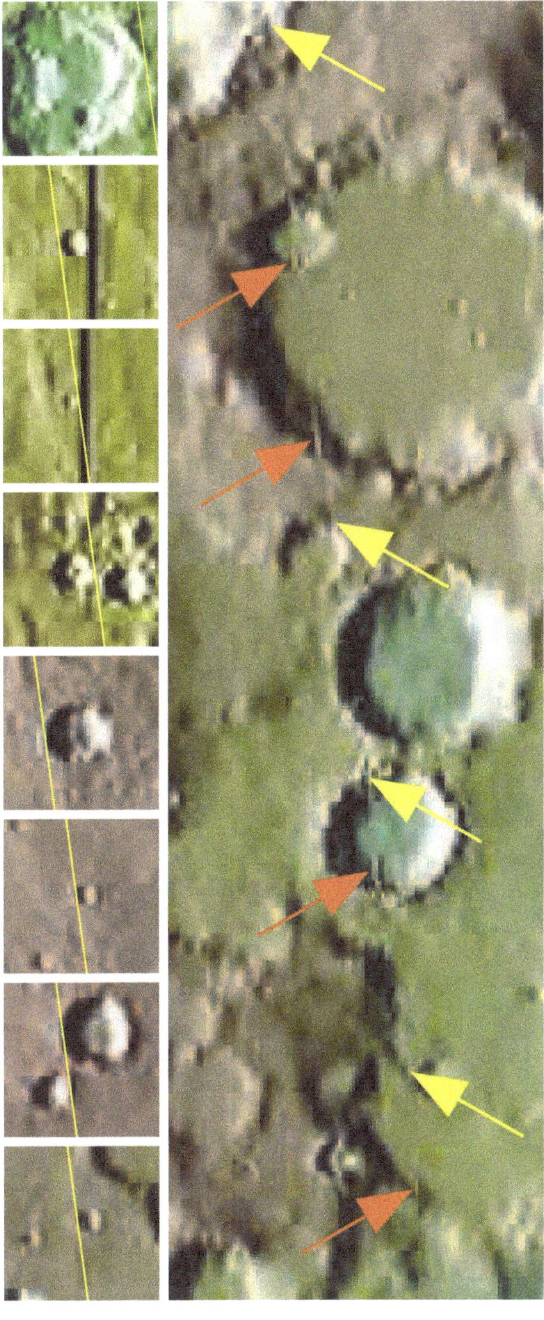

Fig. 6.6: *Large image in the middle shows a series of short grooves and ridges occurring between 51 - 52° E and between 25 - 29° S. A perfectly straight line with a bearing angle of 7.5° in the counter-clockwise direction can be drawn through all the separate grooves (yellow arrows) to connect them. Red arrows point out a series of short ridges which also can be joined by a line which is parallel to the line joining the grooves, but lies about 0.5 km to the west of it. When the ridge line is extrapolated between 30° S and 30° N it is found to align to the perimeters of 9 craters (left column of images). A similar extrapolation of the groove line joining the grooves marked by the yellow arrows aligns to the perimeters of 7 craters shown in the right column of images. USGS Astrogeology.*

extending over an incredible distance of 2223 km from south to north (Fig. 6.7). All of the short grooves and ridges marked by yellow arrows in Fig. 6.7 could be connected with a single straight line with a constant bearing angle of 7.5° in the clockwise direction. The line starts at 109.7433° E 13.2679° S (bottom picture in the left column) and ends at

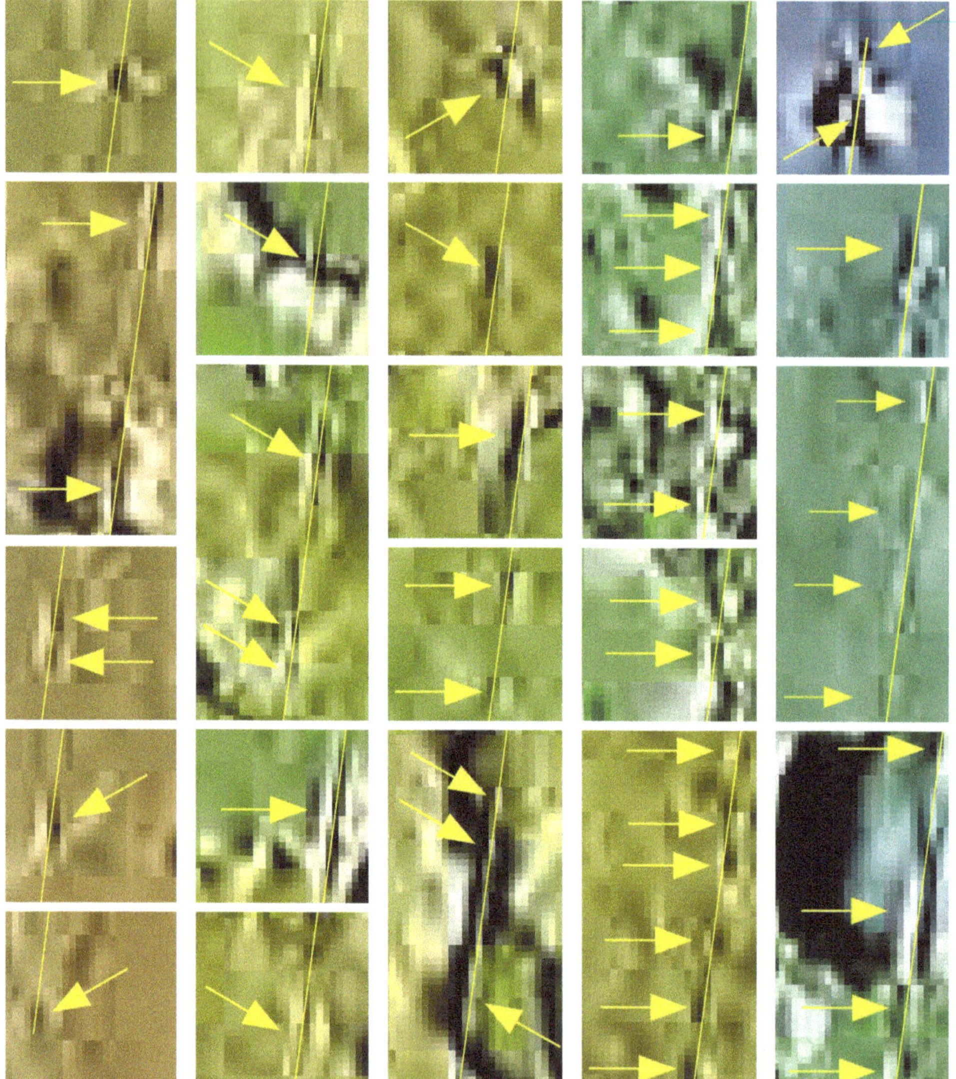

Fig. 6.7: *A groove line composed of multiple short grooves and ridges extending for 2223 km across the surface of Mars. It has a constant bearing angle of 7.5° in the clockwise direction. The pictures are arranged in columns with the groove line starting at 109.7433° E 13.2679° S (bottom picture of left column) and ending at 114.7505° E 23.9179° N (top picture of right column). The pictures travel northwards from bottom to top in each column and from column to column from left to right. The equator lies between the 2 top pictures in the second column from the left. USGS Astrogeology.*

114.7505° E 23.9179° N (top picture in the right column). The pictures travel northwards from bottom to top in each column and from column to column moving from left to right. The equator lies between the 2 top pictures in the second column from the left. Note the series of ridges joined by the line in the bottom picture of the third column. The 2 grooves in the top picture of the right column are found in the Utopia Planitia which is believed by some researchers to be part of a northern ocean in ancient times. The edges of a total of 4 craters were found to be aligned to the groove line when it is extrapolated to cover the entire distance between 30° S and 30° N.

The final long groove line that I will show you has a bearing angle of 7.5° in the counterclockwise direction (Fig. 6.8). Like the previous groove line, it is composed of many intermittent short grooves and ridges, and

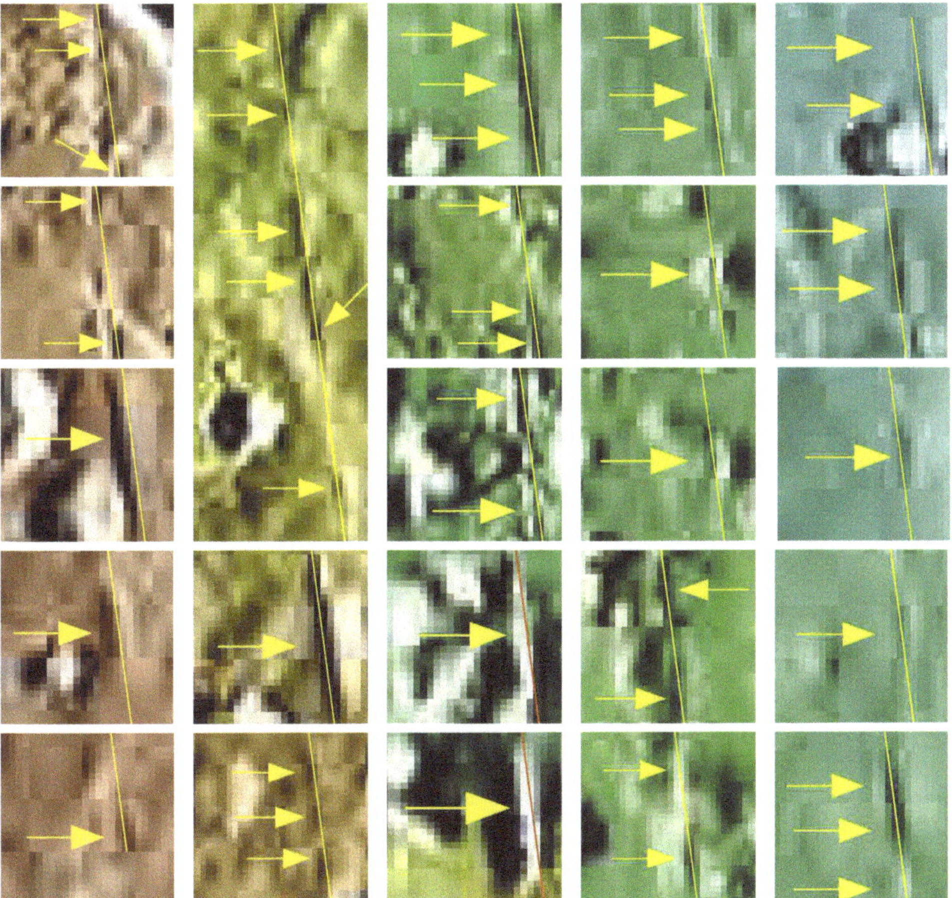

Fig. 6.8: *A groove line composed of multiple short grooves and ridges extending for 2562 km across the surface of Mars. It has a constant bearing angle of 7.5° in the counterclockwise direction. The pictures are arranged in columns with the groove line starting at 125.7993° E 27.3211° S (bottom picture of left column) and ending at 119.9864° E 15.5390° N (top picture of right column). The pictures travel northwards from bottom to top in each column and from column to column from left to right. The equator lies between the 2 bottom pictures in the 2nd column from the left. USGS Astrogeology.*

covers a very long distance, this time an immense 2562 km. It crosses the equator at 122.0576°E and lies between 27.3211° S and 15.5390° N. It is the longest groove line that I came across. Most of the alignments are with grooves or trenches. The ridges appear as linear white regions especially in the bottom 2 pictures in the third column of Fig. 6.8. The edges of a total of 8 craters were found to be aligned to the groove line when it is extrapolated to cover the entire distance between 30° S and 30° N.

Parallel groove lines can be as close as about 0.50 km from one another. Fig. 6.9 shows 2 parallel groove lines which are about 0.65 km apart from one another. They are located at about 122° E, just north of the equator. The line on the left is shown to align to a single groove (top left arrow) while the line on the right aligns to several grooves, 2 of which are pointed out by the yellow arrows on the right. Several more grooves are aligned to each line outside the limits of the image. Each line aligns to the edges of a different set of craters when extrapolated between 30° S and 30° N, the line on the left to 4 craters and the line on the right to 3 craters.

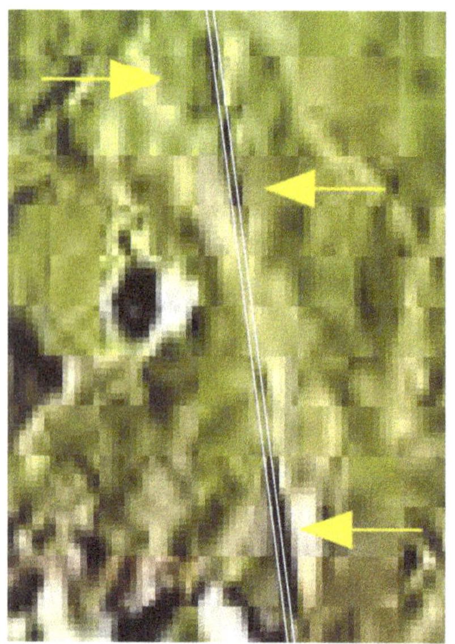

Fig. 6.9: *Parallel groove lines about 0.65 km apart. Each line aligns to a different set of short linear grooves, most of which lie outside the figure. USGS Astrogeology.*

A diagrammatic map of the 61 groove lines which I could identify is shown in Fig. 6.10. The vast majority of these lines connect several short grooves and ridges rather than being a continuous groove. Sometimes the distance between the short grooves and/or ridges making up a line extends more than a hundred kilometers. The groove lines are shown as straight red lines on the diagrammatic map which extends from 0° – 360° in longitude between 30° S and 30° N. Due to the low resolution of the map only about

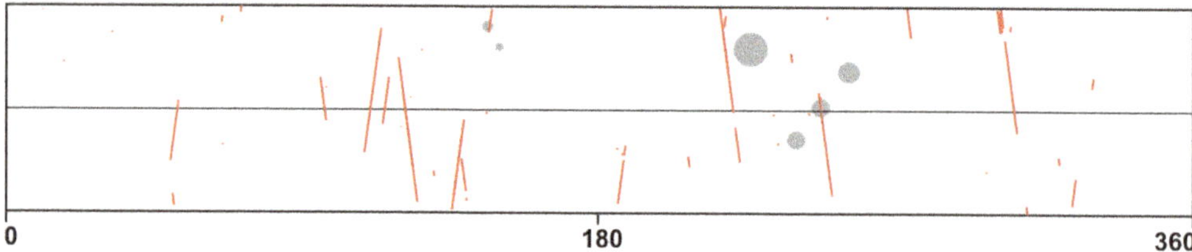

Fig. 6.10: *Location of groove lines across the planet between 30° S and 30° N. Each red line represents a groove line. Grey circles show the locations of the giant mountains.*

half of the lines show up. Any line less than about 200 km appears only as a tiny dot, and some of the lines are so close to one another that they are unresolved. However, this map shows that the groove lines occur at most longitudes and in both the southern and northern hemispheres. Slightly more are in the counterclockwise direction (35) than in the clockwise direction (26) but this difference is not statistically significant. All lines have a clockwise or counterclockwise bearing angle of 7.5°.

Ridge Lines

Besides grooves lines with a bearing of 7.5°, I also discovered several narrow ridge lines which are linear over a considerable distance and have a bearing of 7.5° in the clockwise or counterclockwise direction. One of these has already been shown lying on the surface of Pavonis Mons extending for a distance of 2 latitude degrees. Short linear ridges are sometimes in alignment with a series of short linear grooves and I have included some of these as part of the groove lines discussed in the previous section (e.g., Fig. 6.8). Longer ridge lines have been found which are either in alignment with groove lines or are completely independent of grooves. I found a total of 24 of the longer ridge lines, all having a bearing angle of ± 7.5°. As a group, the ridge lines are much shorter than the groove lines and are mostly intermittent rather than continuous. They vary in length from about 13 km to about 213 km.

Two of the longer ridge lines which I found to be in alignment with groove lines are shown in Fig. 6.11. The ridge line on the left is about 51.6 km long and is found just west of Olympus Mons. The ridge line on the right is about 52.9 km long and occurs in the interior of the Aum Crater.

Several long ridge lines do not align to groove lines. One of these has been shown in Fig. 6.6 (red arrows). The individual ridges which make up this line are very short and spaced at

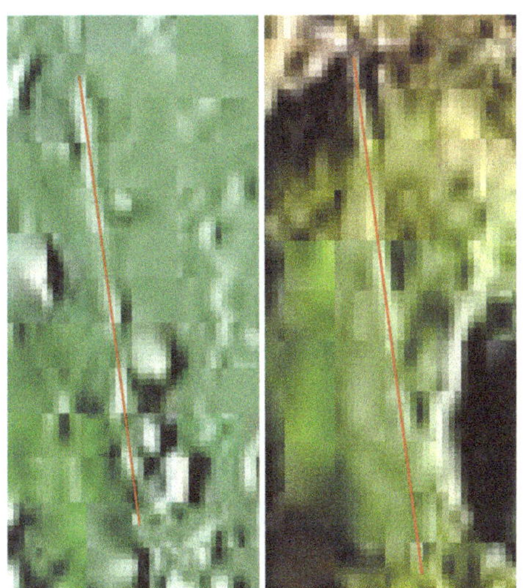

Fig. 6.11: *Examples of longer intermittent ridge lines which are in alignment with groove lines (out of the picture). The ridge line on the left lies just west of Olympus Mons. The ridge line on the right is located inside the Aum Crater. Both ridges have a bearing angle of 7.5° in the counterclockwise direction. USGS Astrogeology.*

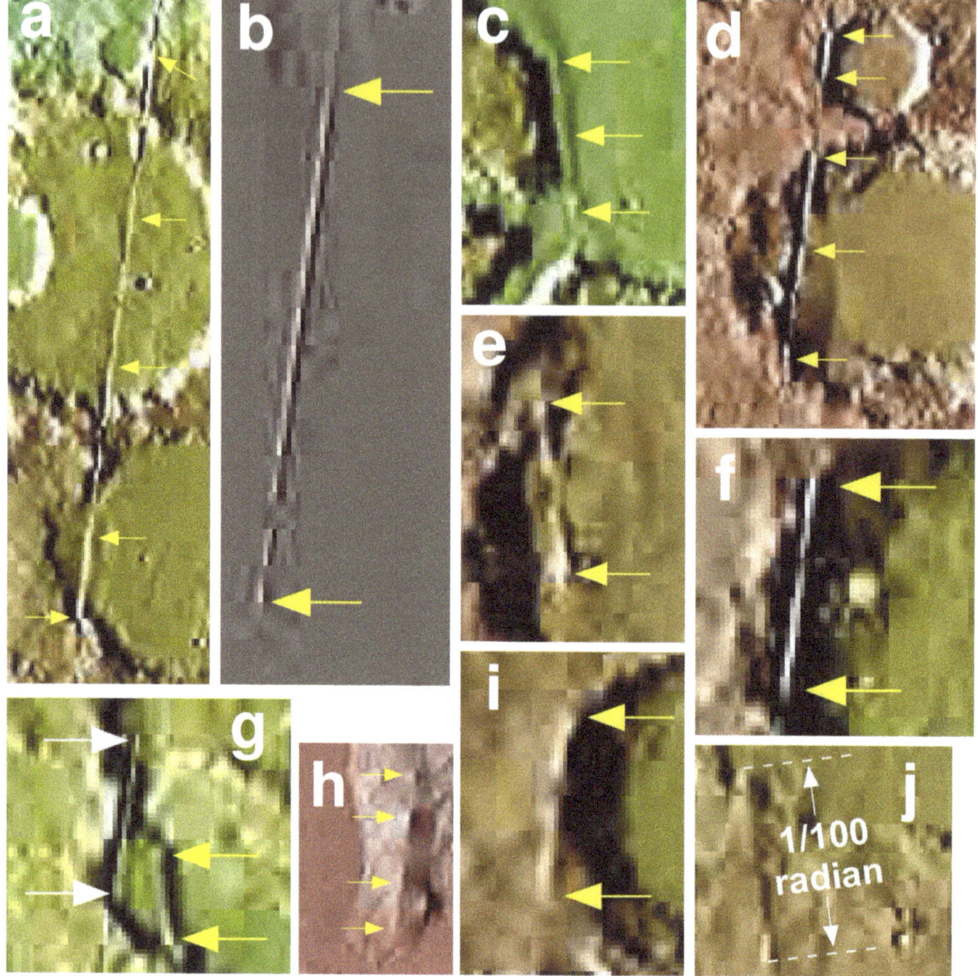

Fig. 6.12: *Ridge lines which do not align to groove lines (except for the left ridge line in picture g - white arrows). Coordinates of the midpoint of each ridge line are:*
a) 188.1102° E 13.2044° S b) 247.3042° E 11.613° S c) 284.9424° E 26.5566° N
d) 137.2824° E 22.0222° S e) 127.475° E 14.6651° S f) 138.8435° E 10.4455° S
g) left 111.6546° E 1.1288° N right 111.7470° E 0.9255° N
h) 249.5228° E 29.1358° S i) 53.6041° E 11.6773° N (j) 0.2746° E 26.1569° S.
The length of the ridge line to the right in picture g is 0.25°. The lengths of the ridge lines in pictures c, e, f and i are each 0.5°. The lengths of the ridge lines in pictures b and d are each 1°°°. USGS Astrogeology.

long intervals. The total length of the line is about 170 km. Another long ridge line is shown in image a of Fig. 6.12. It occurs in the Terra Sirenum region where it traverses the interiors of 2 large craters and is interrupted by only very short breaks. Several grooves lie adjacent on the west side of the ridge line which is about 156 km in total length. A very clear ridge line is shown in image b. It is located to the east of Arsia Mons just west of the Syria Planum. Its total length was measured to be 73.74 km which is very close to the size of 1 sacred degree = 74.09 km. The ridge in image c is located about 17° west of the Sharonov Crater. It is remarkable in that its measured length is 29.69 km in length which is extremely close to the 29.64 km length of 1/2 of a latitude degree. Image d shows a ridge line traversing the western interiors of 2 craters. It is located in the Terra Cimmeria region and its total length was measured to be 74.16 km which is extremely close to the 74.09 km length of 1 sacred degree. The ridge line in image e is located inside the western part of the huge Herschel Crater. Its total length is measured to be 29.93 km which is very close to the 29.64 km length of 1/2 of a latitude degree. The ridge in image f with a measured length of 29.73 km is also approximately 1/2 of a latitude degree. It is located in the Wien Crater at 138.8° E 10.4° S. Image g shows 2 ridge lines, one with a clockwise bearing angle of 7.5° (white arrows) and the other a counterclockwise bearing angle of 7.5° (yellow arrows). The former has a length of 25.54 km and the later, a length of 14.83 km which is very close to the 14.82 km length of 1/4 of a latitude degree. Image h shows a ridge line about 30° south of Pavonis Mons. It is composed of 2 short and 1 long ridge segments, and its total length is about 45 km. The ridge line in image i is located at 53.6° E 11.7° N. With a measured length of 29.56 km, it is very close to 1/2 of a latitude degree in length. The ridge in image j is 1/100 of a radian in length.

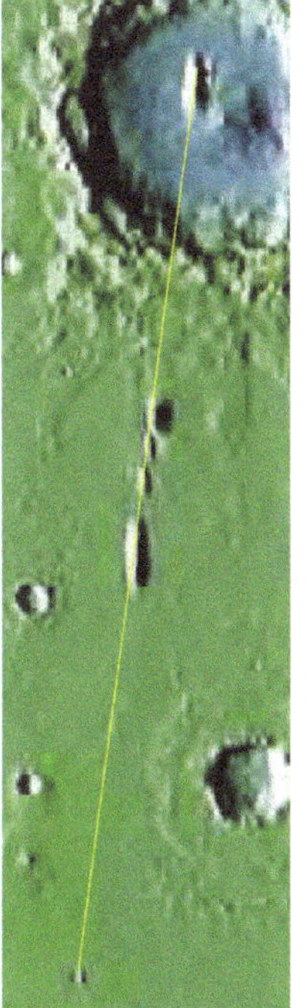

There is also a curious ridge line that is not composed of ridges having bearing angles of 7.5° but instead is made up of several short ridge lines which have bearing angles of 0° (Fig. 6.13). The northern ends of each of the ridges are arranged along a line having a clockwise bearing angle of 7.5°. The most northerly ridge is located inside the Lockyer Crater. This ridge has a longitude of 9.8706°°° E [DPPM] which is very close to $\pi^2 = 9.8696$. The first ridge to the south of the Lockyer Crater has a longitude of 13.0901° E [EMPM] which is very close to $5\varphi^2 = 13.0902$. The next ridge to the south has a longitude of 9.7086°°° E [DPPM] which is very close to $6\varphi = 9.7082$. The third ridge south of the Lockyer Crater has a longitude of 12.1136° E [DPPM] which is close to $7\sqrt{3} = 12.1244$. The longest ridge (the 4th ridge south of

Fig. 6.13: *Ridge line with a clockwise bearing angle of 7.5°. It is composed of 6 short ridges which are oriented in a north-south direction. USGS Astrogeology.*

the Lockyer Crater) has a longitude of 86.9882° W [PCPM] which is close to 32e = 86.9850. The tiny 5th ridge south of the Lockyer Crater has a longitude of 11.3460°° E [EMPM] which is close to asin(1/(φπ)) = 11.3456.

Alignment of Craters to 7.5° in the Absence of Grooves or Ridges

I found a class of craters in which the perimeter of the crater has a sizeable linear segment with a bearing angle of 7.5° in the clockwise or counterclockwise direction yet is not on the path of an extended groove or ridge line. Fig. 6.14 shows a number of examples of these craters at several different longitudes and latitudes which are listed in the caption to the figure.

Fig. 6.15 provides 3 examples where the linear segments from 2 different craters lie on the same line. The first example is shown in Figs. 6.15a and b where the east linear edges of 2 craters lie on the same line having a bearing angle of 7.5° in the clockwise direction. The second example shows that the linear sides of the 2 craters in Figs. 6.15c and d lie on the same line having a bearing angle of 7.5° in the counterclockwise direction. In Fig. 6.15e, the eastern linear sides of 2 craters on opposite sides of the Tithonium Chasma of the Valles Marineris lie on the same line which has a bearing angle of 7.5° in the clockwise direction. Further to the north, this line aligns with the base of a cliff bordering the west side of the Echus Chasma.

The average length of the linear perimeter segments of the 35 craters of this type which I located is about 18 km, with a range from 9.39 km to 42.96 km. Lines extrapolated from the linear segments of these craters to the map limits of 30° S and 30° N were found to be aligned to an average of 4.42 craters (includes the craters with the linear segment). Like some of the ridges, three of the linear segments of the craters had lengths which were virtually identical to the lengths of subunits of latitude degrees. The length of the linear section in the 4th picture of the 3rd row in Fig. 6.14 (image l) is equal to 1/2 of a regular degree. The lengths of the linear section in Fig. 6.14n and of the linear section in the southern crater in Fig. 6.15e are equal to 1/4 of a big degree.

Other Landforms Aligning to 7.5°

Besides craters having linear edges with a bearing angle of 7.5°, there are also instances of other landforms such as calderas, deep trenches, landmasses and cliffs which have linear edge segments with a bearing angle of 7.5°. Fig. 6.16 shows 2 instances where a caldera has a linear

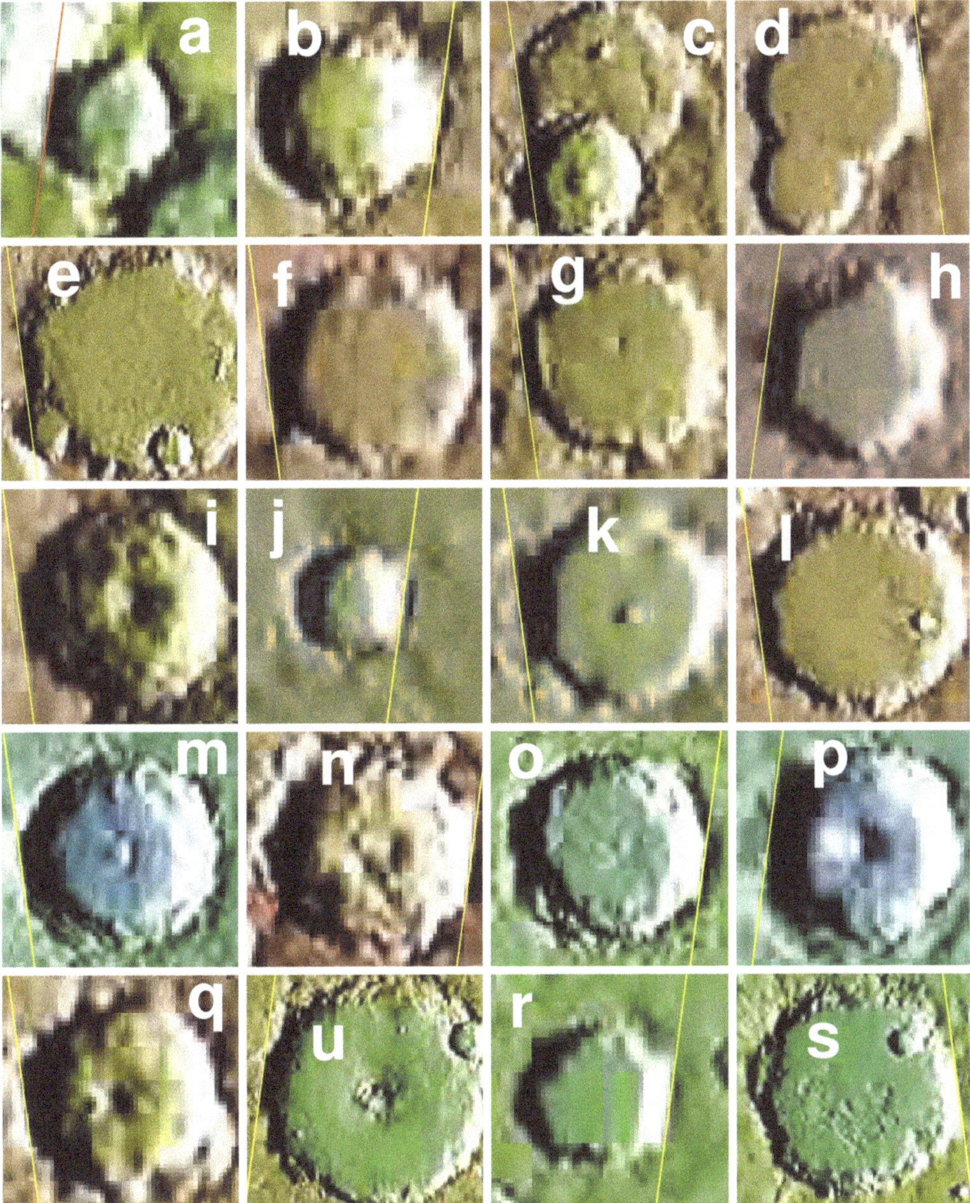

Fig. 6.14: *Craters which have a linear section in their perimeter with a bearing angle of 7.5° in the clockwise or counterclockwise direction. Coordinates of the craters are:*
(a) 1.88° E 9.82° S (b) 6.99° E 28.09° S (c) 17.74° E 11.68° S (d) 22.36° E 23.18° S
(e) 26.05° E 27.86° S (f) 36.66° E 26.48° S (g) 43.10° E 3.21° S (h) 47.49° E 15.55° S
(i) 57.67° E 22.16° S (j) 70.37° E 16.12° S (k) 72.17° E 14.55° S (l) 121.43° E 12.26° S
(m)125.09° E 9.23° N (n)130.94° E 16.78° S (o)138.55° E 15.86° N (p) 148.53° E 4.23° N
(q)173.42° E 26.25° S (u) 273.45° E 21.64° N (r)326.13° E 3.83° S (s) 346.66° E 23.50° S.
USGS Astrogeology.

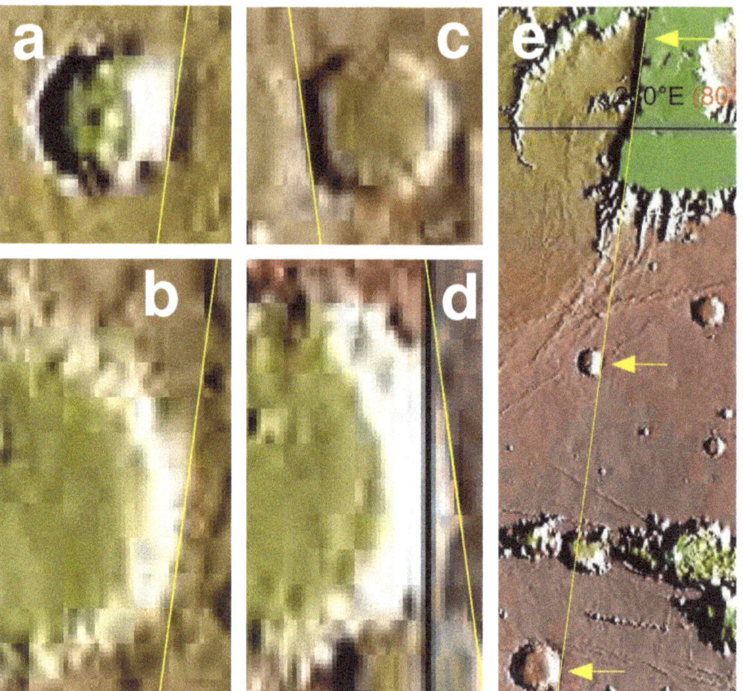

Fig. 6.15: *A line with a clockwise bearing angle of 7.5° aligns with the linear sections of the perimeters of the craters in pictures a (42.37° E 22.41° S) and b (41.37° E 26.89° S). A similar line with a counter-clockwise bearing angle of 7.5° aligns with the linear sections of the perimeters of the craters in pictures c (43.43° E 7.10° S) and d (44.52° E 20.50° S). In picture e, a straight line with a clockwise bearing angle of 7.5° aligns to the linear perimeter sections of a southern crater located at 278.13° E 6.28° S and a more northern crater located at 278.74° E 2.73° S. The same straight line also aligns to the bottom edge of a linear section of the boundary of the Echus Chasma. USGS Astrogeology.*

section of its perimeter with a bearing angle of 7.5°. The western perimeter of the Olympus Mons Caldera has a linear section with a counterclockwise bearing angle of 7.5° (yellow line, left image). Its eastern perimeter also has a straight line section but this is best fit by a line having a bearing angle of 6° in the clockwise direction. The western perimeter of the Apollinaris Mons Caldera has a linear section with a bearing angle of 7.5° in the clockwise direction (white line, right image). This line has a measured length of 29.64 km which is equal to 0.5°.

Three cliffs and a high ridge have been found to have a linear section over several kilometers with a bearing angle of ±7.5°. These are shown in Fig. 6.17. Image a shows the west cliff bordering the Aganippe Fossa just west of Arsia Mons. The cliff is about 32 km long. Image b shows a high ridge inside the Ophir Chasma of the Valles Marineris. It is about 20 km long and is marked by a red line and 2 yellow arrows in the image. Image c shows a cliff in the southern part of the Simud Vallis. The cliff is about 23.5 km long. Image d shows a cliff in the eastern part of the Capri Chasma. The cliff is divided into 2 sections by a tab that juts out slightly

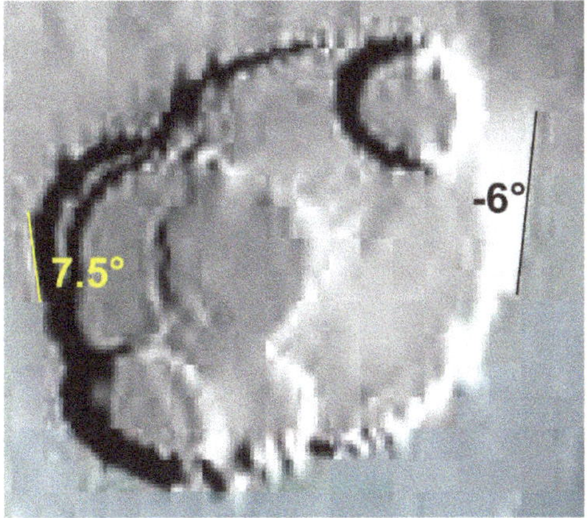

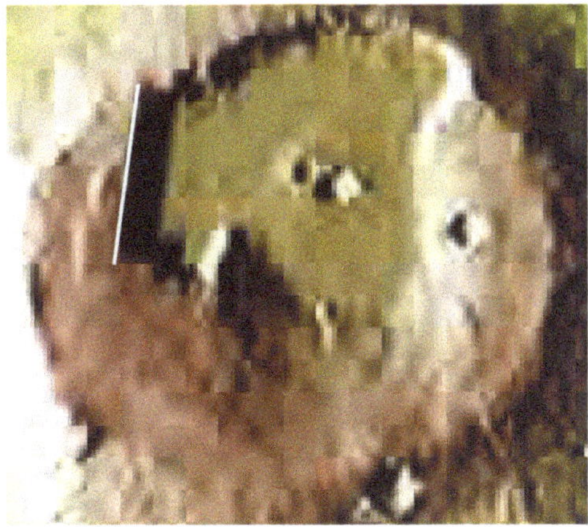

Fig. 6.16: *Left image: west linear segment of Olympus Mons Caldera has a counter-clockwise bearing angle of 7.5° (yellow line) and the east linear side, a clockwise bearing angle of 6° (black line). Right image: the west linear side of the Apollinaris Mons Caldera has a clockwise bearing angle of 7.5° and a length equal to 0.5°. USGS Astrogeology.*

into the Capri Chasma. Each section is longer than 20 km.

Fig. 6.18 shows 2 landmasses whose west linear sides have a bearing angle of 7.5°. The image on the left shows a landmass in the Simud Vallis at about 320° E 10.5° N. Its west side has a bearing angle of 7.5° in the clockwise direction. The image on the right shows a landmass in the

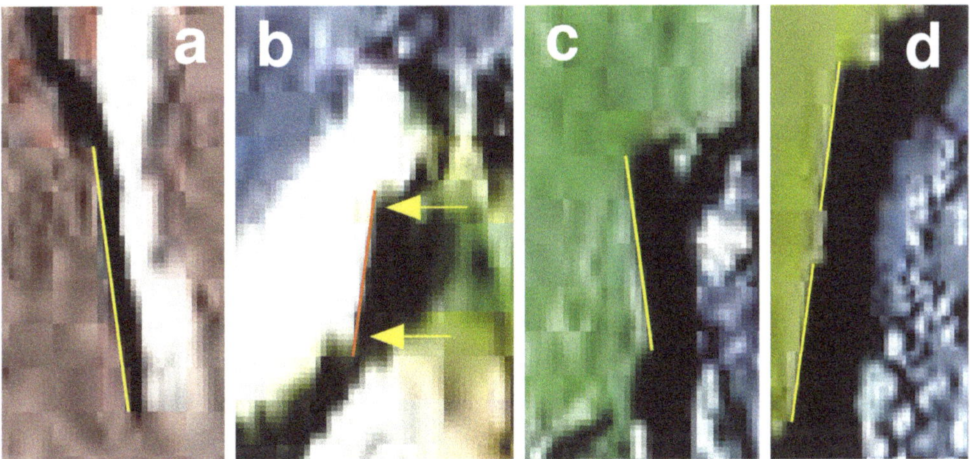

Fig. 6.17: *Linear cliff/ridge edges with a bearing angle of 7.5° in the clockwise or counter-clockwise direction. Image a: the west side of the southern section of the Aganippe Fossa just west of Arsia Mons. Image b: linear high ridge in the Ophir Chasma of the Valles Marineris. Image c: a linear cliff bordering the southern region of the Simud Vallis. Image d: cliff bordering the western perimeter of the eastern section of the Capri Chasma. USGS Astrogeology.*

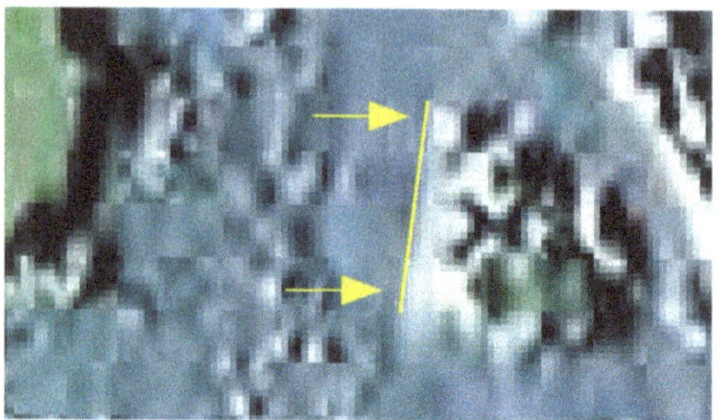

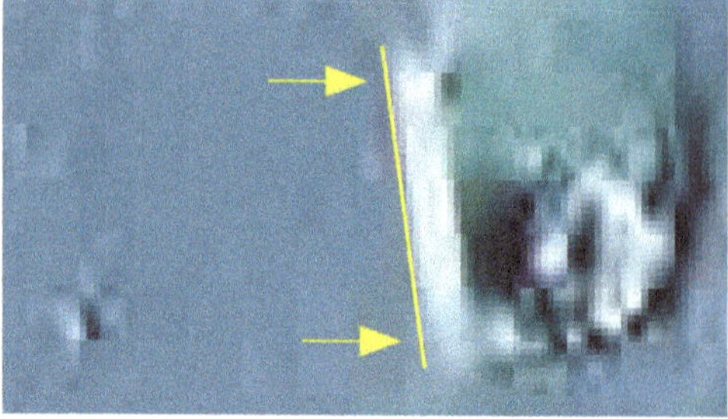

Fig. 6.18: *Left image: an elevated landmass in the Simud Vallis whose west side has a bearing angle of 7.5° in the clockwise direction. Right image: an elevated landmass in the Tiu Valles whose west side has a counterclockwise bearing angle of 7.5° and is 0.5 latitude degrees in length. The crater is the Shawnee Crater. USGS Astrogeology.*

northern regions of the Tiu Valles. Its west side has a bearing angle of 7.5° in the counterclockwise direction. This line has a measured length equal to 29.79 km which is very close to 0.5° (29.64 km).

Linear Structures with a Bearing Angle of 0°

Besides those structures having a bearing angle of 7.5°, there are several linear grooves, ridges, cliffs and sections of crater perimeters with a bearing angle of 0°. The grooves, ridges and cliffs do not occur very frequently, but the craters with linear perimeter sections are very numerous. All of these structures do not extend for great lengths as do some of the structures with a bearing angle of 7.5°, but rather are less than about 30 km in length.

Fig. 6.19 shows examples of grooves and ridges that have a bearing angle of 0° and therefore point to the north and south poles. Hence, one of their purposes may be to help spacecraft orient themselves. Another purpose of the grooves and ridges may be to create sacred geometry numbers with their longitudes. Meaningful longitudes could be found for all of the grooves and ridges in Fig. 6.19 and these are listed in the caption. The groove line in image b has a longitude of 1.00° E (Sharonov Triangle PM). This puts it at exactly 154° E of the Dagger Peak Prime Meridian thereby matching the longitude difference between the Crater Edge and Sharonov Triangle prime meridians, and the Elysium Mons and Sharonov Tower prime meridians. The groove line in image g has a longitude of 18φ° W [DPPM] which is significant since 18 is 1/2 the number of degrees in a pentagram star point.

I found 4 linear cliff structures which have a bearing angle of 0°. These are shown in Fig. 6.20. Meaningful longitudes can be found for all of these straight lines. They are listed in the caption to the figure. A curious

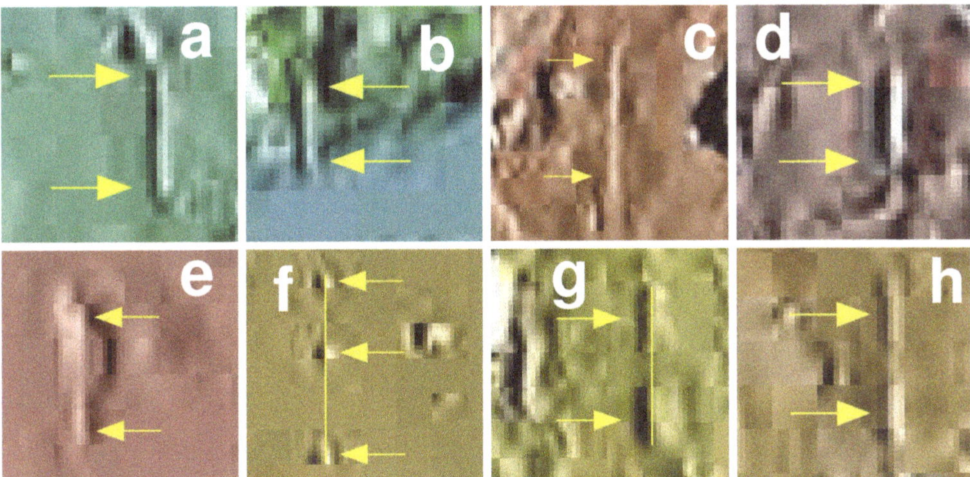

Fig. 6.19: *Groove lines and ridge lines (c and e) which have a bearing angle of 0°. The ridge line in image e is 0.25°°° in length. Meaningful longitudes of the lines are:*
a) 43°° E [STrPM] b) 1° E [STrPM] c) 18.5°° W [STrPM]
d) 13Φ° W [PCPM] e) 5Φ°° W [PCPM] f) 31°° W [CEPM]
g) 18φ° W [DPPM] h) 34√5°°° W [DPPM]. USGS Astrogeology.

arrowhead shape was found in association with the cliff in image d. I found the latitude of the tip of this arrowhead to be e° S, but only if you use planetographic coordinates instead of planetocentric coordinates.

The linear sections of crater perimeters with a bearing angle of 0° mostly do not exceed 15 km. However, they are very plentiful. Fig. 6.21 shows a sample of these structures. Image c shows a crater having linear sections on both its east and west sides. Image e shows the linear base of the crater perimeter rather than the perimeter edge. This particular feature extends for approximately 21 km on crater's east side so it is longer than all of the linear sections shown in the other images. The

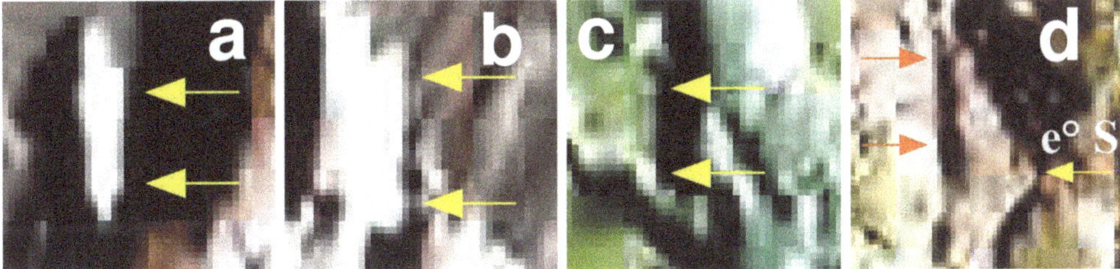

Fig. 6.20: *Straight line cliff edges having a bearing angle of 0°. Meaningful longitudes of the lines are:*
a) 10φ° E [PMPM] b) $(2e^2)$° E [PCPM] c) 2φ°°° W [DPPM] d) 28√3°° W [EMPM]. The latitude of the point of the arrowhead shape in image d is e° S in planetographic coordinates. USGS Astrogeology.

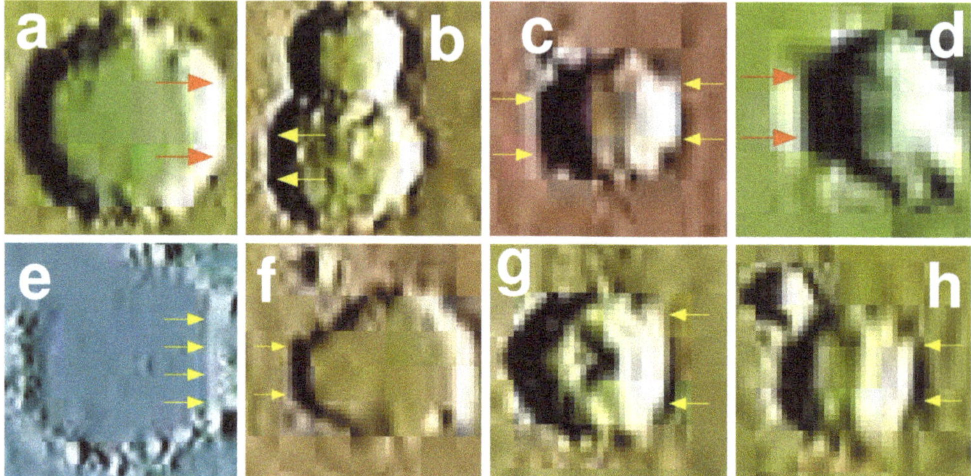

Fig. 6.21: *Crater perimeters with linear sections having a bearing angle of 0°. Coordinates of the craters are:*
(a) 324.15° E 17.98° S (b) 286.17° E 4.97° N (c) 282.27° E 15.95° S (d) 279.61° E 9.18° N (e) 183.11° E 28.58° N (f) 164.88° E 27.98° S (g) 160.53° E 26.17° S (h) 273.17° E 24.58° N. USGS Astrogeology.

crater's northwest side appears to have eroded away. Coordinates of the craters are listed in the caption to Fig. 6.21.

Summary and Conclusions

The surface of Mars is marked by numerous linear groove lines and ridge lines which have a clockwise or counterclockwise bearing angle of 7.5°. Some of these lines are extremely long, running (mostly intermittently) for distances of more than 2500 km. Since they occur in all quadrants of the planet and have constant bearing angles of plus or minus 7.5°, it is astronomically improbable that they are natural formations. There is no natural process that could possibly create rhumb lines (i.e., lines which maintain their bearing angle) over such vast distances. The only sensible conclusion is that they must have been constructed by intelligent beings, but for what purpose?

It is possible that the groove and ridge lines, together with craters having a sizeable linear segment with a bearing angle of ±7.5°, are the remnants of a grid system of lines used by the architects of the planetary topography. To explore this possibility, I constructed a map (see Fig. 6.22) containing all of the lines obtained by extrapolating the ridge and groove lines with a bearing angle of ±7.5° to the 30° S and 30° N boundaries of

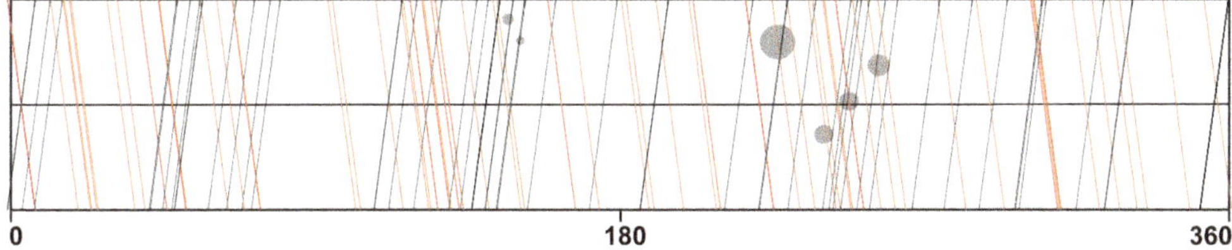

Fig. 6.22: *Extrapolated groove and ridge lines having a clockwise or counterclockwise bearing angle of 7.5°. Also included are lines extrapolated from the linear segments of crater perimeters with a bearing angle of ±7.5°. The lines have been extrapolated to the limits of 30° S and 30° N. USGS Astrogeology.*

my map. I also included the lines extrapolated from the craters having a sizeable linear segment with a bearing angle of ±7.5°. From this map, it can be seen that these lines cover most of the circumference of the planet with a relatively high density. Since the lines have been found in some instances to be as close as 0.5 km apart (e.g., Fig. 6.6), it can be speculated that a very dense meshwork of grid lines may have originally been present to guide the placement of craters and mountains and other topographical features. Most of these lines would then be covered over as items were put into place. To support this view, I found that the majority of the extrapolated lines aligned with the perimeters of several craters. Many of them also passed through the centres of craters. On average, the extrapolated groove lines aligned with the perimeter edges of 3.16 craters, the extrapolated ridge lines with 3.58 craters, and the extrapolated lines from linear-edged craters with 4.42 craters (this count includes the craters with the linear edge).

Another very important function of the ridge lines is to mark out the distance of degrees or fractions of degrees for overhead spacecraft. We have seen that 4 of the ridge lines marked out the distance of 0.5 regular degrees = 29.64 km. In addition, 2 ridge lines marked out the distance of 1 sacred degree = 74.09 km, giving strong support for the sacred degree system. The marking out of specific distances will be covered in more detail in the next chapter.

One purpose of the groove and ridge lines oriented with a bearing angle of 0° may have been to mark meaningful longitudes. The ones that I found all had meaningful longitudes. Another purpose of these groove and ridge lines, as well as craters with linear perimeter segments having a bearing angle of 0°, may have been to provide north – south orientation for overhead spacecraft.

When groove lines and ridge lines with a bearing angle of 0°, are extrapolated to the 30° S and 30° N limits of the map, the extrapolated lines were found to align to the perimeter edges of a combined average of 2.41 craters. The extrapolated lines from linear crater perimeters with a bearing angle of 0° were found to align to an average of 3.56 craters (includes the crater with the linear edge). Hence, the grooves, ridges and linear crater perimeters with a bearing angle of 0° may have been positioned using a grid system consisting of meridian lines, similar to the grid of lines having bearing angles of plus or minus 7.5°. This postulated second grid system may have been used in conjunction with the 7.5° grid system.

This now brings us to a discussion of 7.5. The use of the number 7.5 is very curious as it is not an integer. I puzzled over this for a very long time until it finally dawned on me that if you go to the sacred degree system, 7.5° becomes 6°°°. Now 6 is a much more meaningful number and it can be said to be sacred in many ways. It is the number of sides in a hexagon, one of the polygonal shapes used for several craters as discussed in *Intelligent Mars II*. It also divides evenly into the 60° angles of the equilateral triangle, the 90° angles of the square, all of the angles of the pentagram (36°, 72° and 108°) and the 120° angles of the hexagon. Validation of this theory comes from Fig. 6.16 which acts as a Rosetta Stone for the translation of the value of 7.5. Here we have 2 straight line segments of the caldera perimeter which lie on opposite sides of the Olympus Mons Caldera. On the western side is a straight line segment with a bearing angle of 7.5° (counterclockwise) whose distance I measured to be 14.62 km. This is very close to either 1/4 of a regular degree or 1/5 of a sacred degree, both of which are equal to 14.82 km, well within my error of measurement. On the eastern side of the caldera is a straight line segment with a bearing angle of 6° which I measured to be 29.64 km which is exactly the length of 1/2 of a regular degree. Hence, it is very reasonable to conclude that the bearing angle of the western segment is to be read in sacred degrees (i.e., 6°°°), analogous to the bearing angle of the eastern segment in regular degrees. However, it does not appear that the Martian architects used lines with a bearing angle of 6° to create an entire grid system for aligning objects, but rather only used this bearing angle occasionally to honour the number 6.

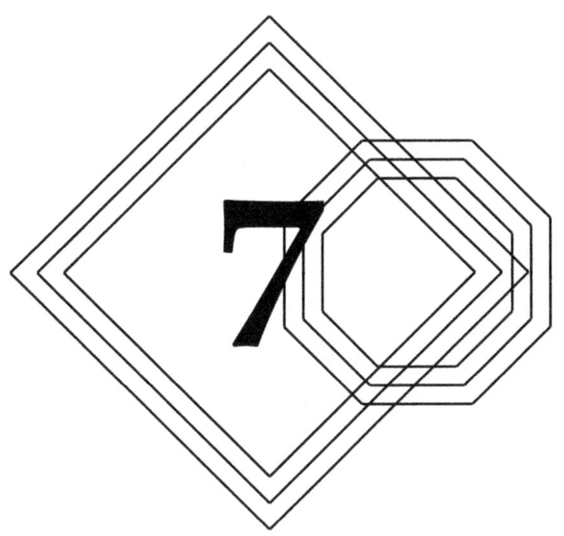

Calibration Markers for Spacecraft

In the previous chapter we have seen several sites on the surface of Mars that mark out the length of whole degrees, 2 degrees, or fractions of regular, big or sacred degrees, and even a fraction of a radian. These sites included ridge lines and linear edges of craters. The degree sizes are measured out in terms of latitude degrees, or degrees of longitude at the equator. Thus 1 regular degree is 59.2747 km, 1 big degree is 66.6840 km and 1 sacred degree is 74.0934 km in length. There are also other markers of degrees on the planet such as the distance between the 2 large craters on Ulysses Mons (*Intelligent Mars II*). This chapter will examine in a systematic manner all of the markers of important lengths that I could find.

The first markers that I wish to examine are those found on the tops of mountains. A faint ridge is located on the northern half of Albor Tholus (Fig. 7.1, left image). It has a clockwise bearing angle of 7.5° and a length of 34.18 km which is very close to the 33.96 km distance covered by 1/100 of a radian. The west side of the caldera on Apollinaris Mons has a measured length of 29.64 km which is the exact length of 0.5° (Fig. 7.1, right image). It has a clockwise bearing angle of 7.5° as noted in the previous chapter. Moving on to Olympus Mons, there is a linear segment of the western perimeter of the caldera which has a measured length of 14.64 km that is very close to the 14.82 km length of 0.25° or 0.20°°° (Fig. 7.2). The east side of the perimeter of the caldera has a linear section with

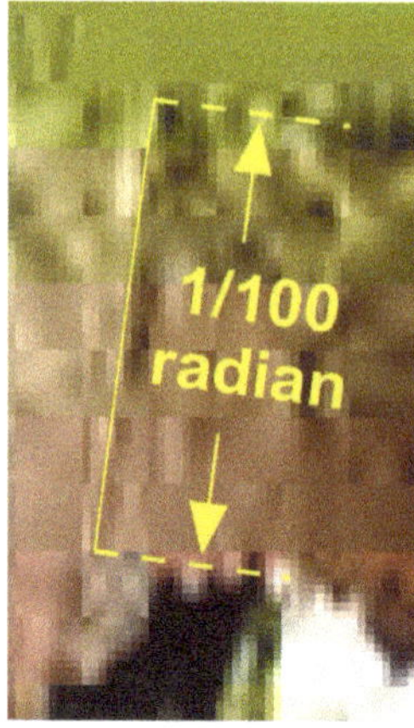

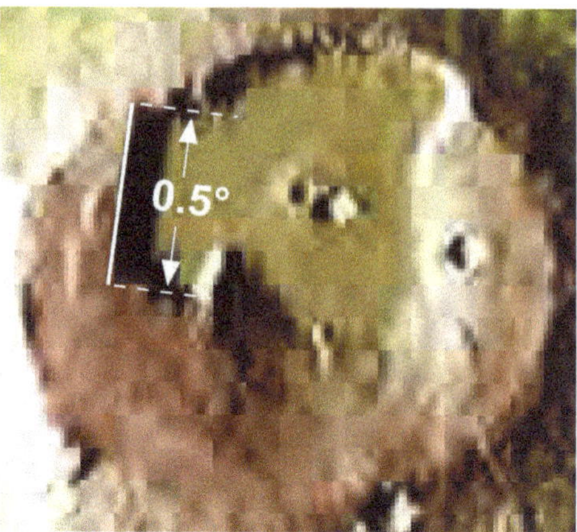

Fig. 7.1: *Left image shows a ridge line on the northern half of Albor Tholus which is 1/100 of a radian in length. Right image shows the west linear side of the Apollinaris Caldera with a length equal to 1/2 latitude degree. USGS Astrogeology.*

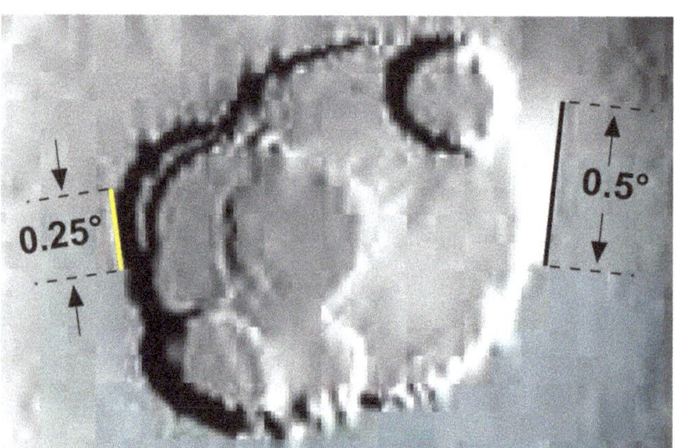

Fig. 7.2: *West linear side of Olympus Mons Caldera has a length of 1/4 latitude degree while the east linear side has a length equal to 1/2 latitude degree. USGS Astrogeology.*

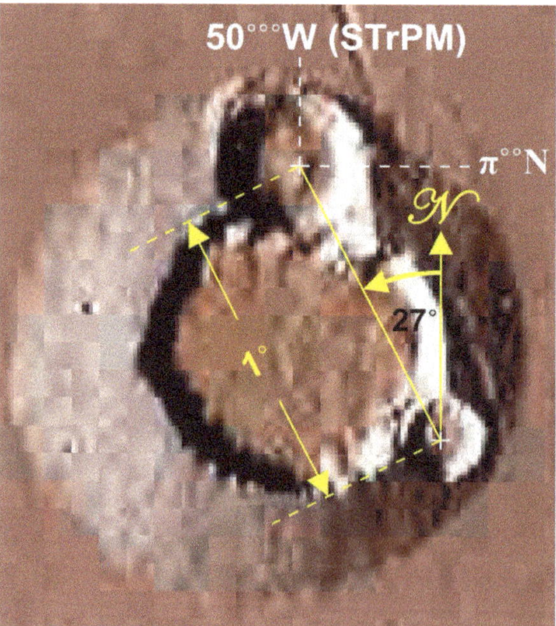

Fig. 7.3: *The distance between the 2 large craters on Ulysses Tholus is 1 latitude degree. The imaginary line connecting the centres of the craters has a bearing angle of 27° in the counterclockwise direction. The centre of the northern crater is 50°°° W of the Sharonov Triangle Prime Meridian and has a latitude of π°° N. USGS Astrogeology.*

a measured length equal to 0.5°, the same length as for the west side of the Apollinaris Caldera. The west linear segment has a counterclockwise bearing angle of 7.5° and the east linear side has a clockwise bearing angle of 6° as noted in the previous chapter. As was shown in *Intelligent Mars II*, the distance between the centres of the 2 large craters on Ulysses Tholus has a measured length of 59.29 km which is very close to the 59.27 km length of 1 latitude degree (Fig. 7.3). The bearing angle of the imaginary line connecting the 2 craters has a measured value of 27.0253° which is close to 27°, the value of 1/4 of the angle size between the star points of a pentagram. The centre of the northern crater is 50°°° W of the Sharonov Triangle Prime Meridian and its latitude is π°° N.

The ridge on Pavonis Mons has a measured length of 118.46 km which is very close to the value of 118.55 km, the length of 2° (Fig. 7.4). This is not the only calibration mark provided by this ridge. If we look closely, we see that the ridge is divided into sections with short breaks in between adjacent sections. A number of important lengths are marked out by these sections. Going from north to south, the length between the start of the first section and the start of the third section is measured to be 37.21 km which is very close to 37.05 km, the distance covered by 0.5°°°. The distance covered from

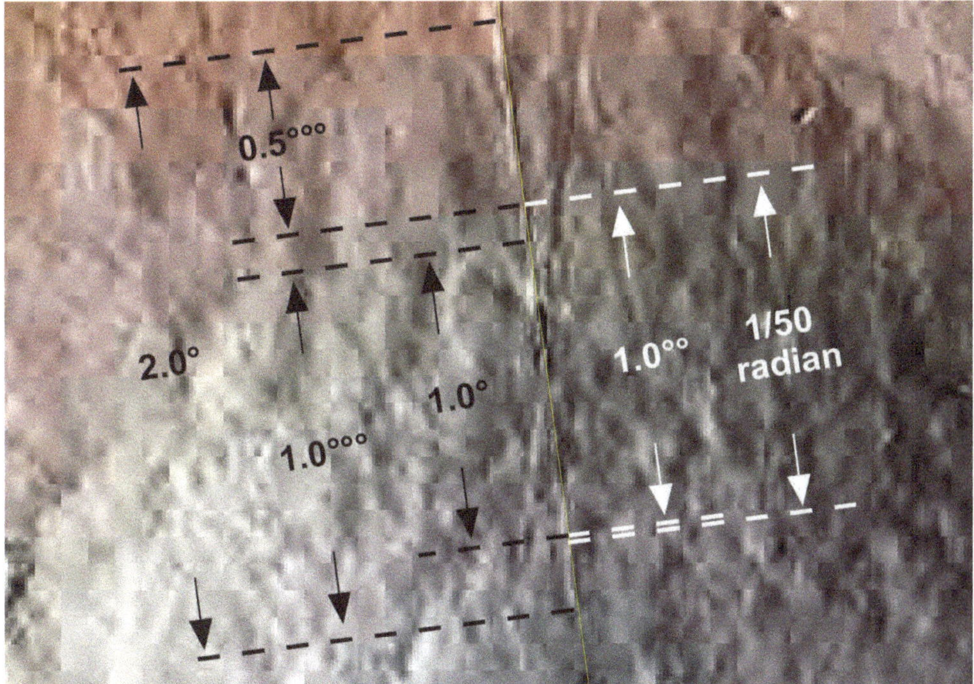

Fig. 7.4: *The ridge on Pavonis Mons is composed of several sections which provide a measure of all the fundamental degree systems that I have found to be used on Mars. USGS Astrogeology.*

the end of the third ridge section to the end of the final section is measured to be 73.86 km which is very close to the length of 74.09 km for 1°°°. The distance covered from the end of the third ridge section to the end of the second to last ridge section is measured to be 59.08 km which is very close to the value of 59.27 km for the length of 1.0°. The distance covered from the start of the third section to the end of the second to last section is 66.46 km, which is very close to the distance of 66.68 km covered by 1.0°°. The distance covered from the start of the third section to the beginning of the final section is 67.81 km. which is very close to the length of 1/50 of a radian (67.92 km). Finally, the distance from the beginning of the final section to the south end of the ridge on the southern wall of the Pavonis Mons Caldera (not shown) is 118.41 km which gives another measure of the length of 2° (118.55 km). Thus with a single ridge composed of multiple sections, the Martian architects manage to give a measure of regular, big and sacred degrees as well as of the size of 1/50 of a radian. This is a very strong confirmation of the existence of these degree systems used for coordinate measurements on Mars as well as the use of radians.

Before leaving the section on calibration markers located on the mountains, it should be mentioned that the distances between each of the mountains are sacred geometry formulae which could be used to gauge distances especially from great altitudes. These formulae have been worked out in *Intelligent Mars I.* Thus the distance between Olympus Mons and Pavonis Mons is $R/\sqrt{5}$ km. The distance between Albor Tholus and Hecates Tholus is $R/(2\sqrt{5})$ km. Since these distance formulae contain the equatorial radius R, these distances are immediately convertible to radians simply by setting $R = 1$. This works out because R km is the distance covered by 1 radian. Thus $R/\sqrt{5}$ km is the distance covered by $1/\sqrt{5}$ radians. However, when the distance formula uses the northern polar radius R′, the value of R′ has to be substituted by the value of 0.994114 to convert the formula into units of radians since R′ is smaller than R.

Groove and Ridge Markers Not on Mountains

A number of groove and ridge lines (all having bearing angles of ±7.5°) that are not located on mountains also mark out special distances. Fig. 7.5 shows 5 of these groove lines. Image c shows a groove line marking out the distance of 0.5 big degrees while the other images show groove lines with lengths of 0.5 regular degrees. The lines are distributed across the planet approximately 45 – 95 longitude degrees apart from one another.

We have already seen ridge lines on Albor Tholus and Pavonis Mons which mark out various distances in units of degrees and radians. Fig. 7.6 shows 4 more ridge lines which mark out the distance of 0.5 regular

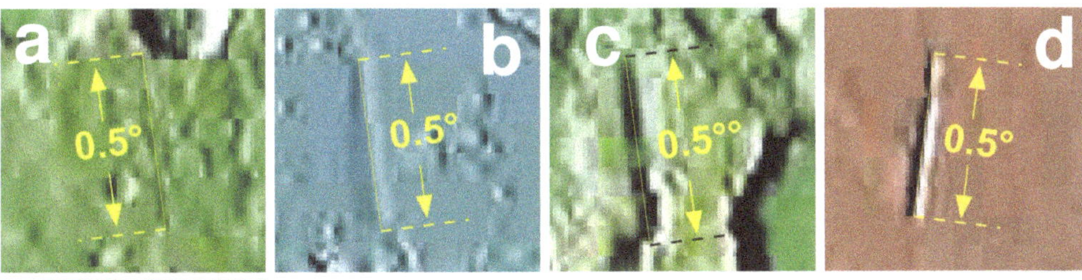

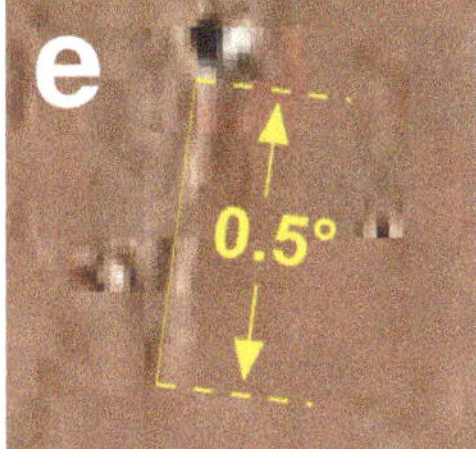

Fig. 7.5: *Groove lines that mark half degree lengths. All degrees are regular degrees except for image c which is in big degrees. Approximate coordinates are: (a) 32.7° E 22.7° N (b) 127.1°E 17.8° N (c) 185.9° E 11.8°S (d) 233.1° E 1.4° S (e) 297.3° E 18.4° S. USGS Astrogeology.*

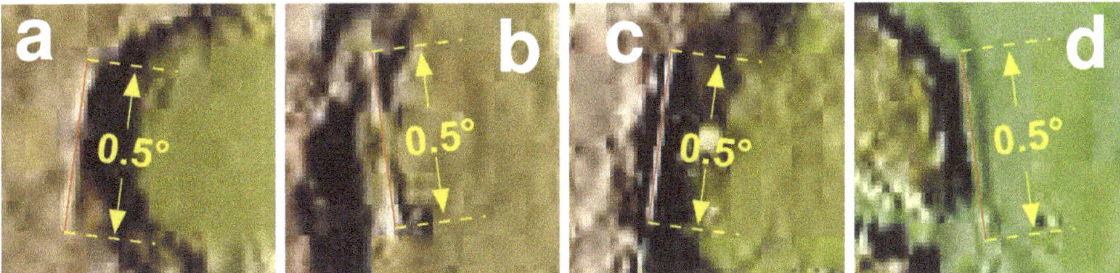

Fig. 7.6: *Ridge lines that mark half degree lengths. All degrees are regular degrees. Approximate coordinates are: (a) 53.6° E 11.7° N (b) 127.5° E 14.7°S (c) 138.8° E 10.4° S (d) 284.9°E 26.6° N. USGS Astrogeology.*

degrees. Like the groove lines, the ridge lines are spread quite far apart, thus providing spacecraft with opportunities to calibrate distance at several different locations around the planet.

Fig. 7.7 shows a remarkable ridge line located east of Arsia Mons just west of the Syria Planum. Like the ridge line on Pavonis Mons, this ridge line acts as a measuring ruler to mark out the lengths of whole or 0.5 degrees in multiple degree systems. At first glance it looks like a continuous line, but closer examination reveals that it is composed of segments which provide markings for the different measured distances. The measured length of the entire length of the ridge line is 73.74 km which is very close to the value of 74.09 km for 1 sacred degree. A big degree (66.68 km) is marked out by the distance (66.51 km) measured from the beginning of the first section (north) to the point where the fourth section shifts its course westwards by one pixel. The measured distance to the end of the third section is 59.27 km which is exactly the distance covered by 1°. The measured length of the first section of the

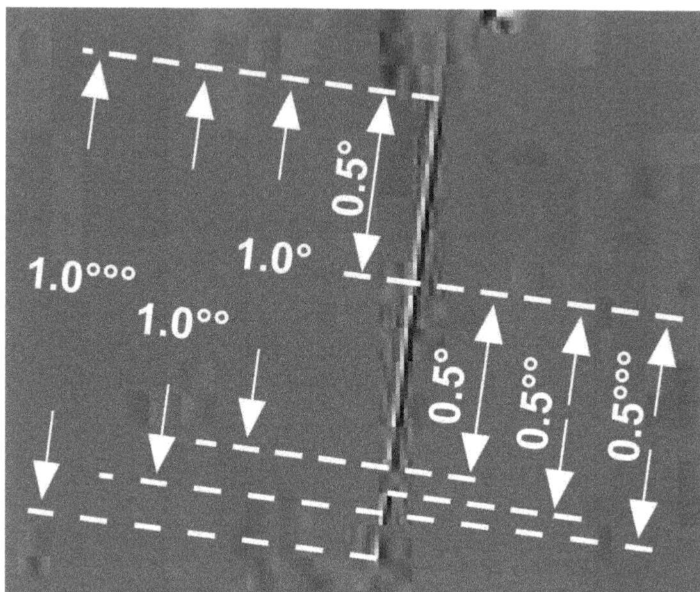

Fig. 7.7: *Ridge line east of Arsia Mons at about 247.3° E 11.6° S. It is composed of several segments which mark out different lengths corresponding to whole degrees or 0.5 degrees in the regular, big and sacred degree systems. USGS Astrogeology.*

ridge line is 29.67 km which is very close to the value of 29.64 km for 0.5°. The measured length from the end of the first section to the end of the third section (29.61 km) provides another measure of 0.5°. Finally, the measured length from the end of the first section to (a) the beginning of the fourth section (33.56 km) is close to the half value (33.34 km) of a big degree, and (b) the shift point of the fourth section (36.85 km) is close to the half value (37.05 km) of a sacred degree.

Another important line which acts like a calibrated ruler is shown in Fig. 7.8. Here a series of white lines run adjacent to a groove line. The white lines most likely form a ridge rather than simply the eastern slope of the trench or groove since the white lines are interrupted whereas the trench is continuous. The segments of the line mark out the unit distances of 3 different degree systems. The distance from the start of the northern segment to the start of the last segment is measured to be 74.16 km which is very close to the 74.09 km covered by 1 sacred degree. The second to last southern segment is composed of 2 parts which are approximately equal in length. The northern part is 2 to 3 pixels wide while the southern part is very faint being only 1 pixel wide and its midline is displaced west. The distance from the start of the northern segment at the top of the picture to the junction point of the 2 parts of the second to last southern segment is measured to be 67.96 km which is very close to the 67.92 km distance covered by 1/50th of a radian. The third to last southern segment is displaced westwards by one pixel near its southern end. The distance from the start of the northern segment to the point of westward

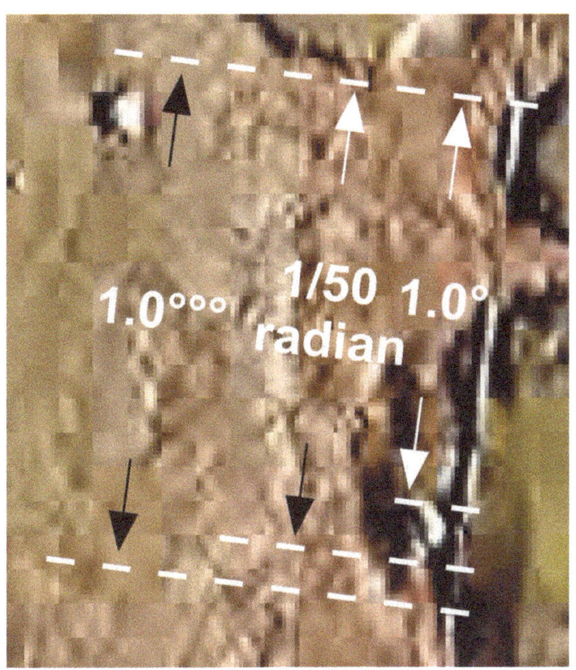

Fig. 7.8: *Ridge line located at about 137.3° E 22.0° S which is approximately 16 degrees south of the Gale Crater. It is composed of several segments which mark out different lengths corresponding to the units of 3 different systems. The distances of a regular degree and a sacred degree are provided as well as the distance covered by 1/50th of a radian. USGS Astrogeology.*

displacement of the third to last southern segment is measured to be 59.26 km which is very close to the 59.27 km distance covered by 1 degree in the 360 degree system.

A third off-mountain ridge line which marks out important distances is shown in Fig. 7.9. This line is composed of multiple segments and is located just west of the southern half of Olympus Mons. The measured distance from the start of the northern segment to the start of the third segment is 33.54 km which is very close to the distance of 33.34 km covered by 1/2 of a big degree. The measured distance from the start of the third

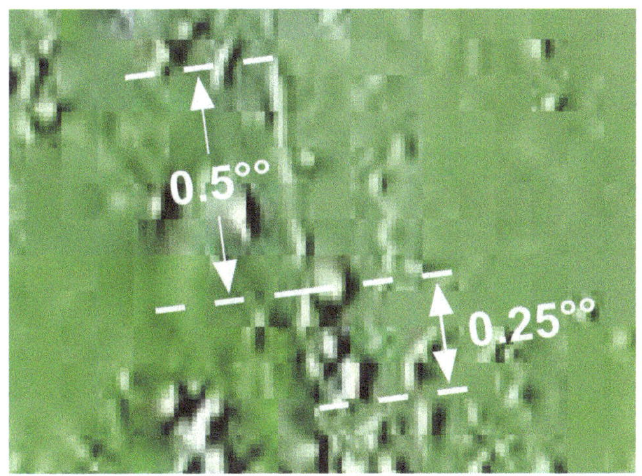

Fig. 7.9: *Ridge line located at about 218.7° E 16.2° N, just west of the southern half of Olympus Mons. It is composed of several segments of varying size. The final southern segment is very short. This ridge line marks out the distances covered by 1/2 and 1/4 big degrees. USGS Astrogeology.*

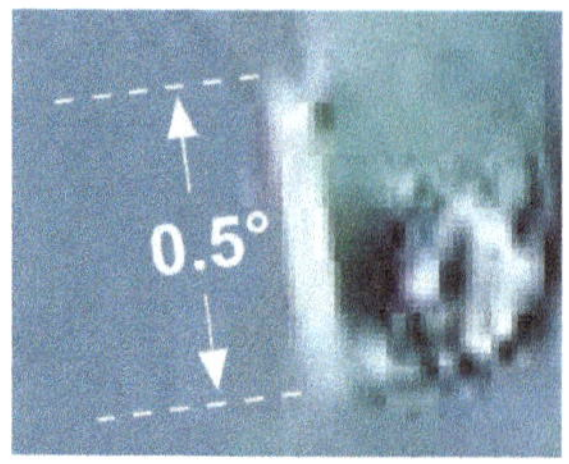

Fig. 7.10: *The bottom west edge of an elevated landmass in the Tiu Valles has a length of 0.5°. The cliff edge is located at approximately 328.2° E 22.5° N. USGS Astrogeology.*

segment to the start of the tiny final southern segment is 16.81 km which is very close to the distance of 16.67 km covered by 1/4 of a big degree.

The bottom western edge of a linear cliff bordering a landmass in the Tiu Valles covers a measured distance of 29.79 km (Fig. 7.10). This is very close to the 29.64 km length of 0.5 degrees in a 360 degree system.

Three linear crater edges with lengths of fractions of degrees were shown in the previous chapter.

Summary and Conclusions

There are several markers of the distance covered by a fraction or multiple of a degree (or fraction of a radian) on the surface of Mars. The lengths are measured out in terms of latitude degrees (or degrees of longitude at the equator) in a planetocentric system which assumes the planet to be a sphere with a radius equal to the equatorial radius R. All 3 major degree systems are used where 1 degree in the 360 degree system is 59.2747 km, 1 big degree is 66.6840 km and 1 sacred degree is 74.0934 km in length. The size of 1/50 of a radian is also used in two of the markers where the distance covered is equal to 67.9238 km.

Table 7.1 shows how frequently a marker of a given size is used for grooves, ridges, cliffs and crater edges. The distance most often used is 0.5° which lends huge support to the possibility that the Martians used a 720 degree system as well as a 360 degree system. The presence of

Table 7.1: *Frequency of occurrence of markers on the surface of Mars*

	Unit			
System	**2**	**1**	**0.5**	**0.25**
Regular Degree	2	4	14	3
Big Degree		2	3	3
1/50 Radian		2	2	
Sacred Degree		3	2	1

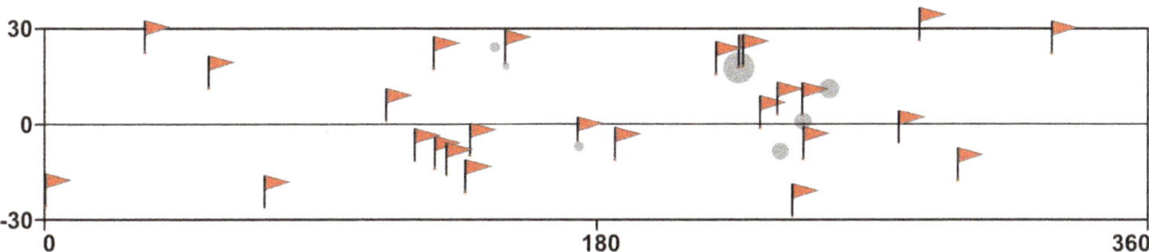

Fig. 7.11: *Map of markers on the surface of Mars which provide overhead spacecraft with a measure of degrees, or fractions or multiples of degrees, in several different degree systems. Although the markers are distributed around the planet, they are most concentrated in the regions of the large mountains.*

markers with lengths in terms of whole or fractions of big and sacred degrees also lends support for the use of the big and sacred degree systems on Mars. The finding of markers in terms of a fraction of a radian was unexpected, and it points to the use of this system for coordinate mapping and/or distance measuring on Mars as well as the various degree systems. One radian covers the distance of the length of R, the equatorial radius of Mars, which is equal to 3396.19 km. It appears that the Martians used the unit of 1/50 of a radian or smaller as a practical unit of measurement in the radian system.

These markers are distributed all around the planetary surface of Mars between 30° S and 30° N (Fig. 7.11). However, they are most concentrated in the areas of the large mountains. There are 8 of them located near Elysium Mons, 2 located near Apollinaris Mons and 8 located near Olympus Mons and the Tharsis Montes. Since these markers could be used to judge altitude and distance by overhead spacecraft, it is reasonable to assume that they were constructed for viewing by the pilots of these vehicles. They seem to be arranged along 2 flight paths, a northern one aligned with Elysium Mons and Olympus Mons, and a more southern one close to the equator which is aligned with Apollinaris Mons and the Tharsis Montes. This suggests that their landing sites were located near or on the major mountains, and that perhaps small space vehicles might have entered biospheres within the mountains which were judged to be artificially constructed in *Intelligent Mars I*. Mysterious dark spots on the flank of Arsia Mons were seen in Mars Odyssey THEMIS images. The spots were speculated to be entrances to extremely deep caves (https://www.planetary.org/blogs/emillakdawalla /2007/0984.html). There were 7 of these, and even a very high resolution image of one of them by the HiRISE camera could not determine a bottom for the hole. The holes are over 100 meters in diameter which is an adequate size for a small spacecraft to enter.

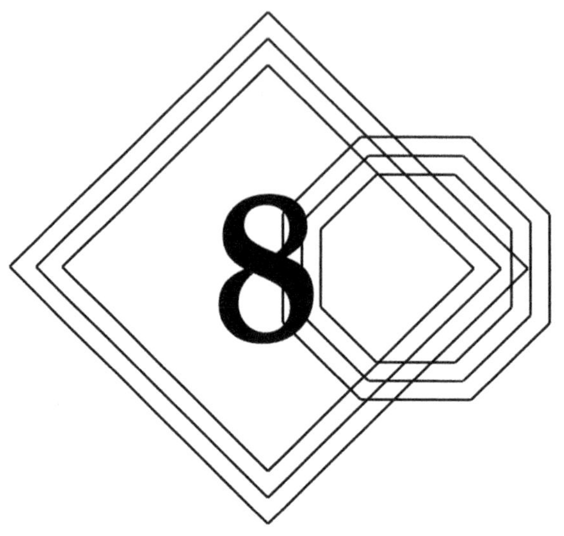

8 Large Scale Artifacts

To say that there are several artificially constructed objects on Mars would actually be an understatement. There are a huge number of artifacts on Mars! The reason why most people haven't observed any artifacts at all is due to 4 main reasons. The first and foremost is the likelihood that NASA has been busy covering them up either by eliminating or altering them in photos, or by explaining them away as natural objects such as they have persisted in doing for the famous face observed in the Cydonia region of Mars. They seem to be deliberately presenting a myth to the public that perceives Mars to be a bleak and barren planet that has never supported any life forms either in the distant or recent past. The second reason is that the Martian architects have cleverly disguised many of the artifacts to look like natural terrain such as mountains and craters. The third reason is our own psychology which prevents us from seeing the obvious, i.e., inattentional blindness. The fourth reason is a lack of knowledge of the measurement systems and the mind-set of the Martian architects. For example, people are not familiar with the prime meridians or with the degree systems used by the Martian architects, so they do not notice when a site has special coordinates. Nor are they familiar with the concept of *participatory* sacred geometry (see *Intelligent Mars II*) where the observer has to be a co-creator to uncover the hidden sacred geometric shapes that are present within the planetary topography such as the virtual squares and octagons emanating from the

Aum Crater, or the Vitruvian Martian fitted to the giant mountains.

What I hope to focus on in this chapter are large scale artifacts which have multi-kilometer dimensions. Since all of the artifacts which I will be discussing are from observations made from MOLA maps with a resolution of about 0.67 km, smaller features cannot be seen. The artificiality of objects or markings has to be established in many cases on the basis of shapes, bearing angles and/or coordinates which are highly unlikely to have occurred naturally. Other artifacts can actually be identified as symbols or as structures which served a particular function. The advantage of using the MOLA maps for this purpose is that NASA has failed to remove many artificial constructs from this source, at least partially, leaving enough data for anyone willing to spend the time to look and decipher the information. Either NASA underestimated our capacity to do so or they simply did not want to invest a whole lot of resources in sanitizing the data thoroughly. There is also the possibility that NASA did not even notice some of them. I will not discuss the artificial mountains of Mars in this chapter since they have been presented already in the first book of this series. The same goes for artificially constructed craters which were discussed in the second book and in the earlier chapters of this book.

Sunken Enclosure Between Biblis Tholus & Olympus Mons

Other than the mountains, the craters, and the Pentagram and Pentagon pyramids, for the longest time I had not noticed any other kinds of artifacts on the MOLA maps. Even the mountains seemed at first to me to be artificial only in their placement, and that they were simply formed by volcanic eruptions which were somehow engineered to occur at specific locations. I never entertained any thought that they must have been deliberately shaped and constructed out of enormous volumes of liquefied rock rather than from natural lava flows. It was not until my analysis of the artificiality of the mountains and craters was well underway that my whole outlook started to change regarding the presence of other artifacts. I suddenly noticed an area of land at a lower elevation than the surrounding terrain which had a perimeter that was composed of straight lines.

This unusual landform is located between Olympus Mons and Biblis Tholus at about 233.7° E 8.2° N (Fig 8.1). At first, I had the impression that it was rectangular in shape because the northern part of the enclosure seemed to be bounded by linear sides which formed right angles with each other. However, upon closer examination, I decided that the shape was more irregular, having at least 7 sides instead of 4, and that the construct was not simply an eroded rectangle. I fit straight lines to each of the linear portions of the perimeter of the region and measured their

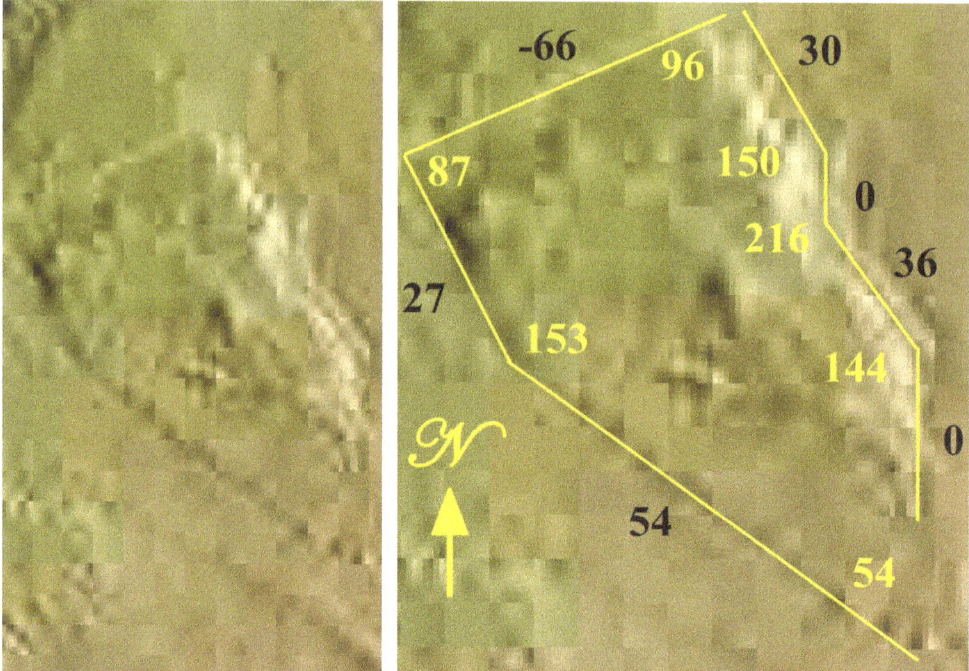

Fig. 8.1: *Left image: a large enclosure located between Olympus Mons and Biblis Tholus. It is irregular in shape with several linear sides. The eastern sides appear to be composed of a double wall. Right image: the same enclosure enlarged and fitted with straight lines. The black numbers are the bearing angles of the fitted sides. The yellow numbers are the interior angle sizes of the enclosure. The southern section of the east side was left incomplete as there are no markings in this region. USGS Astrogeology.*

bearing angles. I was astounded to find that the bearing angles were evenly divisible by 3 for all sides except for 2 portions of the east side which lay in a precise north-south direction and therefore had bearing angles of 0°. The bearing angles for 4 of the sides are evenly divisible by 6. Also 3 of the bearing angles (27°, 54° and 36°) are associated with the pentagram. The value of 27° is one-quarter of the size of the 108° angle between the star points of the pentagram, and 54° is one-half of 108°. The size of the angle of a pentagram star point is 36°. Parts of the eastern side were actually bounded by a double wall rather than a single ridge or edge. The southern section of the east side, having a bearing angle of 0°, could only be fit for its northern part as there are no markings for its southern part which may have been intended as an entry way for the enclosure. The values of all of the internal angles of the enclosure are also divisible by 3. Note that 153° is the complement of 27° and that 216° is twice the value of 108°. The internal angle of 54° at the bottom of the image is the angle that would have been produced if the southern section of the east side was extrapolated to join the southwest side. Hence, a number of the internal angles also reflect the pentagram. The length of the region is about 80 km. The entire structure looks like it is in decay suggesting that it is very ancient. I could see no evidence that it had been deliberately damaged.

Irregular Enclosure in the Daedalia Planum

Southwest of the enclosure just discussed, there is another curious enclosure. It is found in the Daedalia Planum at the eastern end of the Mangala Fossa (Fig. 8.2). Its approximate coordinates are 118.7° E 16.2° S and it has a very complex geometric shape with parts of the perimeter on both sides of the Mangala Fossa. The linear portions of the perimeter at the northern end of the enclosure have bearing angles of 60° in the counterclockwise direction and 54° in the clockwise direction. There are also 2 linear parts of what seems to be the perimeter of the same enclosure at the northwest corner which have bearing angles of 45° in the clockwise direction and 0°. The interior edge of the western border of the enclosure has a bearing angle of 30° in the counterclockwise direction.

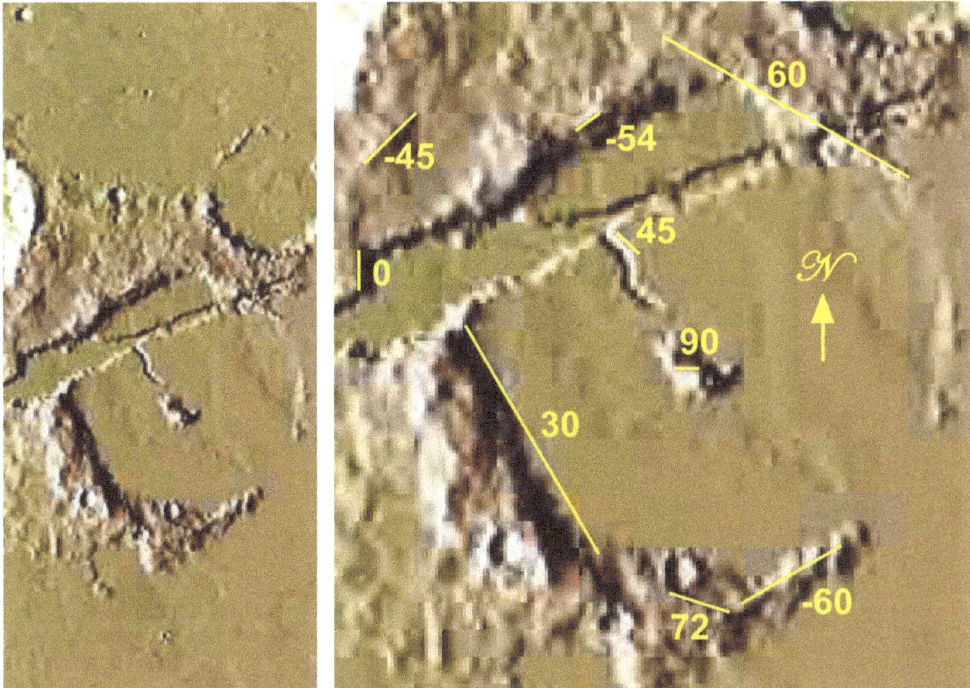

Fig. 8.2: *Left image: a large enclosure located in the Daedalia Planum. It is very irregular in shape with several linear portions in its perimeter. Right image: the same enclosure enlarged and fitted with straight lines. Yellow numbers are the bearing angles of the linear portions of the perimeter and of structures inside the enclosure. USGS Astrogeology.*

Linear segments of the southern perimeter have bearing angles of 72° in the counterclockwise direction and 60° in the clockwise direction.

A portion of a southward offshoot of the Mangala Fossa within the northern part of the enclosure has a bearing angle of 45° in the counterclockwise direction. A part of the structure in the middle of the enclosure has a bearing angle of 90°. The purpose of the enclosure is difficult to assess. The very meaningful values of the bearing angles of linear parts suggest that the architect was honouring sacred geometry and therefore point to artificiality in the construction of the enclosure. The bearing angle value of 60° is associated with the equilateral triangle. The angle value of 30° is 1/2 of this value. The bearing angle value of 45° is associated with the square since its diagonal creates angles of this size. The bearing angles of 54° and 72° honour the pentagram since the former value is half the size of the angle between 2 adjacent star points, and the latter value is the size of the base angles of the star points in a pentagram. The angles of 0° and 90° create a means to determine the 4 cardinal points of the compass for an overhead spacecraft. The enclosure is about 87 km (east to west) by 93 km (north to south).

The Major 3rd Enclosure

About 500 km east of the Hibes Montes there is a rectangular enclosure (181.4453° E 2.8877° N) with a small square-shaped crater placed near its southern corner (Fig. 8.3). The northwest side aligns with the edge of the enclosure close to the corners and follows a line of lower terrain in the middle region where some higher terrain penetrates into the enclosure. The bearing angles of its sides (36° and -54°) are associated with the pentagram since 36° is the size of the angle of a star point and 54° is one-half the size of the angle between 2 star points. The rectangle therefore symbolizes the golden mean φ which is integral to the structure of the pentagram. The length of the rectangle is about 39.29 km and the width is about 31.43 km. The ratio of the length to the width is 1.25 which is the same as the interval value for the major 3rd interval in the chromatic scale. I have, therefore, called this enclosure the Major 3rd Enclosure. The richness in symbolism of this structure becomes even more apparent when the coordinates of its vertices are determined. The western vertex has a latitude of 3.0000° N which celebrates the number 3. The longitude of its eastern vertex is 30.0004°° E or 27.0004°°° E (Dagger Peak PM). Once again the number 3 is celebrated since 27 is 3^3 but also √3 is honoured since the tan(30) is 1/√3. In addition, the number 27 is 1/4 the size of the angle between the star points of a pentagram. The latitude of the eastern vertex is also of interest since it is 10.8750 degrees north of the Arsia Mons

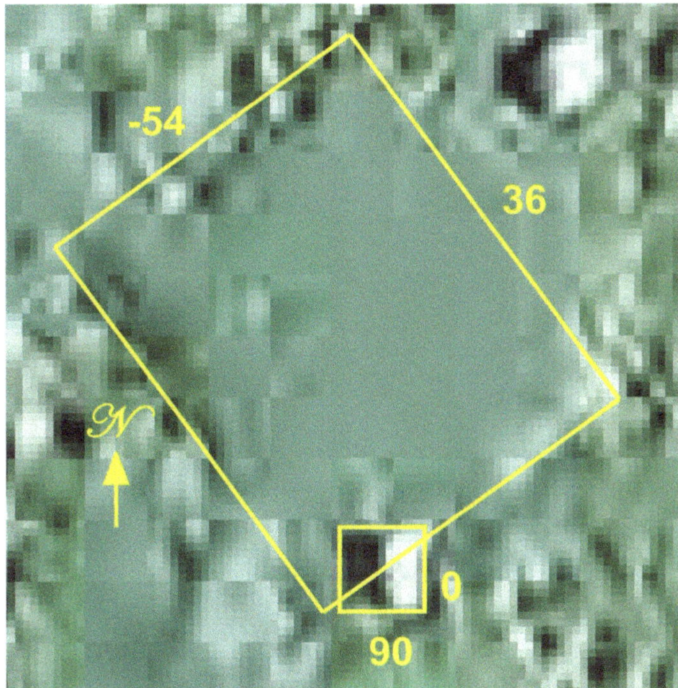

Fig. 8.3: *Rectangular enclosure about 500 km east of the Hibes Montes. The bearing angles of its sides relate to the pentagram. A small square crater sits on its southern side. USGS Astrogeology.*

Prime Latitude. The value of 10.8750 is very close to 4e = 10.8731.

The centre of the square that sits on the southern side of the enclosure has a latitude of 2.2459°° N which is very close to √5 = 2.2361. It is also 9.4455° N of the Arsia Mons Prime Latitude which is only about 1 min of a degree more than the value for 3π = 9.4248 degrees. In terms of sacred degrees, it is 8.5010°°° N [AMPL] which is very close to 8.5000. The longitude of the centre of the square is 29.6952° E (Dagger Peak PM) which is very close to the value of 21√2 = 29.6985.

Since a square is associated with the number 4 (4 equal sides) and the number √2 (the diagonal of a square is √2 times the length of a side), the square together with the rectangle surrounding the enclosure celebrate all of the basic irrational numbers (π, φ, e, √2, √3 and √5) and the integers 3 and 4. This is a very remarkable achievement with such simple geometric figures. The enclosure even celebrates the musical note of a major 3rd.

The Triple Region

An interesting artificial construct can be seen just west of the Aureum Chaos at about 329.5° E 4.0° S (Fig. 8.4). Close examination indicates that this construct consists of a middle enclosure and 2 large craters which sit at either end. The middle region has a lower elevation than the

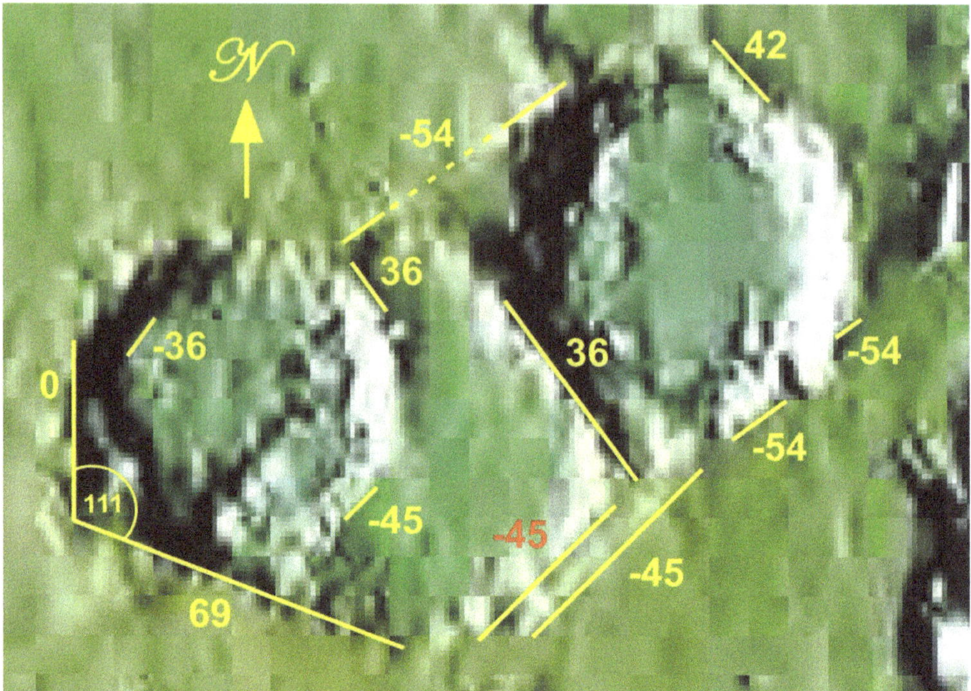

Fig. 8.4: *The Triple Region enclosure has a middle section bounded on either side by a crater. The perimeters of all 3 sections have straight line segments with meaningful bearing angles. The 3 sections are tied together by alignments to common straight lines. USGS Astrogeology.*

surrounding terrain but it is not as deep as the 2 craters. The southwest side of the middle enclosure is continuous with the southwest side of the western crater and has a bearing angle of 69° in the counterclockwise direction. The southeast side of the middle enclosure appears to have 2 levels, and the higher level extends for a short distance along the southeast perimeter of the eastern crater. Both levels of the southeast side have a bearing angle of 45° in the clockwise direction.

The long linear wall between the middle enclosure and the eastern crater has a bearing angle of 36° in the counterclockwise direction and is parallel to a short length of the eastern perimeter of the western crater. The northwest walls of both the middle enclosure and the eastern crater are on a common line with a bearing angle of 54° in the clockwise direction. This line is parallel to 2 short linear sections of the southeast perimeter on the opposite side of the eastern crater.

A section of the northeast side of the eastern crater is linear and has a bearing angle of 42° in the counterclockwise direction. A section of the bottom edge of the northwest wall of the western crater has a bearing angle of -36°, and the southeast perimeter of the crater has a linear section with a bearing angle of -45°. Finally, parts of the western perimeter of the western crater lie on a common line of bearing angle 0° which forms an angle of 111 degrees with the southwest side of the crater. Note that other than the bearing angle of 0°, all bearing angles and this internal angle are evenly divisible by 3. The whole complex is about 105 km in length.

The bearing angles of 36° and 54° of the Triple Region are associated with the pentagram and hence, with the value of φ, as pointed out in the discussion on the Major 3rd Enclosure. Also the bearing angle of 45° is associated with $\sqrt{2}$ since both the cosine and sine of 45° are equal to $1/\sqrt{2}$.

The Square Root of 5

While measuring the sizes of several northern craters, I came across a very faint image of what looked to be the symbol for $\sqrt{5}$ (Fig. 8.5). It is followed by a small "u" type symbol which is a little more distinct. The $\sqrt{5}$ symbol is so faint that I hesitated to put it in this book. I was concerned that it could simply be a mapping artifact. However, when I discovered that the western edge of the $\sqrt{5}$ symbol was located at $\sqrt{5}$° W (Pavonis Caldera Prime Meridian), I became more convinced that my interpretation of the faint design has a high probability of being correct. Of course, this has vast implications. It would mean that the symbols for

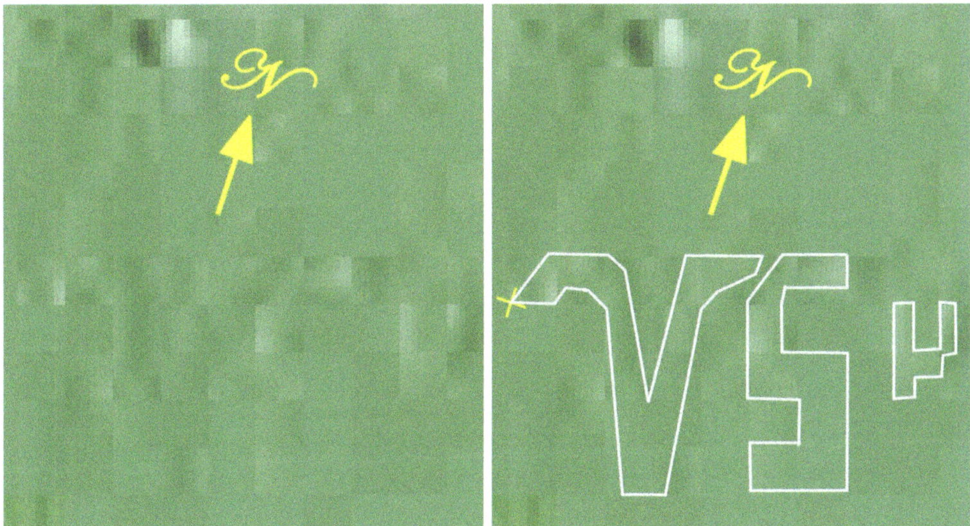

Fig. 8.5: *The figure on the left is an untouched view of a faint image of what appears to be a $\sqrt{5}$ symbol on the surface of Mars. It is located at about 245.7° E 55.4° N. The figure on the right is my outline superimposed on the image. The yellow cross at the left side of the outlined image marks the longitude of $\sqrt{5}$° W [PCPM]. USGS Astrogeology.*

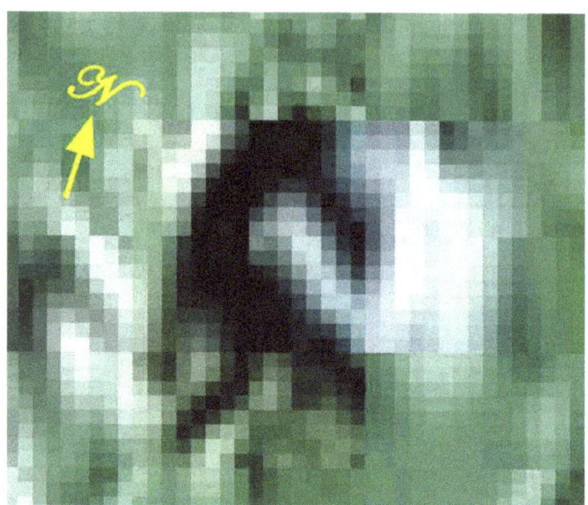

Fig. 8.6: *Square root sign inside a small crater located at 305.8926° E 30.8804° N. The left side of the symbol is very clear. The right side is positioned on a white background but is delineated by a light greyish band in the crater wall. The symbol is tilted by about 18° in the counterclockwise direction from being aligned with a meridian line. USGS Astrogeology.*

both the square root and for the number 5 are exceedingly ancient, billions of years, rather than a few hundred years as we currently believe. The √ symbol was first used in print by the German mathematician Christoff Rudolff in 1525. However, it may have been used by Arabic mathematicians earlier than this. The symbol for the number 5 is believed to have developed from the Hindu numerals which looked very different from the Arabic number symbols used today. The "u" type symbol may have been an ancient symbol for degrees.

Another finding lent further strong support to my interpretation of this faint image. I discovered that the image is rotated so that its vertical dimension is very close to 18° in the counterclockwise direction from a meridian line. The number 18 is one-half of the value for the number of degrees in the star point of a pentagram. A pentagram encodes the value for the golden mean in several ways as discussed in *Intelligent Mars I*, and the golden mean is intimately connected to the value for √5 since it is equal to (√5 +1)/2. The height of the symbol is about 27 km.

I found another possible occurrence of the √ symbol in a crater located at 305.8926° E 30.8804° N (Fig. 8.6). This time the shape is very clear and is a perfect representation of the square root symbol. The left side of the symbol is well delineated by the dark background of the crater's interior. The right side of the symbol tends to blend in with the white background of the crater wall, but it is nevertheless delineated by a light greyish band on its eastern edge. The symbol's vertical bearing is rotated counterclockwise with respect to due north. It is difficult to measure the bearing angle exactly but 18 degrees does seem to be a very credible value. The symbol is about 10 km in length.

A third instance of the √ symbol will be seen in Fig. 12.2d of Chapter 12.

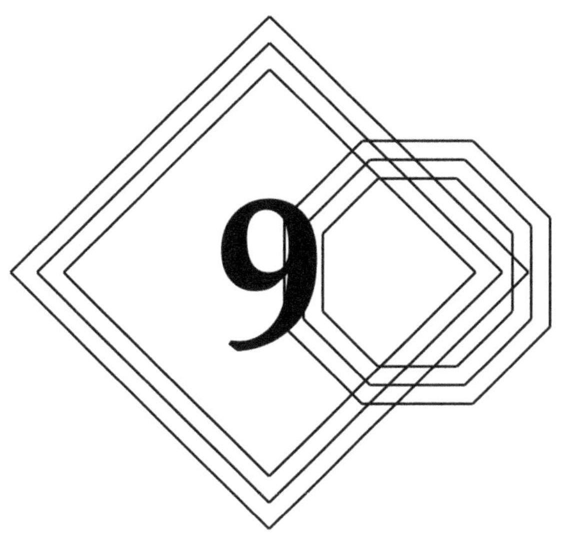

Special Monuments and Landmarks Devoted to Irrational Numbers

Further evidence for artificiality in the Martian landscape is to be found in structures and landforms which pay homage to the irrational numbers of π, φ, e, √2, √3, and √5. It appears that these were the most important irrational numbers in Martian sacred geometry, showing up very frequently in distance measurements and coordinate values of mountains, craters and other landforms. This chapter will draw attention to special structures and landforms which can be considered to be monuments since their primary purpose appears to be the honouring of a pure irrational number in a latitude coordinate. A number of craters which were found to have a pure irrational number in a latitude or longitude coordinate were presented in *Intelligent Mars II* so, for the most part, they will not be discussed here.

There are several factors which have to be taken into account in order to reveal the presence of pure irrational numbers in the landscape of Mars. Coordinates can be coded in one of several degree systems, the main ones being the 360 degree system, the 320 degree system and the 288 degree system. Longitude degrees can be referenced to one of 8 prime meridians and latitude degrees to one of 2 prime latitudes. Also, latitudes can be measured either in terms of planetocentric coordinates (assumes the planet to be a perfect sphere with a radius equal to the equatorial radius of Mars) or in terms of planetographic coordinates (treats the planet as an oblate spheroid). Another element to consider is the discernment of the particular feature of a planetary construct which was intended to be the focus of attention. For example, it could be the centre of a crater, the tip of an arrow formation or the end of a landform.

Monuments at the Tops of Long Ramps

I have discovered the presence of 2 long ramps which terminate at the latitude of a pure irrational number. The first of these is associated with the Pentagram Pyramid. Fig. 9.1 (left image) shows the outline of the Pentagram Pyramid with its centre located at π° N in planetographic coordinates as was reported in *Intelligent Mars I*. What was not reported was the presence of what appears to be a ramp-like structure which penetrates into the Pentagram Pyramid at the base of the northeast star

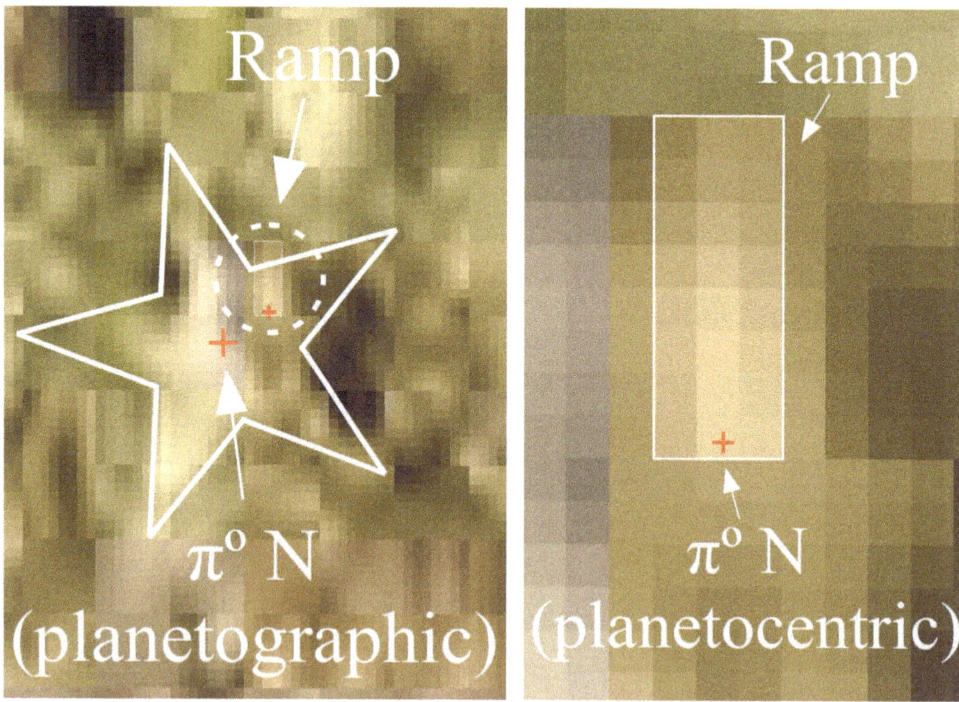

Fig. 9.1: *Left image: a ramp can be seen at the base of the northeast star point of the Pentagram Pyramid (see inside dashed circle). The centre of the Pentagram Pyramid is located at the latitude of 3.1046° N = π° N (planetographic). Right image: a magnification of the ramp shows it to be a rectangular structure in which the pixel colour gets brighter gradually from north to south. The brightest pixel is at the southern end of the ramp structure which suggests that it is the highest point. The red cross shows that the latitude of 3.1416° N = π° N (planetocentric) occurs within this pixel. USGS Astrogeology.*

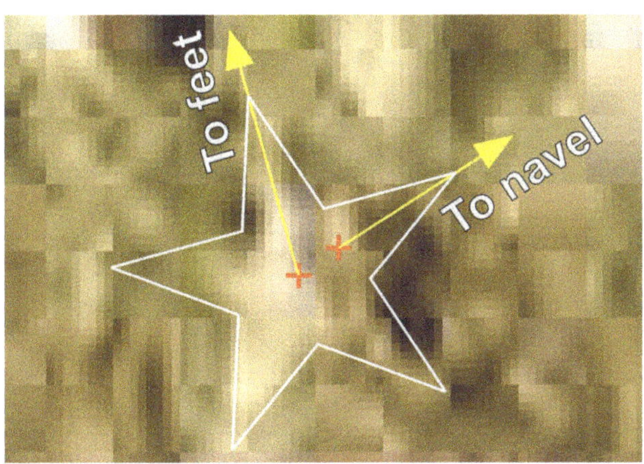

Fig. 9.2: *A straight line from the centre of the Pentagram Pyramid passes through the northwest star point to the feet of the Vitruvian Martian. Another straight line from the location of the marker for the planetocentric latitude of $\pi°$ N on the ramp passes through the northeast star point of the Pentagram Pyramid to the navel of the Vitruvian Martian. USGS Astrogeology.*

point (see inside dashed circle, left image of Fig. 9.1). An enlargement of the ramp-like structure is shown in the right image of Fig. 9.1. It reveals that pixels within the ramp (white rectangle) show a steady increase in brightness as they go from north to south. The red cross inside the white rectangle is located at a latitude of exactly $\pi°$ N in planetocentric coordinates. It is positioned inside the brightest pixel, the pixel which would correspond to the highest point on the ramp-like structure. I located its longitude dimension slightly to the east in this pixel since the eastern column of pixels that compose the ramp are brighter than the western column, suggesting that the ramp extends more into the eastern column than the western column. One would expect that some sort of structure like an obelisk is located at the position of the red cross, but its dimensions would be too small to detect with the MOLA map resolution.

Besides the presence of markers for $\pi°$ N in both the planetocentric and planetographic coordinate systems, what is truly incredible is that these markers are used in conjunction with the northern star points to locate the size and position of the radius of the circle which fits the Vitruvian Martian (Fig. 9.2). In my original calculation of the Vitruvian Martian presented in *Intelligent Mars I*, I used the measurements which I made from Da Vinci's drawing. These measurements were compatible with the NW star point of the Pentagram Pyramid pointing to the bottom pair of feet of the Vitruvian Martian and the NE star point pointing to the navel of the Vitruvian Martian. However, with a more recent geometric model of the Vitruvian Martian which perfectly integrates the circle, square, equilateral triangle and pentagram fitting the humanoid shape, I discovered that the navel of this model is situated 4.66 km east of the sight line from the NE star point of the Pentagram Pyramid. This puzzled me for a long time since I had every reason to believe that the new model was the correct one used by the Martian architects. It then occurred to me that if the ramp tower was

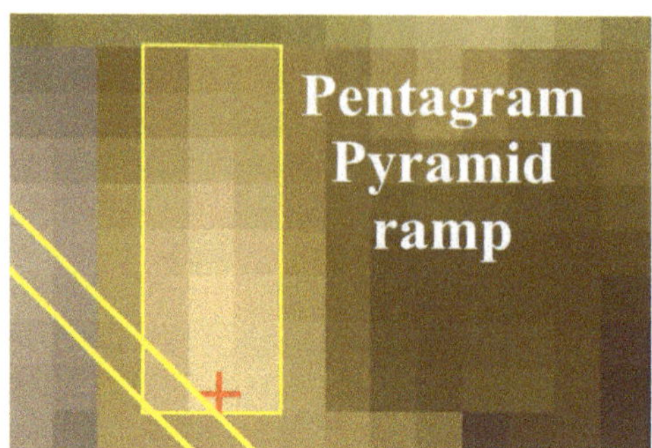

Fig. 9.3: *The northeast sides of 2 expanded squares pass very close to the location of the peak of the ramp located in the Pentagram Pyramid. The lower square has an interval ratio of 27π:1 with respect to the fundamental square fitting the Aum Crater. The upper square has an interval ratio of 60√2:1. USGS Astrogeology.*

located 166 meters (= 200 Martian meters) west of the point where it would be aligned with the NE star point and the centre of the Pentagram Pyramid, it would now sight the new model location of the Vitruvian Martian navel by lining up with the tip of the NE star point. The lines of sight to the feet and navel shown in Fig. 9.2 use the new model of the Vitruvian Martian. All images of the Vitruvian Martian used throughout this book are based on my new model (to be presented elsewhere).

Another remarkable feature of the location of the peak on the ramp in the Pentagram Pyramid is that it is very close to the northeast sides of 2 expanded squares from the Aum Crater (Fig. 9.3). These squares belong to the special set of expanded squares in which their interval ratios with respect to the fundamental square fitting the Aum Crater contain an irrational number. In Chapter 2, if you recall, one such square had an interval ratio of $4.5e^2$:1 and the other an interval ratio of 76√2:1. The squares in Fig. 9.3 have interval ratios of 27π:1 (lower square) and 60√2:1 (upper square). These numbers are remarkable in that 27 is 1/4 the size of the angles between the star points of a pentagram and 60 is the size of the angles in an equilateral triangle. The irrational number π is the ratio between the circumference and diameter of a circle. The irrational number √2 is a factor in the size of a square's diagonal. The Pentagram Pyramid is used to mark out the fit of the Vitruvian Martian to a pentagram, an equilateral triangle and a circle. The Vitruvian Martian also fits a square (*Intelligent Mars I*). In addition, the latitude of the peak of the ramp is π° N. To bring all of this together in one location is sheer genius! It strongly suggests that the Martian architects intended to convey the concept that the Vitruvian Martian is created from irrational numbers as well as the integers of harmonics. The distance of the lower square to the red cross on the ramp in Fig. 9.3 is 0.87 km and the distance of the upper square is 0.23 km.

The second ramp that I discovered is located inside the southeast crater on Ulysses Tholus (Fig. 9.4). This ramp is the same length as the one associated with the Pentagram Pyramid, namely about 5 km, but it is narrower, being 2 pixels wide instead of 3. The western column of pixels is

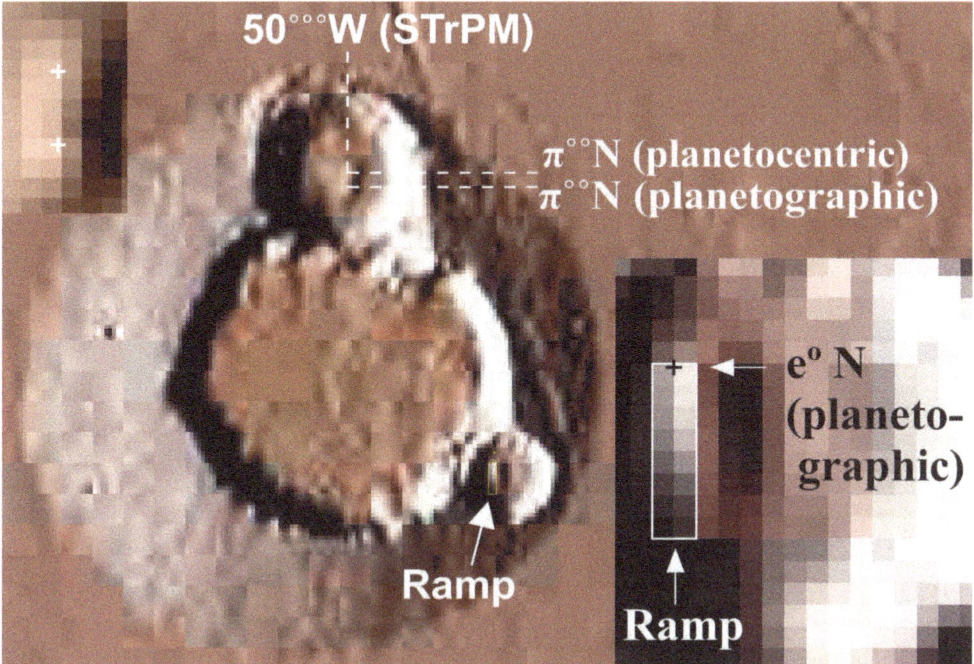

Fig. 9.4: *A ramp can be seen inside the southeast crater on Ulysses Tholus. The pixels of the ramp get progressively brighter from south to north (see inset at bottom right). The latitude of 2.6862° N = e° N (planetographic) occurs at the northern end of the ramp which is where the brightest pixels are located (black cross). The northern crater of Ulysses Tholus contains 2 peaks. The northern peak occurs at the centre of the crater and its latitude is 3.5343° N = π°° N in planetocentric coordinates. The brightest pixel of the southern peak is at the latitude of 3.4927° N = π°° N in planetographic coordinates. Both peaks are at the longitude of 238.6047° E which is 50°°° W [STrPM], and are shown at greater magnification in the inset at the top left. USGS Astrogeology.*

less bright than the eastern column, suggesting that the ramp extends only partially into the western column. The black cross at the northern end of the ramp (see inset bottom right) is located at the latitude of e° N in planetographic coordinates. Despite an intense search, I could not locate a corresponding marker for the planetocentric latitude of e anywhere nearby. However, I discovered that the northeast sides of 2 expanded octagons with interval ratios containing irrational numbers pass close by the peak of the ramp marked by the black cross (Fig. 9.5). The closest octagon has an interval ratio of 68√5:1, and it passes within 0.30 km of the black cross. The second has an interval ratio of 56e:1, and it passes within 2.69 km of the black cross. Note that 56 = 7x8 and that the 8-sided octagon occurs in the 8th octave. Incredibly, the northeast side of this e-dimension octagon passes through the latitude of e° N planetocentric at a point where a perpendicular line from its side intersects the latitude of e° N

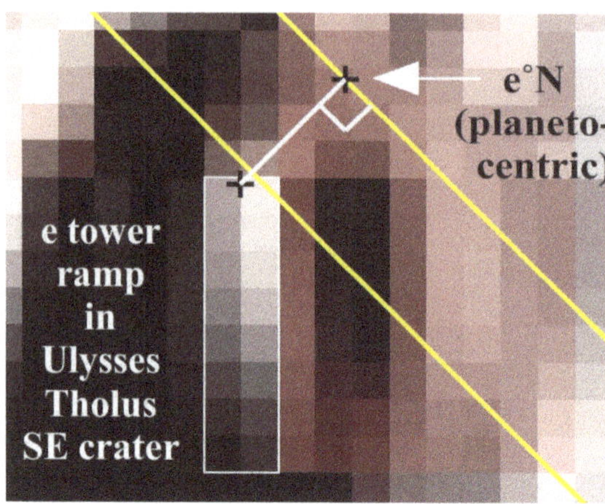

Fig. 9.5: *The northeast side of the lower octagon has an interval ratio of 68√5:1 and passes within 0.30 km from the cross marking the latitude of e° N planetographic at the top of the ramp. The perpendicular from the northeast side of the upper octagon (which has an interval ratio of 56e:1) passes through the cross on the ramp. It does so from the point where it marks the latitude of e° N planetocentric on the upper octagon at the upper cross. USGS Astrogeology.*

planetographic on the ramp. It does so at a location just slightly west of the midline of the ramp and this is where I elected to place the cross marking e° N planetographic. Hence, the marker for e° N planetocentric is virtual rather than physical, and I placed a second black cross at the top of the perpendicular line to locate it. The longitude of the black cross at the top of the ramp is 8.1498° W [PCPM] which is very close to the value of 3e = 8.1548. This fits in very well with the "e" theme of the ramp. The numbers 3 and e have been found to be associated with the torus fitting the Janssen Crater (see Fig. 9.10 below). The longitude of the upper cross is 55.1556°° W [STrPM] which is very close to 39√2 = 55.1543.

Going back to Fig. 9.4, the northern crater of Ulysses Mons was discovered to have 2 sites which mark the latitude of π°° N. Near the centre of the crater are 2 peaks or mounds. The more northern peak is located at the actual centre of the crater and it marks the latitude of π°° N in planetocentric coordinates. The southern mound has a latitude of π°° N in planetographic coordinates. An enlargement of the mounds is shown in an inset at the top left corner of Fig. 9.4. Together they form a monument paying tribute to π in big degrees.

Pavonis Mons and φ

The survey centre of Pavonis Mons has been previously reported in *Intelligent Mars I* to have a latitude of φ° N in planetographic coordinates. The value of φ appears once again in the latitude coordinate of (φ/π)° N for the centre of the Pavonis Mons Caldera. In *Intelligent Mars I*, the possibility was discussed that the Pavonis Mons Caldera, in conjunction with a rectangular area extending northeast from the caldera (Fig. 9.6), may have been intended to be a symbol for the planet Mars. Our present astrological symbol for Mars, which consists of a circle with an arrow, may have evolved from this Pavonis Mons symbol. It was also noted that the rectangular length is approximately π times the width and that the rectangle has a bearing angle of about 9° in the clockwise direction. In *Intelligent Mars II*, it was further speculated that the caldera and rectangle may also have been the original symbol for the golden ratio which eventually got converted into φ. In my search for monuments to the irrational numbers, I decided to examine the rectangle coming off the Pavonis Mons Caldera more fully. I created a rectangle with a length equal to π times its width and rotated it 9° clockwise. I then proceeded to size and fit this rectangle to the rectangle on the MOLA map.

What I discovered was that the middle of the north side of the fitted rectangle is at the exact latitude of 1.6180° N = φ° N in planetocentric coordinates (Fig. 9.6). The longitude (247.4189° E) of this site turns out to be (π/10)° E [PMPM] thus showing that it honours the values of both φ and π. Incredibly, the latitude of the northeast corner of the rectangle is 1.5989° N = φ° N in planetographic coordinates, the same latitude as for

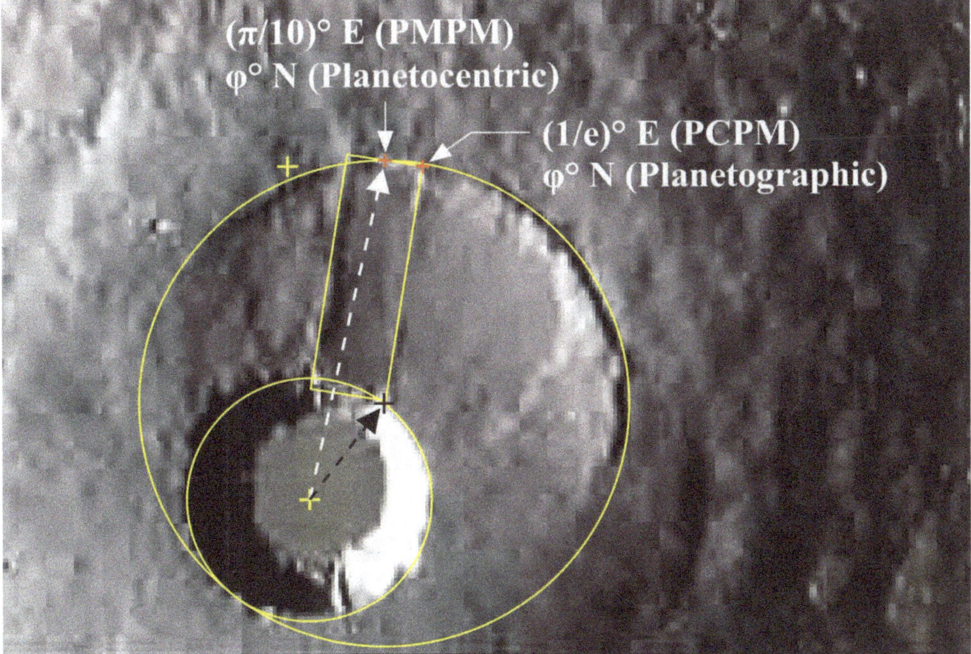

Fig. 9.6: *Pavonis Mons Caldera with a rectangle extending from its northeastern perimeter. The middle of the top side of a fitted rectangle is located at (π/10)° E [PMPM] φ° N (planetocentric). The east top corner of the rectangle is located at (1/e)° E [PCPM] φ° N (planetographic). The rectangle's length is π times its width, and its bearing angle is -9°. Upper yellow cross marks the Pavonis Mons survey centre (φ° N planetographic). Lower yellow cross marks the caldera centre at (φ/π)° N. The large circle is twice the small circle's size and is centred on the small circle's perimeter at the black cross. USGS Astrogeology.*

the Pavonis Mons survey centre. Its longitude is 247.5403° E which is (1/e)° E [PCPM] so this corner of the rectangle honours both φ and e. This gives us 3 sites which are a pure representation of the value of φ. Since the value of φ = 1.6180 is found everywhere in nature, it can be considered to be a fertility number or a number representing creation. Hence, it looks like there is a third symbol intended to be represented by the Pavonis Mons Caldera and its associated rectangle – a phallic symbol! And not just an ordinary phallic symbol, but the *primordial* phallic symbol since φ is a factor in one of the "words" uttered by the Aum sound as was pointed out in Chapter 3 (i.e., for the octagon passing through the centre of the pentagram pyramid). But wait! There is also the partial perimeter of an ancient shallow caldera passing through the middle of the north side of the fitted rectangle. This strongly suggests that the ancient caldera represents the *primordial womb* symbol! It turns out that this older caldera is exactly twice the size of the current caldera, and that its southwest perimeter lines up exactly with the southwest perimeter of the smaller caldera. The pair of calderas is inclined (black dashed arrow joining the 2 centres) at an angle of exactly -37.7612° = -atan(√3/√5)° from due north. This permits the perimeter of the older caldera to be at the latitude of φ° N at the middle of the north side of the rectangle. The diameter of the smaller caldera is very close to π/4 latitude degrees and the larger caldera, π/2 latitude degrees. The bearing angle of the white dashed arrow to the rectangle's north side midpoint is -12.5950° which is very close to 12.5664° = 4π°. So the two calderas, together with the rectangle, honour especially the irrational numbers of φ and π, and also e, √3 and √5. A square fitted within either of the circles would give a measure of √2.

The inescapable conclusion from this sophisticated sacred geometry construct is that *women* are from Mars, not just men as we have been led to think by the academics of our male-dominated culture. They have focused solely on the phallic aspect of the symbol for Mars and have conveniently omitted the womb aspect. There is strong evidence that detailed knowledge of Mars has existed on planet Earth for thousands of years because the megalithic yard is equal to 1/1000 of the equatorial radius of Mars divided by 2^{12} to a high degree of accuracy (see Chapter 12, *Intelligent Mars I*). Hence, it is highly likely that knowledge of the Pavonis Mons Caldera and its associated rectangle and partial caldera also existed, and that the womb aspect of the components of this symbol has been deliberately left out to promote male domination.

Further tributes to φ, π and e

The southern end of one of the markers of the length of 0.5° (see Chapter

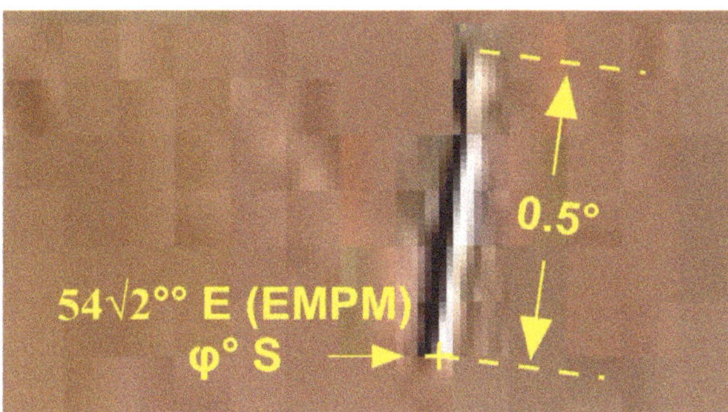

Fig. 9.7: *Groove line just northwest of Arsia Mons with a length of 0.5° or 29.64 km. Its southern end has a latitude of φ° S and a longitude of 54√2°° E [EMPM]. The number 54 honours the pentagram and √2 honours the square. USGS Astrogeology.*

7, Fig. 7.5d) was found to have very interesting coordinates (Fig. 9.7). This marker consists of a linear groove located just northwest of Arsia Mons. The marker's southern end has a latitude of φ° S so it honours the golden ratio as a pure irrational number of latitude degrees. The longitude of this location is 54√2°° E [EMPM] thus honouring the pentagram and the square. The value 54 is 1/2 the number of degrees between the star points of a pentagram and √2 is a factor in the diagonal of a square.

The western perimeter of the huge Schiaparelli Crater has a curious indent that is shaped like an arrow (Fig. 9.8). The tip of the arrow (yellow cross) was found to have a longitude of 108°°° W [DPPM] and a latitude of π°° S in planetographic coordinates. The number 108 is the number of degrees between the star points of a pentagram. Hence, even though the longitude is in terms of sacred degrees, it is the numerical value of 108 which is being honoured, not the magnitude of 108 sacred degrees. The latitude is a pure irrational number. It is highly unlikely that this location would have been used for navigational purposes by overhead spacecraft. Its purpose is most likely to pay homage to π and to the pentagram, and hence, probably φ, since φ is embedded in the pentagram.

One of the most outstanding monuments to π occurs in the western perimeter of the Ophir Chasma of the Valles Marineris (Fig. 9.9). Here we see a giant arrow-shaped wedge projecting westwards from the southern part of the west side of the Ophir Chasma. The bearing angle of the

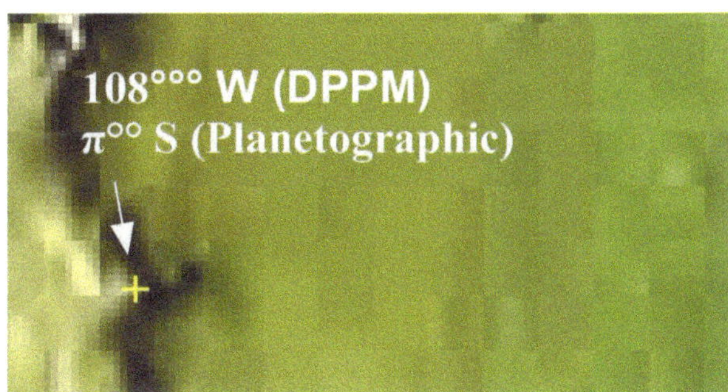

Fig. 9.8: *An arrow-shaped projection into the west side of the giant Schiaparelli Crater. The tip of the arrow (yellow cross) has the coordinates of 13.1047° E 3.4929° S. Its longitude is equal to 108°°° W [DPPM] and its latitude is equal to π°° S (planetographic). USGS Astrogeology.*

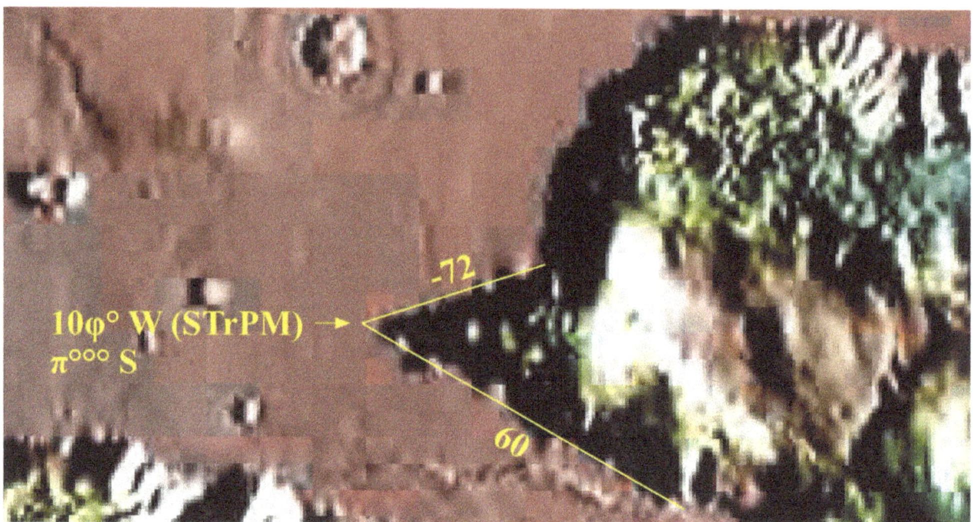

Fig. 9.9: *The tip of an arrow-shaped projection from the west side of the Ophir Chasma of the Valles Marineris marks the latitude of π°°° S. It is located 10φ° west of the Sharonov Triangle Prime Meridian. The bearing angles of the sides of the arrow shape are also very meaningful, being 72° clockwise and 60° counterclockwise. USGS Astrogeology.*

northern side of the arrow is -72° while the bearing angle of the southern side is 60°. Note that 72 is the size of the angles at the base of a star point of a pentagram and that 60 is the size of the angles of an equilateral triangle. The intersection of the sides forming the tip of the arrow has the coordinates of 284.9244° E 3.9270° S. The longitude translates into 10φ° W [STrPM] and the latitude into π°°° S. The latitude is a pure irrational number when measured in sacred degrees while the longitude is a simple multiple of φ. It is highly unlikely that the arrow construct could have been naturally formed since it has very long linear sides (about 38 km for the northern side and about 67 km for the southern side) and the sides have highly meaningful bearing angles. This has vast implications for the genesis of the Valles Marineris which will be examined in the next chapter.

The Janssen Crater is a huge monument to the value of e (Fig. 9.10). The centre of 30 squares which fit the perimeter and/or structures lying outside the crater perimeter occurs at e° N. The sides of the squares are e latitude degrees in length (see *Intelligent Mars II*). Another instance of a monument to e was shown in Fig. 6.20 of Chapter 6. It is an arrow shape whose tip has a latitude of e° S (planetographic).

Tributes to √2, √3 and √5

There are also sites that pay tribute to the irrational numbers of √2, √3 and

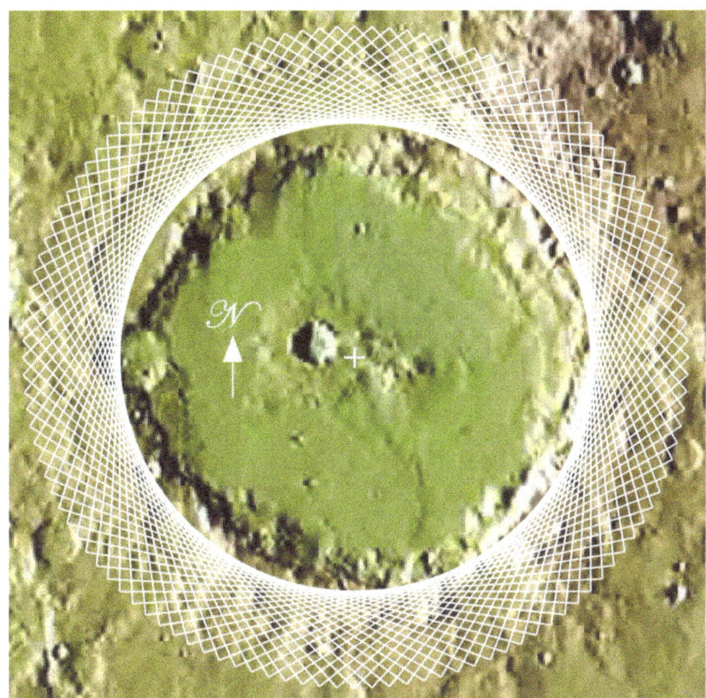

Fig. 9.10: *The Janssen Crater is a monument to the irrational number e. A wheel or torus is created by superimposing 30 squares whose sides have bearing angles which increase in steps of 3°. Each square has a side length of e latitude degrees and the squares are aligned to linear portions of the Janssen Crater perimeter and/or to structures lying outside the crater perimeter. The centre of the squares is located at 37.5495° E e° N. USGS Astrogeology.*

√5. To start with, there is a crater reported in *Intelligent Mars II* which can be fit to an octagonal shape. It was found that the northern ends of the east and west sides of the fitted octagon are at a latitude of √2° S. Upon further examination of this crater (Fig. 9.11), I discovered that the centre of a small red square inside the crater also has a latitude of √2° S. Its longitude

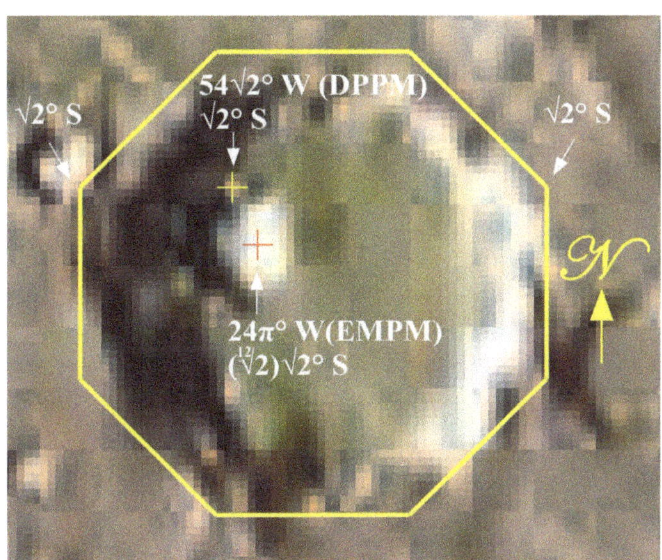

Fig. 9.11: *An octagon-shaped crater which creates a monument to √2. The northern vertices of the east and west sides have a latitude of √2° S. The center of a small red-coloured region which is square-shaped with a side length of 2 pixels has the coordinates of 54√2° W [DPPM] √2° S (yellow cross). The centre of a larger white area (red cross) has the remarkable coordinates of 24π° W [EMPM] ($\sqrt[12]{2}$)√2° S. USGS Astrogeology.*

likewise honours √2 being 54√2° W [DPPM], and pays tribute to the pentagram as well since the value of 54 is 1/2 the number of degrees between the star points of a pentagram. The side length of this square is only 2 pixels so it likely represents an obelisk or platform to mark this location. The interior of the octagon-shaped crater also contains a curious white-coloured area which is too far from the centre of the crater to be its central peak. There are 2 possibilities for the latitude of the centre of this white area which are too close to each other to be able to differentiate them. One possible value is 1.5000° S and the other is 1.4983° S. The former is just slightly south of the later. Both numbers represent the interval ratio for the perfect 5th in a chromatic scale. The former is the chromatic scale harmonic value since it is the ratio of 2 integers, namely 3:2. The latter number is the interval ratio for a perfect 5th in the equal temperament tuning system used in western music today. This system divides the octave into 12 equal intervals, so that the interval ratio for the perfect 5th in this system would equal the 12th root of 2 raised to the 7th power (i.e., the perfect 5th is the 7th note in the equal temperament chromatic scale). Amazingly, the interval ratio for the perfect 5th in the equal temperament system is also equal to the 12th root of 2 multiplied by √2. This is the value that I chose to show in Fig. 9.11 since it has a parallelism to the other instances where √2 appears in this crater, and serves to drive home the message that this crater honours √2 in a big way. The longitude of this location is also meaningful being 24π° W [EMPM].

There are 2 sites which honour √2 in big degrees. The site in the left image of Fig. 9.12 occurs about 400 km northwest of the Gale Crater. It appears to be a raised area which has an arrow shape pointing due east. The coordinates of the red cross in the image are 130.5008° E 1.5723° S. The longitude is 15.7081°° W [DMPM] which is extremely close to both 5π = 15.7080 and to $6\varphi^2$ = 15.7082, and may be honouring the equation of $\pi = 6\varphi^2/5$ which differs from π by only 0.000048. Its latitude is √2°° S in planetographic coordinates. Hence, its coordinates are quite extraordinary in that they celebrate not only the irrational numbers of π, φ and √2, but also the integers 5 and 6 which likely refer to the pentagram and hexagram. The site in the right image of Fig. 9.12 is a small raised area inside a crater located in the Meridiani Planum about 185 km west of the MER Opportunity landing site. Its conventional coordinates are 351.3702° E 1.5723° S. The longitude is 16π = 50.2655° E [STrPM], and the latitude is √2°° S in planetographic coordinates like the previous site.

There is a north-pointing arrow-shaped plateau in a crater located very close to the western edge of the Zephyria Planum (Fig. 9.13 left image). The arrow shows up as a white-coloured area in the eastern half of the crater. The coordinates of the arrow tip are 152.0695° E 1.7321° S. Both coordinates

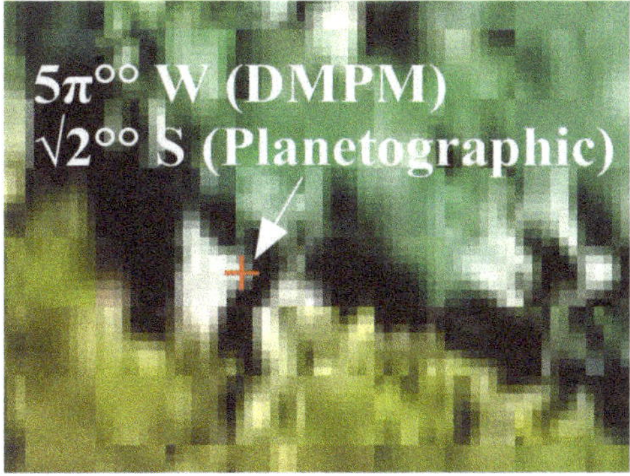

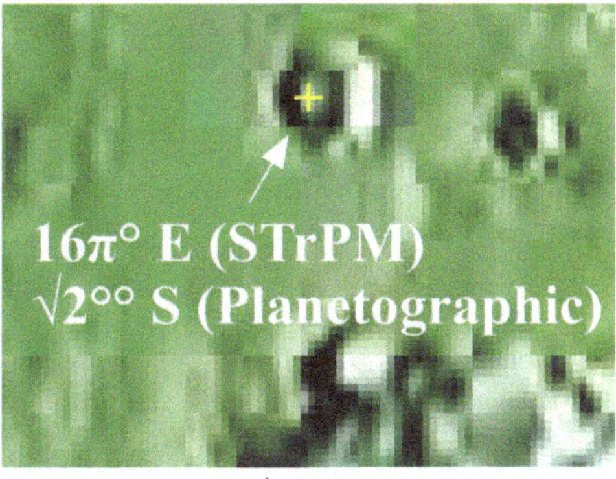

Fig. 9.12: *Two landmarks with exactly the same latitude of √2°° S in planetographic coordinates. The red cross in the left image marks the tip of an east-pointing arrow-shaped landform which has a longitude of 5π°° W [DMPM]. The yellow cross in the right image indicates the position of a raised area in a crater which might be the central peak of the crater. Its longitude is 16π° E [STrPM]. USGS Astrogeology.*

honour √3 since the longitude coordinate is 2√3°° E [DMPM] and the latitude coordinate is √3° S. I also discovered a site which honours √5 (Fig. 9.13 right image). It is the tip of a west-pointing arrow-shaped crater located on the northern edge of the Noctis Labyrinthus. Its conventional coordinates are 255.1724° E 2.5156° S. The longitude coordinate translates into 108° E [EMPM]. The latitude coordinate is √5°° S. The value of 108 is the number of degrees between the star points of a pentagram.

In conclusion, monuments to all 6 of the major irrational numbers which were considered very important by the Martian architects have been discovered on the planetary surface of Mars. Some of these sites are truly sophisticated. They could only have been created by a highly intelligent civilization and not by natural causes.

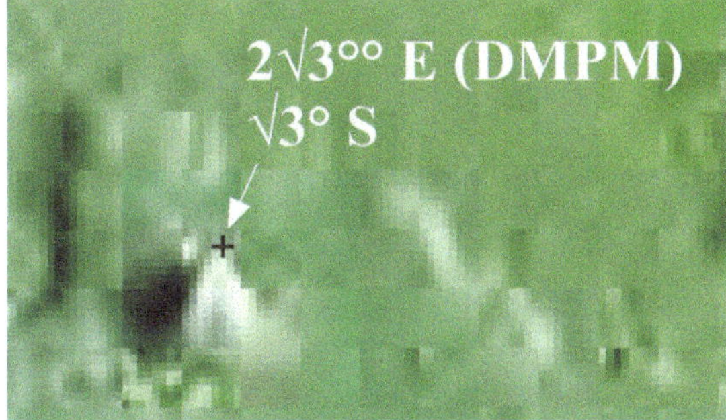

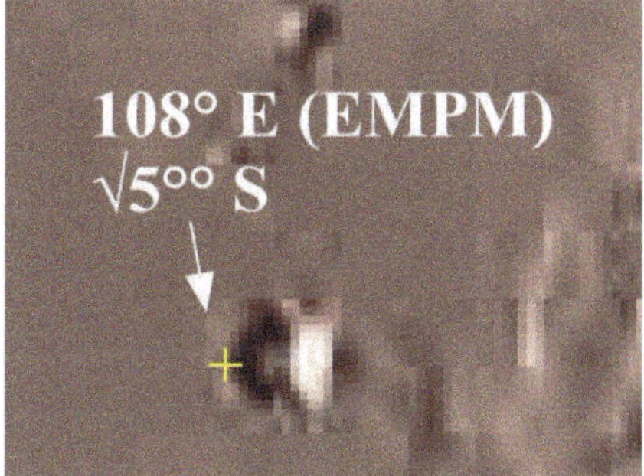

Fig. 9.13: *Left image: an arrow-shaped plateau inside a crater near the Zephyria Planum. The tip (black cross) of the north-pointing arrow has the coordinates of 2√3°° E [DMPM] √3° S. Right image: an arrow-shaped crater pointing due west. The arrow tip (yellow cross) has the coordinates of 108° E [EMPM] √5°° S. USGS Astrogeology.*

Enormous Scale Artifacts

The chapter on large scale artifacts has shown some large artifacts which range from about 10 km to more than 100 km in length. Larger artifacts than these can be seen in the giant mountains of Mars which are several hundred kilometers in diameter. Now I am going to examine 2 of the largest features on Mars, namely the Hellas Basin and the Valles Marineris. Their dimensions cover thousands of kilometers. It seems preposterous to even think that these landmarks could in any way be artificial. But we thought that of the mountains too. So let's check them out.

The Hellas Basin

The Hellas Basin is one of the most striking features of the Martian landscape. It is about 2100 km wide and about 8 km deep[1], and is currently believed to have been created by a huge object striking the planet as much as 4 billion years ago. This is the conventional thinking. But having established that objects as large as Olympus Mons must have been artificially created, I decided that I should ignore academia and examine the crater for myself.

I started by looking for straight lines in the perimeter, but since the perimeter is actually a band of material hundreds of kilometers wide there was no well demarcated line for me to follow. I then decided to use the border between the blue-green area and the purple interior region of Hellas on my MOLA maps. This is where the walls of the crater make a steep plunge downwards and it is reasonably well delineated. When I examined the southern part of the line I noticed that it was parallel to the 55° S line of latitude over a great distance (Fig. 10.1). This meant that it has a bearing angle of 90°. When I measured the latitude of this part of the border, I found it to be situated at 54° S. East of this was another long linear portion of the purple border which I measured to have a bearing angle of 60° in the clockwise direction. The east side of the Basin runs in a straight north-south direction for about 5 degrees of latitude so it has a bearing angle of 0°. Moving to the southwest side of the crater (Fig. 10.2), there is yet another long linear portion of the border. It has a bearing angle of 54° in the counterclockwise direction. West of this is a shorter linear region with a bearing angle of 36° in the counterclockwise direction.

To sum up so far, the bearing angles of linear sections of the border between blue-green and purple regions surrounding the Hellas Basin are 0, 36, 54, 60 and 90 degrees. And the southern border was located at a

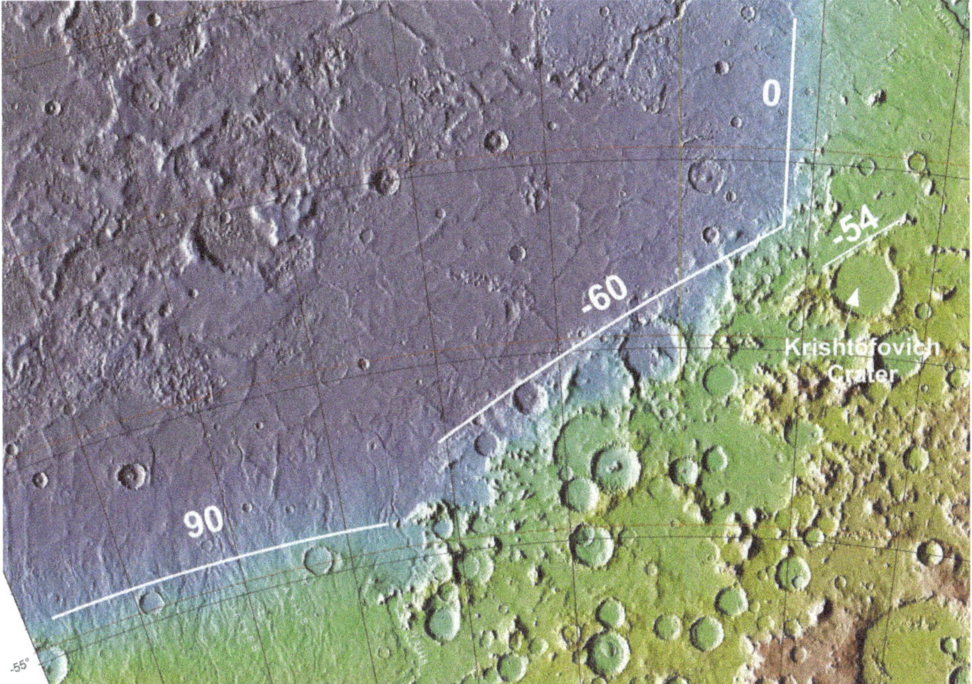

Fig. 10.1: *South and east regions of the Hellas Basin. The border between blue-green and purple locates a sudden drop in elevation. It is linear over long sections, and has interesting bearing angles. The line with a 90° bearing angle is over 500 km long. Also shown is a linear cliff with a bearing angle of 54 degrees in the clockwise direction just beyond the northwest edge of the Krishtofovich Crater. USGS Astrogeology.*

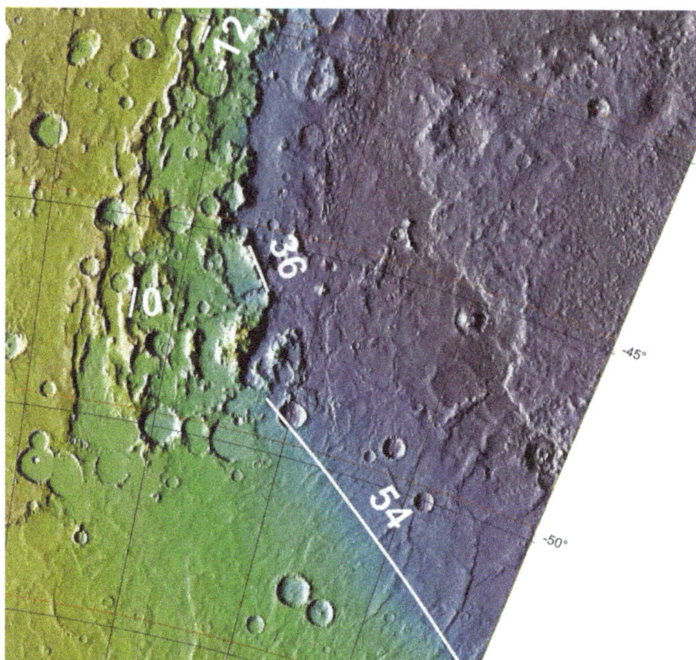

Fig. 10.2: *West side of the Hellas Basin. The border between blue-green and purple has 2 linear regions, a long one with a bearing angle of 54° and a short one with a bearing angle of 36°, both being in the counterclockwise direction. Also marked out are a cliff with a bearing angle of 0° and a ridge with a clockwise bearing angle of 12°. USGS Astrogeology.*

latitude of 54°! These numbers are associated with the pentagram and the golden mean φ (i.e., 36° and 54°), the equilateral triangle and √3 (i.e., 60°), and the compass (i.e., 0° and 90°). This is more than a smoking gun. It is very hard to avoid the astounding conclusion that the gigantic Hellas Basin has also been artificially constructed! This is an engineering feat well beyond our wildest imagination!

If we examine the region of the basin between the purple border and outside rim where the basin wall slopes down more gently we can find other linear structures. There is a long linear cliff just beyond the northwest edge of the Krishtofovich crater located in the southeast corner of the Hellas Basin (Fig. 10.1). The edge of the cliff has a bearing angle of 54° in the clockwise direction. On the west side of the Hellas Basin (Fig. 10.2), I found a shorter cliff with a bearing angle of 0° which points to the poles of the planet. Further north on the west side, there is a ridge with a bearing angle of 12° in the clockwise direction. So once again we see an association with the pentagram (54°) and the compass (0°). The number 12 divides evenly into all 3 of the angles of the pentagram and is discussed in the chapter on the Martian meter in *Intelligent Mars I*.

There is more evidence for artificiality in the northwest corner and in the interior depths of the Hellas Basin. In the northwest corner, the border between the blue-green and purple regions has 2 linear sections, one with

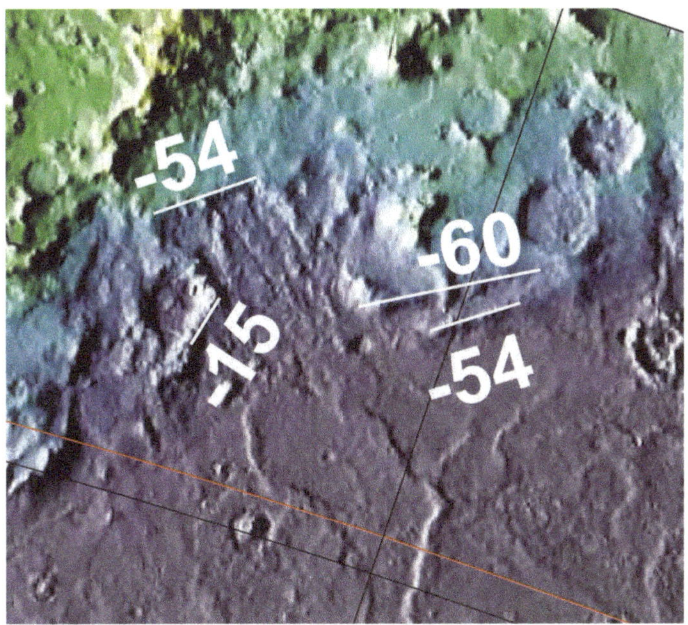

Fig 10.3: *Northwest corner of Hellas Basin. The border between blue-green and purple has a short linear region with a bearing angle of 54° and another with a bearing angle of 60°. Also marked out is the linear eastern edge of a small crater with a bearing angle of 15° and a ridge with a bearing angle of 54°. All of these bearing angles are in the clockwise direction. USGS Astrogeology.*

a clockwise bearing angle of 54° and the other with a clockwise bearing angle of 60° (Fig. 10.3). Further into the basin there is the eastern linear edge of a small crater with a clockwise bearing angle of 15°, and a linear ridge with a bearing angle of 54° in the clockwise direction. In an area just southeast of the centre of the Hellas Basin interior I was able to measure the bearing angles of 3 more edges of linear cliffs (Fig. 10.4). These had bearing angles of 15° and 39° in the clockwise direction and 48° in the counterclockwise direction. All of these numbers are evenly divisible by 3.

In summary, there is ample evidence for the artificiality of one of the

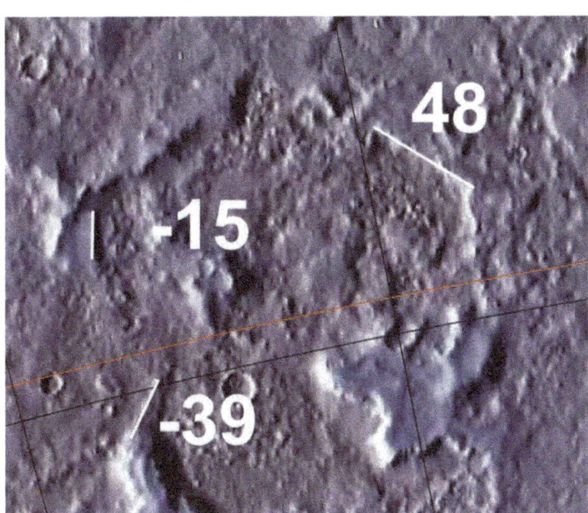

Fig. 10.4: *Interior of Hellas Basin just southeast of its centre. Linear cliffs were found with bearing angles of 15° and 39° in the clockwise direction, and 48° in the counterclockwise direction. Faint black grid lines cross at 75° E 45° S on the right side of the figure. USGS Astrogeology.*

largest landforms on the planet. We must not take for granted that any object on Mars is necessarily natural, no matter how large. Could this possibly include the Valles Marineris too?

The Valles Marineris

The Valles Marineris formation is among the most magnificent sites on Mars, extending east-west about 5000 km across the planet just south of the equator. It plunges to depths as low as 9 km below the surrounding surface, and the Melas Chasma portion is believed to contain the lowest terrain on Mars. Current theory (Tanaka and Golombek, 1989)[2] suggests that the Valles Marineris has been formed chiefly from tensional fracturing of the planetary surface. Because it is so vast a system, it has been divided into various sections to help identify where you are in relation to the whole. These regions are generally called chasmata and chaoses although the system starts in the west with a labyrinth. Some of the regions are isolated from the main system such as the Hebes and Juventae chasmata which lie to the north of the main channel.

It was outrageous for me to think that, in addition to the giant mountains and the Hellas Basin, the Valles Marineris might be the product of intelligent engineering. Nevertheless, I decided to examine this area very closely for signs of artificiality. I searched for linear sections of the perimeter and of the edges of internal structures in the various chasmata and other regions that make up the Valles Marineris. Having located several of these, I then proceeded to measure their bearing angles to see if standard values were present.

I will start by presenting my results from the west end of the Valles Marineris and then work eastwards. The Noctis Labyrinthus at the west end of this system consists of huge irregular blocks of terrain surrounded by deep canyons. I came across the top edge of a linear cliff and found that it has a bearing angle of 7.5° in the counterclockwise direction (Fig. 10.5). Just north of this is another similar cliff with the same bearing angle. Interestingly, I was able to replace the 2 lines which I placed at the top edges of the cliffs with a single line. Thus these 2 cliffs are perfectly aligned to each other. About 70 km east of this I found another cliff with the same bearing angle of +7.5°. When I fit a line to this cliff, I found that if I extended the line northwards, it was perfectly aligned to the bottom edge of another cliff, and to the top edge of the cliff adjacent to it, further to the north.

More linear cliffs with a bearing angle of 7.5° were found in the southeastern portion of the Noctis Labyrinthus (Fig. 10.6, lower right image). One cliff (lower left) has a counterclockwise bearing angle of 7.5°.

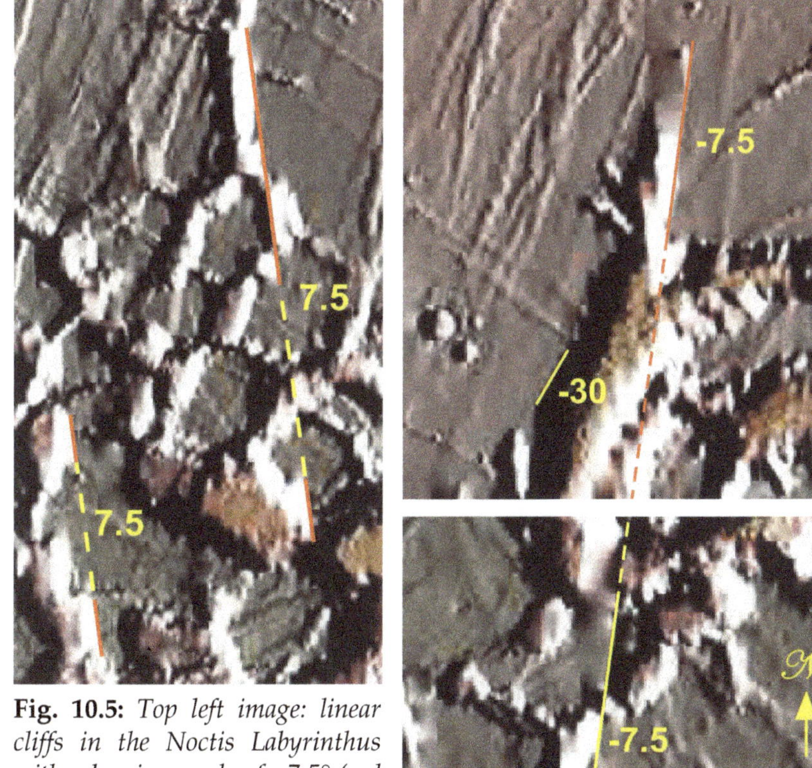

Fig. 10.5: *Top left image: linear cliffs in the Noctis Labyrinthus with a bearing angle of +7.5° (red lines). Individual cliffs are aligned with each other across unaligned terrain (dashed yellow lines). USGS Astrogeology.*

Fig. 10.6: *Linear cliffs in the Noctis Labyrinthus which are east (upper right image) and southeast (lower right image) of those in Fig. 10.5. All have a clockwise or counterclockwise bearing angle of 7.5° except for a cliff with a clockwise bearing angle of 30° (upper image) and a more eastern cliff with a clockwise bearing angle of 45° (lower image). The cliffs joined by the dashed line in the 2 images are about 300 km apart. The inset in the lower right corner shows a ridge with a bearing angle of 7.5° in the clockwise direction. It occurs about 140 km east of the dashed line. USGS Astrogeology.*

To the east and northeast of this is a pair of adjacent cliffs which are aligned to a common line which has a clockwise bearing angle of 7.5°. These two cliffs were also perfectly aligned with the top edge of a long cliff approximately 300 km away to the north (upper right image). East and south of this line is a pair of adjacent cliffs with a clockwise bearing angle of 7.5° (lower image, bottom middle area). A ridge with a bearing angle of 7.5° in the clockwise direction was found at the eastern edge of the Noctis Labyrinthus (Fig. 10.6, inset bottom right corner). The straight line cliffs and ridge with a bearing angle of ±7.5° immediately reminded me of the groove and ridge lines discussed in Chapter 6 which had bearing angles of ±7.5°. In that chapter, I also showed examples of cliffs and linear edges of landforms in the middle and eastern regions of the Valles Marineris having that same numerical value of bearing angle (see Figs. 6.17 and 6.18). Now here are multiple structures at great distances from each other in the Noctis Labyrinthus portion of the Valles Marineris which are aligned to the bearing angles of ±7.5°. This is very strong additional evidence that grid lines with bearing angles of ±7.5° were used in the planning and execution of the layout of the Martian landscape.

There are plenty of examples of straight line top or bottom edges of cliffs in the Noctis Labyrinthus region with bearing angles other than 7.5°. In Fig. 10.6 there is an example of a long linear top edge of a cliff with a clockwise bearing angle of 45° (lower right image) and a shorter cliff with a clockwise bearing angle of 30° (upper right image). Several more linear structures are shown in the left image of Fig. 10.7 which occur in the southeastern portion of the Noctis Labyrinthus just west of the previous figure. There is one example each having a bearing angle of 24°, 54° or -60°, and 2 linear structures with a bearing angle of 0°. I found another

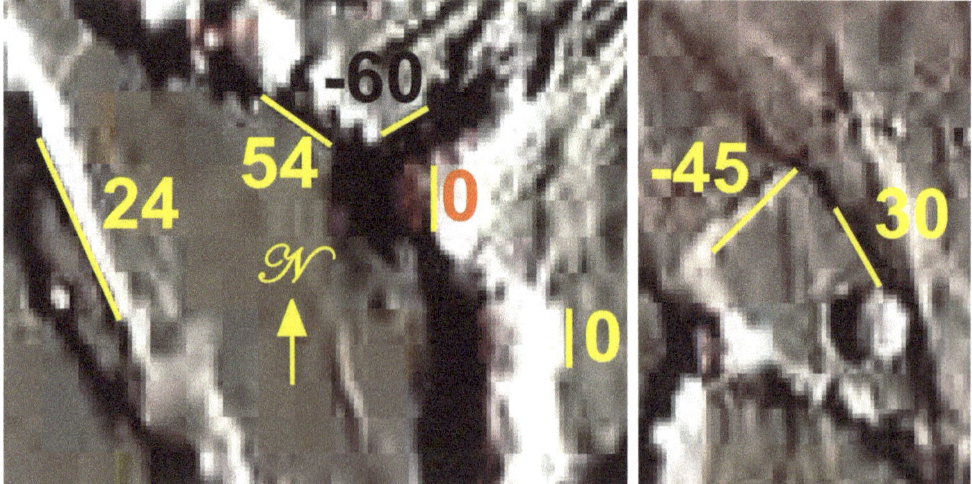

Fig. 10.7: *Left image: linear cliffs in the Noctis Labyrinthus in a region between those in the previous 2 figures. Bearing angles of the tops or bottoms of linear cliffs are 0°, 24°, 54° and -60°. Right image: linear cliffs in the northwestern region of the Noctis Labyrinthus with bearing angles of -45° and 30°. USGS Astrogeology.*

cliff with a bearing angle of -45° and also one with a bearing angle of 30° in the northwestern part of the Noctis Labyrinthus (Fig. 10.7, right image).

Further east, in the Ius Chasma region of the Valles Marineris, I came across a cluster of cliffs and other structures with straight line edges (Fig. 10.8). In the interior of the Ius Chasma there are 2 cliffs which have linear bottom edges, one with a bearing angle of 42° in the clockwise direction and another with a bearing angle of 42° in the counterclockwise direction. In addition, there are several structures with a bearing angle of 0° which point to the north and south poles. On the northern border of the chasma there is a top edge of a cliff with a clockwise bearing angle of 30° and another with a bearing angle of 90°. On the southern border of the chasma there is a cliff whose bottom edge has a clockwise bearing angle of 30° and another cliff whose top edge has a counterclockwise bearing angle of 60°. About 200 km east of this region in the Louros Valles area there is a pair of cliffs whose bottom edges are aligned with each other and have a common bearing angle of 36° in the counterclockwise direction (Fig. 10.9).

When I moved to the Melas Chasma which contains the deepest part of

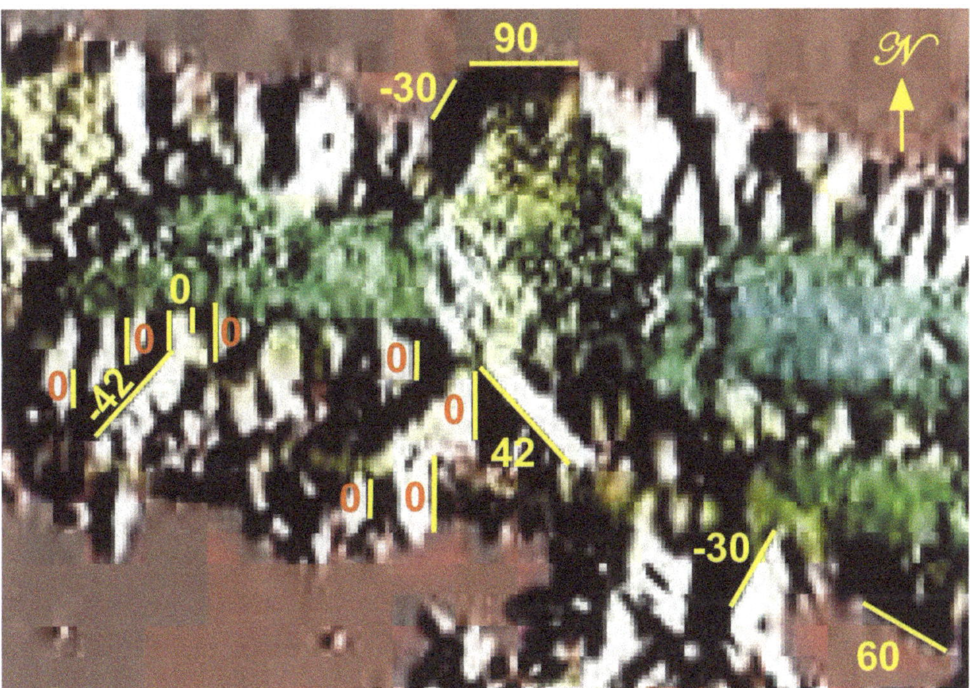

Fig. 10.8: *The Ius Chasma has many straight line cliffs and ridges. Most are oriented north-south with a bearing angle of 0°. There is a matched pair of lines with a bearing angle of 42°, one clockwise and the other counterclockwise. There are also lines with -30°, 60° and 90° bearing angles. USGS Astrogeology.*

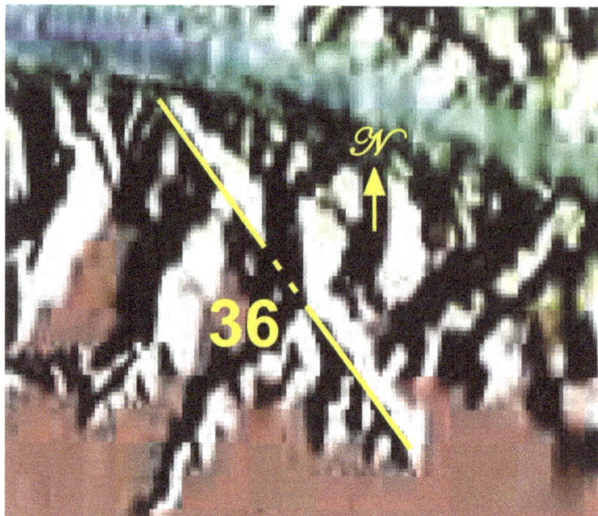

Fig. 10.9: *The base edges of 2 long cliffs in the Louros Valles which come off the southern side of the Ius Chasma. They are linear and can be fit with a common line having a bearing angle of 36° in the counter-clockwise direction. USGS Astrogeology.*

the Valles Marineris, the first thing that I noticed was a long cliff on its southern border with a counterclockwise bearing angle of 36° (Fig. 10.10).

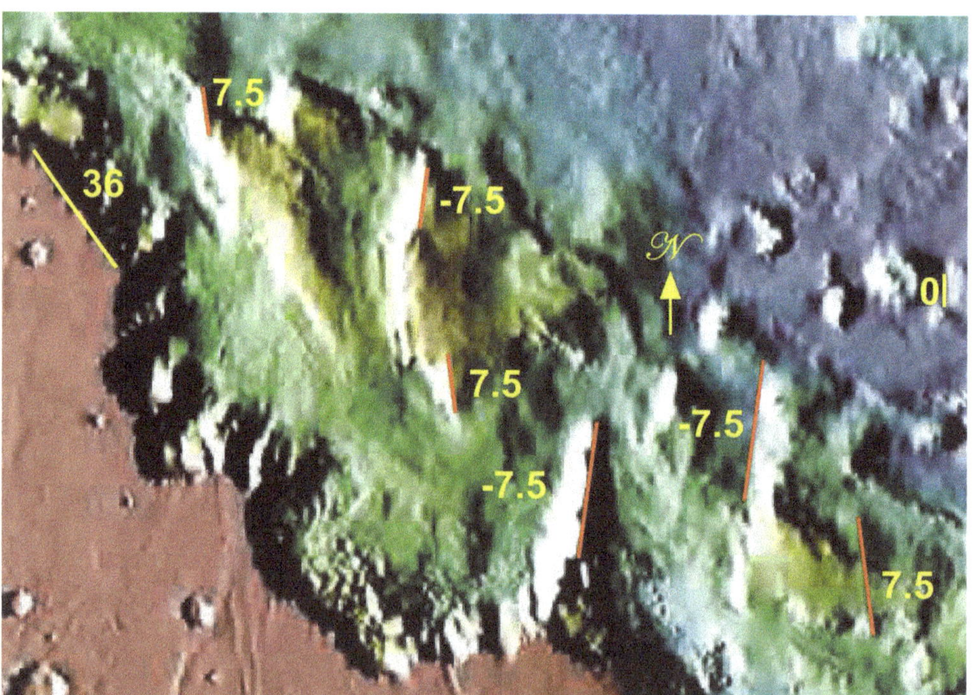

Fig. 10.10: *Several cliffs and ridges in the Melas Chasma are linear and have a bearing angle of 7.5°. There are 3 instances in the clockwise direction and 3 in the counterclockwise direction The top of a long cliff (left) has a bearing angle of +36°, and the base edge of a short cliff (right) has a bearing angle of 0°. USGS Astrogeology.*

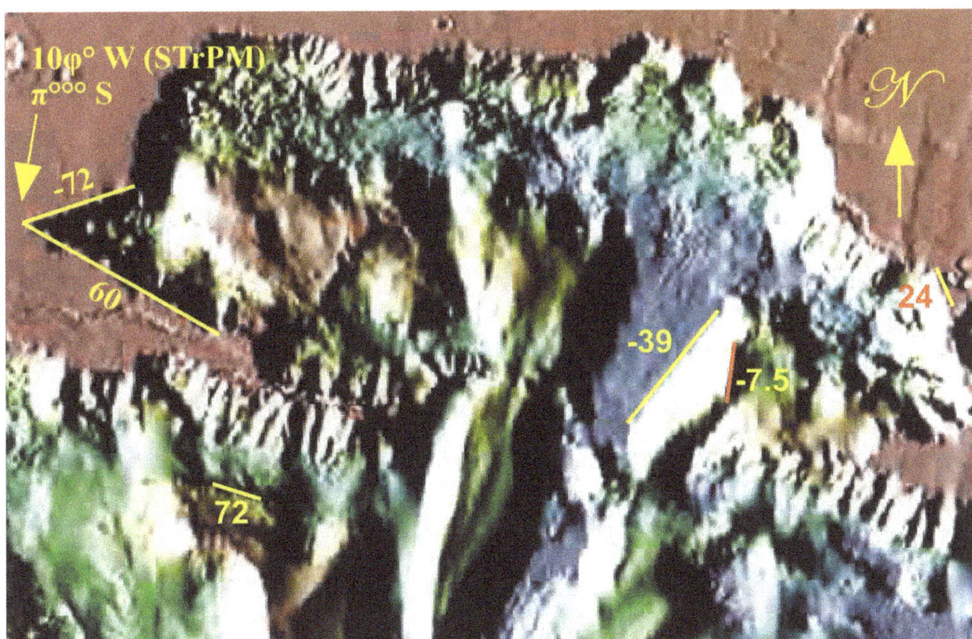

Fig. 10.11: *The Condor Chasma (bottom third of image) and Ophir Chasma (upper two thirds of image) lie to the north of the Melas Chasma. A cliff in the Condor Chasma has a counterclockwise bearing angle of 72°. Cliffs with linear edges around the perimeter of the Ophir Chasma have bearing angles of 60°, -72° and 24°. The bottom edge of a ridge inside the Ophir Chasma has a bearing angle of 39°, and a section of the top of the ridge, a bearing angle of 7.5° (both clockwise). USGS Astrogeology.*

Just east of this are linear ridges, cliffs and the bottom edge of a cliff all having a bearing angle of 7.5°, 3 in the clockwise direction and another 3 in the counterclockwise direction. None of these showed alignments with landforms further north. To the east of these is the bottom edge of a cliff with a bearing angle of 0°.

North of the Melas Chasma, Fig. 10.11 shows linear features in the northern part of the Condor Chasma (bottom 1/3 of figure) and in the perimeter of the Ophir Chasma (upper 2/3 of figure). A linear cliff in the northern part of the Condor Mensa (which is inside the Condor Chasma) has a bearing angle of 72° in the counterclockwise direction. A long linear bottom edge of a ridge with a bearing angle of 39° in the clockwise direction occurs in the central part of the Ophir Chasma. A section of the top of the ridge has a clockwise bearing angle of 7.5°. There is also a linear top edge of a cliff having a counterclockwise bearing angle of 24°. The cliff forms part of the perimeter of the eastern part of Ophir Chasma. The large arrow tip formation located in the western part of the Ophir Chasma has been discussed in the previous chapter (see Fig. 9.9). The southern cliff edge of the arrow tip has a bearing angle of 60° in the counterclockwise direction and the northern cliff edge, a bearing angle of 72° in the clockwise direction. The size and bearing angles of the sides of the arrow tip strongly suggest artificiality. This is virtually confirmed by the coordinates of the tip of the arrow (10φ° W [STrPM] π°°° S). This is very similar to what was found for the Denning Crater where an arrow tip protruding from its western perimeter marked the latitude of 18° S (see *Intelligent Mars II*). The encoding of π in sacred degrees lends much support for the use of the 288 degree coordinate system on Mars.

The top edges of cliffs in the Hebes Chasma to the northwest of the Ophir Chasma have counterclockwise bearing angles of 36° and 84° (Fig. 10.12, left image). Northeast of the Ophir Chasma is the Juventae Chasma (Fig. 10.12, right image). In its northern part are found 2 linear cliff edges, one with a bearing angle of 54° and the other with a bearing angle of 36°, both in the clockwise direction.

Moving back to the main channel of the Valles Marineris, as we travel eastwards out of the Melas Chasma we come to the Coprates Chasma (Fig. 10.13). In the middle of the channel (left picture) there is a linear ridge at the south end of a crater which has a counterclockwise bearing angle of 60°. The perimeter of the northern side of the channel has 2 linear

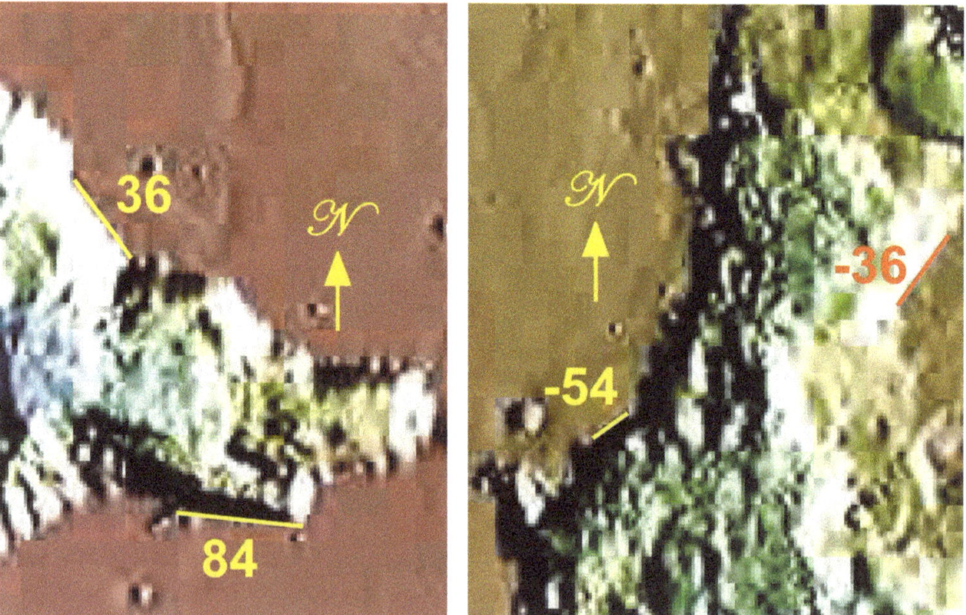

Fig. 10.12: *The perimeter of the eastern part of the Hebes Chasma (left) has straight line segments with counterclockwise bearing angles of 36° and 84°. The perimeter of the northern part of the Juventae Chasma (right) has straight line segments with clockwise bearing angles of 54° and 36°. USGS Astrogeology.*

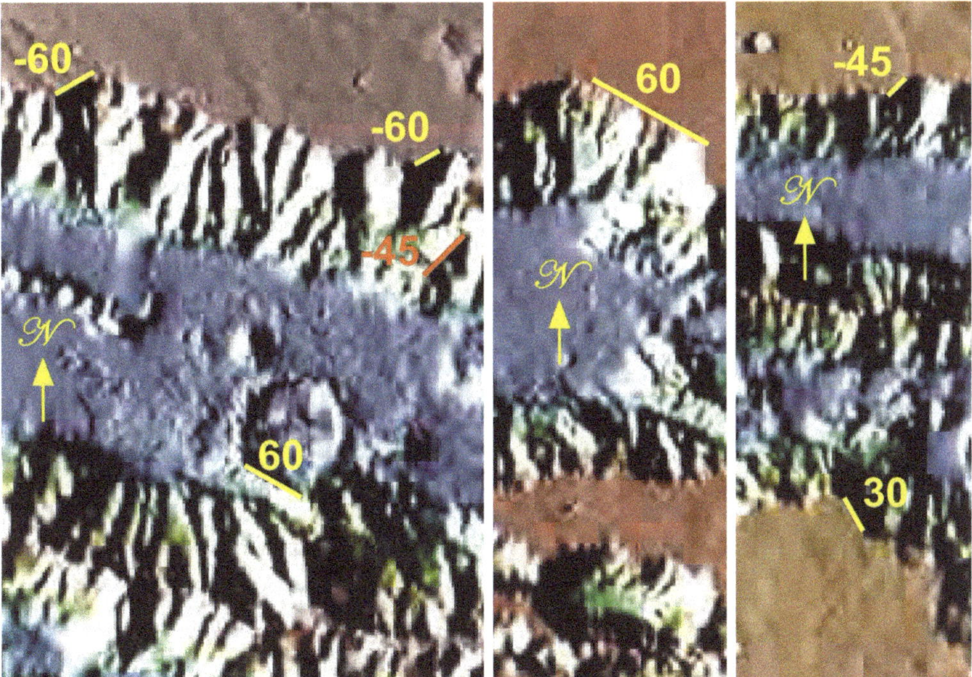

Fig. 10.13: *Left image: a crater in the west end of the Coprates Chasma has a linear perimeter section with a counterclockwise bearing angle of 60°. The north side of the chasma has 2 cliff edges with bearing angles of 60°, and a section in the wall with a bearing angle of 45°, all clockwise. Centre image: middle of the Coprates Chasma has a long linear cliff with a counterclockwise bearing angle of 60°. Right image: the perimeter of the east end of the chasma has a short linear section with a counterclockwise bearing angle of 30° and another with a clockwise bearing angle of 45°. USGS Astrogeology.*

segments each with a clockwise bearing angle of 60°. On the side of the northern wall of the channel there is a ridge with a bearing angle of 45° in the clockwise direction. Further east (middle picture), there is a long linear section of the perimeter of the northern side of the channel with a counterclockwise bearing angle of 60°. In the east end of the Coprates Chasma (right picture) there is a linear section of the southern cliff with a counterclockwise bearing angle of 30° and a linear section of the northern cliff with a clockwise bearing angle of 45°.

Travelling further east brings us to the Eos Chasma where there are a couple of ridges with a bearing angle of 0° inside the channel, and a long cliff edge with a counterclockwise bearing angle of 30° on the southern border of the channel (Fig. 10.14 left). There is also the bottom edge of a cliff on the east side of the Eos Chasma with a counterclockwise bearing angle of 30° (Fig. 10.14 right). The northern part of the channel containing the Eos Chasma is called the Capri Chasma. It is divided from the Eos

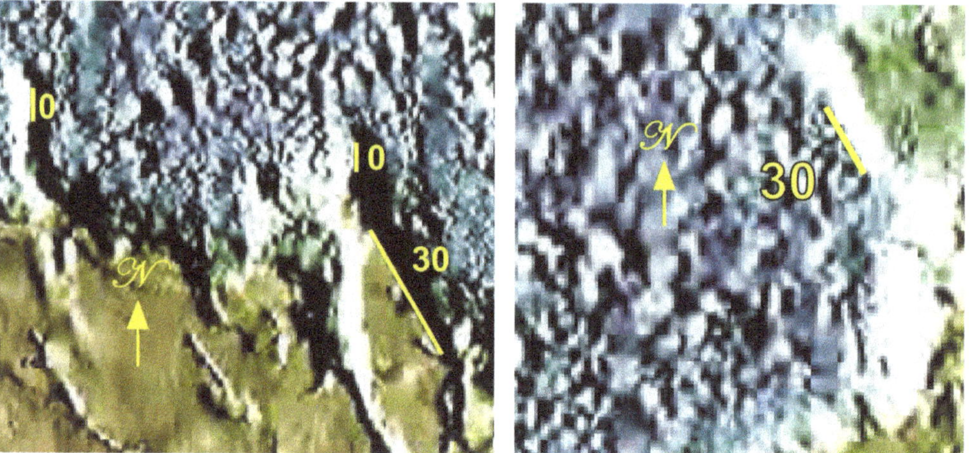

Fig. 10.14: *The west side of the southern part of the Eos Chasma (left) contains interior ridges with linear portions having a bearing angle of 0°. A long linear cliff with a bearing angle of +30° forms part of the southern perimeter. The base of a cliff on the east side of the Eos Chasma (right) also has a bearing angle of +30°. USGS Astrogeology.*

Chasma by the Capri Mensa. The northeast part of the Capri Mensa and the part of the Capri Chasma just north of this area is home to several straight line ridges and cliff edges (Fig. 10.15, left image). Two of these ridges and the top edge of a cliff have bearing angles of plus or minus 7.5°. There are also 2 ridges with clockwise bearing angles of 18°, 3 ridges with bearing angles of 0° and a single ridge with a clockwise bearing angle of 36°. The bottom edge of a cliff has a clockwise bearing angle of 30° and northeast of this is a landform with a linear section having a counterclockwise bearing angle of 54°. About 700 km northeast of here at the far eastern end of the Capri Chasma (Fig. 10.15, right image) there is a linear cliff in 2 sections which are separated by a tab. The tab projects eastward into the chasma. The 2 sections are aligned to each other and have a clockwise bearing angle of 7.5°. This has been shown previously in Fig. 6.17d.

In the Ganges Chasma which lies to the north of the Capri Chasma, there is a curious arrow-like formation in its southwest corner (Fig. 10.16). In the interior of the shaft of the arrow there is a linear interface between a black region and a white region which has a bearing angle of 15° in the clockwise direction. The head of the arrow is composed of 2 straight lines with bearing angles of 7.5° (counterclockwise) and 69° (clockwise). From previous experience with arrow-like formations, I decided that I should measure the coordinates of the tip of the arrowhead. These turned out to be 307.6724° E 9.0731° S. Its latitude can also be expressed as 0.8653°° S [AMPL] which is equal to $(e/\pi)^{\circ\circ}$ S [AMPL], a sacred formula which is

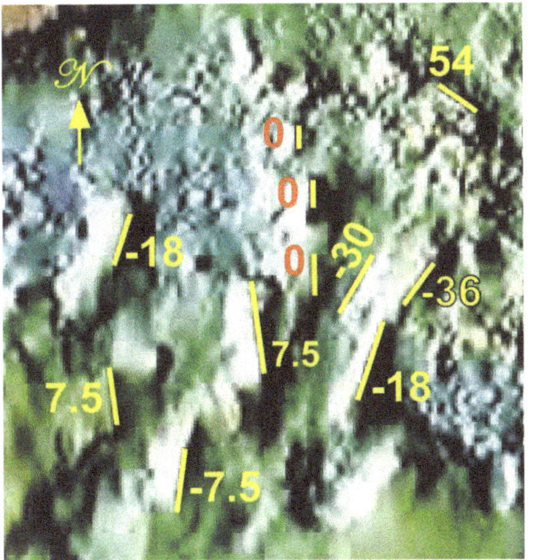

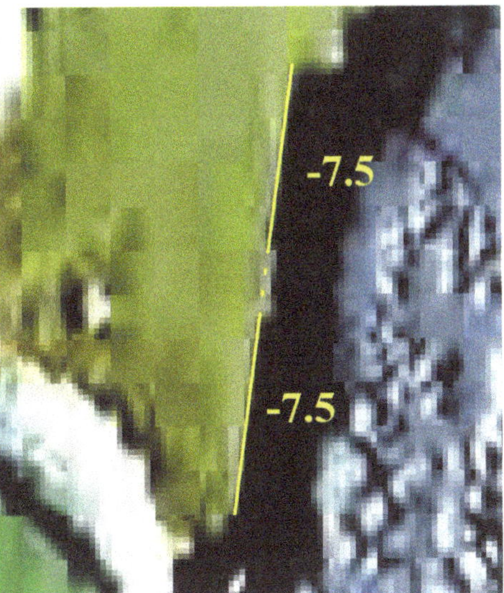

Fig. 10.15: *Left image: the northern part of the Capri Mensa with the Capri Chasma to the north of this. This region is host to 3 linear structures with a bearing angle of plus or minus 7.5°, 3 with a bearing angle of 0° and 2 with a bearing angle of -18°. There are also linear structures with bearing angles of -30°, -36° and 54°. Right image: at the far eastern end of the Capri Chasma there is cliff with 2 linear sections separated by a tab that projects eastward. Both sections have a clockwise bearing angle of 7.5° and are aligned with each other. USGS Astrogeology.*

very meaningful although not as simple as the $\pi^{\circ\circ\circ}$ S which was observed for the tip of the arrow in the Ophir Chasma. The longitude is equal to 160.5000° E (Elysium Mons PM) or 6.5000° E (Sharonov Tower PM). Perhaps the latter number was intended to be 13 degrees in a 720 degree system. The number 13 is often used to represent a leader with 12 followers, and is a Fibonacci number.

Our journey down the main channel now resumes with the Hydraotes

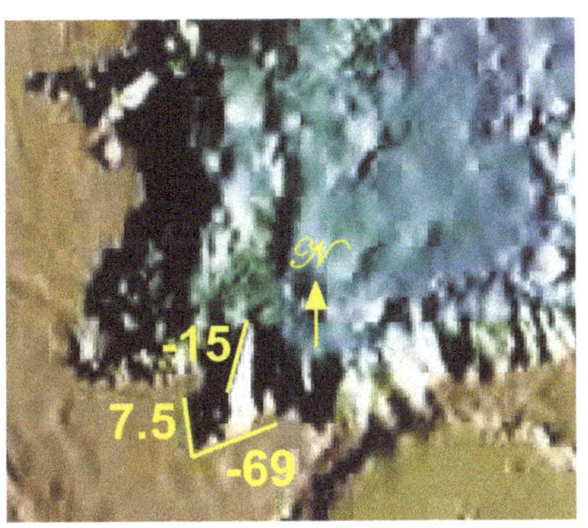

Fig. 10.16: *There is a southwest-pointing arrow formation in the southwest part of the Ganges Chasma. The tip of the arrow has a latitude of 9.0731° S which is equal to $(e/\pi)^{\circ\circ}$ S [AMPL]. Its longitude is 6.5000° E (Sharonov Tower PM). The middle of the arrow shaft has a bearing angle of 15° clockwise. USGS Astrogeology.*

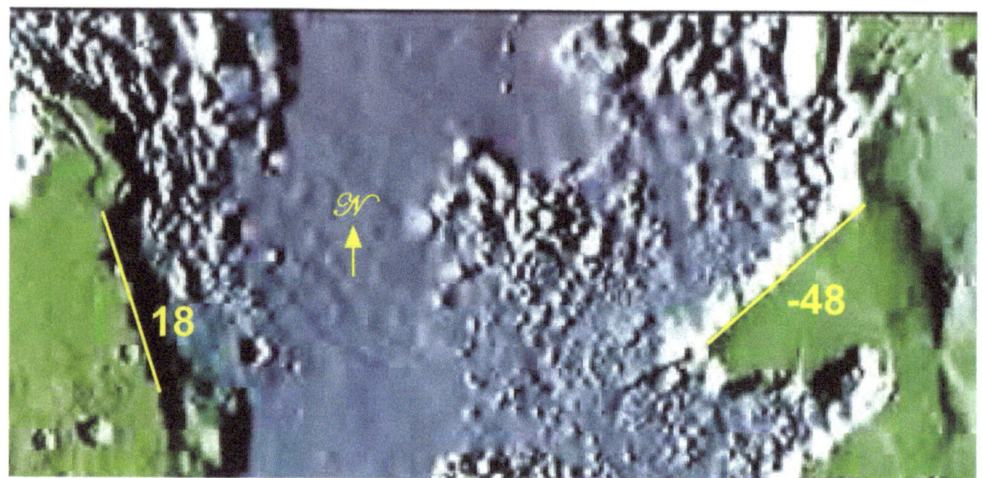

Fig. 10.17: *The mouth of the Hydraotes Chaos has long linear sections in its perimeter. The section on the west side has a counterclockwise bearing angle of 18° while on the east side there is a linear section with a clockwise bearing angle of 48°. USGS Astrogeology.*

Chaos which is at the eastern end of the Valles Marineris. Its mouth opens up with 2 long linear cliffs, one on the west side with a counterclockwise bearing angle of 18° and the other on the east side with a clockwise

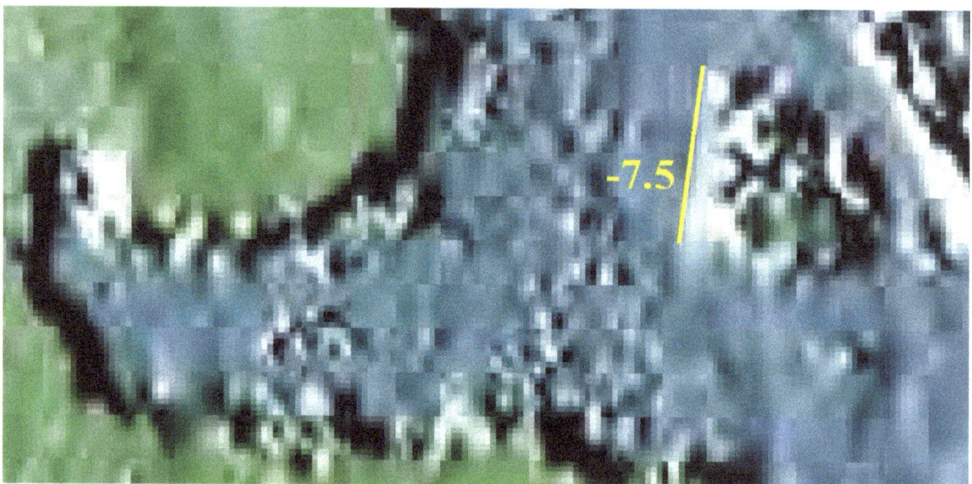

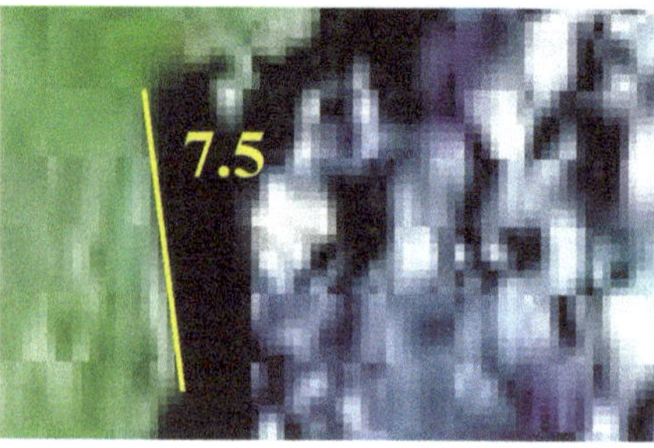

Fig. 10.18: *Left image: a linear cliff in the southwest part of the Simud Vallis with a counterclockwise bearing angle of 7.5°. Upper image: an elevated landmass further north in the Simud Vallis whose west side has a bearing angle of 7.5° in the clockwise direction. This was shown previously in Fig. 6.18. USGS Astrogeology.*

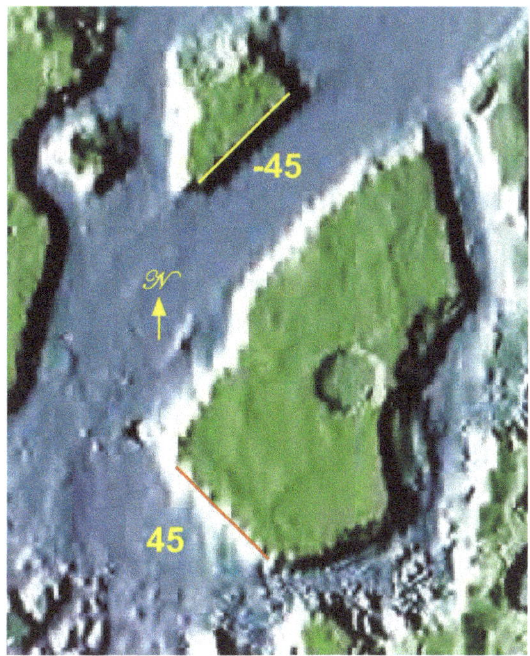

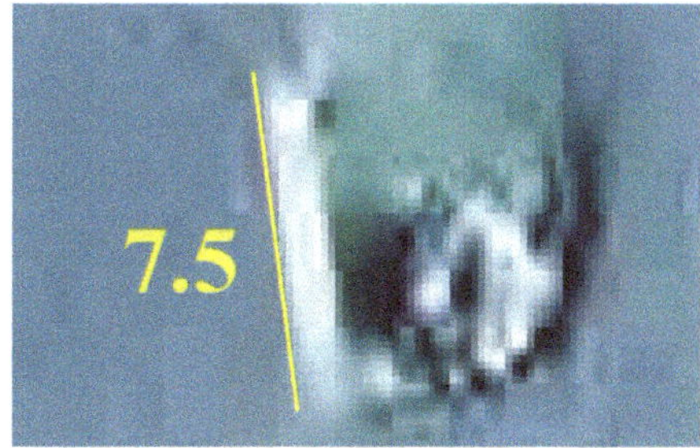

Fig. 10.19: *Left image: just north of the Hydraotes Chaos, there are 2 "islands" in the east part of the Tiu Valles which have linear cliffs. The southern one has a counterclockwise bearing angle of 45° and the northern one, 45° in the clockwise direction. Top image: an elevated landmass in the Tiu Valles whose west side has a bearing angle of 7.5° in the counterclockwise direction. USGS Astrogeology.*

bearing angle of 48° (Fig. 10.17). Proceeding north from the northwest end of the Hydraotes Chaos, we enter the Simud Vallis. On its western perimeter there is a linear cliff with a counterclockwise bearing angle of 7.5° (Fig. 10.18, lower image). Further north of this is a landmass in the Simud Vallis whose west side has a clockwise bearing angle of 7.5° along its base (Fig. 10.18, upper image).

If we return to the Hydraotes Chaos and proceed northward from its northeast end rather than its northwest end which leads to the Simud Vallis, we enter into the Tiu Valles and immediately come to 2 "islands" (Fig. 10.19 left). The larger island has a long linear cliff with a bearing angle of 45° in the counterclockwise direction while the smaller island has a bearing angle of 45° in the clockwise direction. The Tiu Valles wanders northward for a great distance. About 1000 km north of the "islands", we come to another "island" upon which rests the Shawnee Crater. The base of the linear west side of this "island" has a bearing angle of 7.5° in the counterclockwise direction. This was previously shown in the right image of Fig. 6.18 to illustrate landforms with bearing angles of 7.5° and in Fig. 7.10 to illustrate that this linear feature covers a distance of 0.5 latitude degrees. With these 3 islands in the Tiu Valles, we have now completed our long journey along the Valles Marineris having covered over 5000 km of absolutely stunning terrain.

Analysis of Valles Marineris Data

The bearing angles of the numerous straight lines found in various landforms in the Valles Marineris have many repeating values and were not evenly distributed among all possible values between 0° and ±90°. It is now time to examine these findings more systematically to uncover any patterns that may exist and thereby shed some light on the case for artificiality. My findings can be summarized as follows:

1. A total of 82 straight line landforms have been found in the various regions that make up the Valles Marineris. These straight lines coincide with either the top or bottom edges of cliffs, or linear ridges. They stretch for as long as 67 km. Perfectly straight lines are rarely found in nature, particularly for long distances, and especially for repeating bearing angle values when the landforms are hundreds of kilometers apart.

2. Ignoring rotational direction, the frequency of occurrence of bearing angle magnitudes found in the Valles Marineris is shown in Table 10.1. Using the binomial expansion equation under the assumption of 1 degree of resolution, giving 91 possible values between 0° and 90°, it was found that when the number of repeated occurrences of a particular bearing angle exceeds 3 for a group of 82 lines, statistical significance is reached (i.e., the result is unlikely to be due to chance). The chance of obtaining 4 repeated occurrences under random conditions is 1 in 77. The magnitudes which had 4 or more repeated occurrences are 0°, 7.5°, 30°, 36°, 45° and 60°. The chance of obtaining the 5 repeated occurrences of 36° is 1 in 461. The chance of obtaining the 6 repeated occurrences of 45° is 1 in 3311. The chance of obtaining the 7 repeated values of 60° is 1 in 27,934. The chance of obtaining the

Table 10.1: *Frequency of occurrence of bearing angle magnitudes for linear landforms in the Valles Marineris.*

Bearing Angle (°)	# of events	Bearing Angle (°)	# of events	Bearing Angle (°)	# of events
0	16	36	5	60	7
7.5	22	39	1	69	1
15	1	42	2	72	2
18	3	45	6	84	1
24	2	48	1	90	1
30	8	54	3		

8 repeated values of 30° is 1 in 271,869. The chance of obtaining the 16 repeated occurrences for 0° is less than 1 in 100 trillion. Finally, the chance of obtaining the 22 occurrences for 7.5° is less than 1 in 100 billion billion. Hence, this result could not have happened by nature acting randomly. It had to have been deliberately engineered by intelligent beings.

3. Except for 0° and 7.5°, all of the bearing angles are evenly divisible by 3.
4. Most of the bearing angles can be grouped into those that represent the pentagram (18°, 36°, 54° and 72°), the equilateral triangle (30°, 60°), the square (45°), the compass (0°, 45°, and 90°) and multiples of 12 (24°, 36°, 48°, 60°, 72°, and 84°). Only 7.5°, 15°, 39°, 42° and 69° do not belong to these groups.
5. From the analysis presented in Chapter 6, it appears likely that a grid of lines with bearing angles of 7.5° in the counterclockwise and clockwise directions was one of the techniques used to position various features of the landscape on Mars. The cliffs in the Valles Marineris having a bearing angle of ±7.5° were possibly constructed using this grid line information. Another grid of meridian lines having bearing angles of 0° was likely used to position the 16 landforms with a 0° bearing angle. These were the 2 most common bearing angles encountered.
6. The tips of arrow-like outcroppings were used in 2 instances to mark very significant latitudes.

From all of this, I can only conclude that like the giant mountains and many of the craters on Mars, both the Hellas Basin and the Valles Marineris did not come about by natural geological forces, but were artificially created to simulate natural terrain. This astounding conclusion is eerily similar to an exceedingly humorous novel by Douglas Adams entitled *The Hitchhikers Guide to the Galaxy*[3] where the character Slartibartfast boasts about his award winning creation of the Earth's Norwegian Fjords. Slartibartfast takes the main character, Arthur Dent, to a hyperspace factory on Magrathea which is devoted to the construction of whole planets. He shows Arthur the second version of the planet Earth being built to replace Arthur's former home planet which had been blasted out of existence by the Vogons to make way for a hyperspace highway. Was Douglas Adams prescient or did he have access to privileged information about Mars and technologies far beyond our own? Did he know that the Valles Marineris were engineered into existence? Did he suspect that the planet Mars is an artificially constructed planet? Or is it simply that his mind had opened up to possibilities that were previously out of reach to the human imagination?

In 1964, the Soviet astronomer Nikolai Kardashev postulated 3 levels of advanced technology that civilizations could achieve, based on the amount of energy that they could bring under their control. A Type I civilization could capture all the energy available for its planet. A Type II civilization could capture all of the energy radiated by its star. A Type III civilization could capture and use all the energy of an entire galaxy. By these criteria, it would seem that the creators of the Martian terrain and perhaps the planet itself would need to have been at least a Type II, and more likely, a Type III civilization to accomplish the feat of manufacturing planets. A Type III civilization might have the capability of creating the heavier elements which traditional science now ascribes to supernovas.

References

1. *Bleamaster, Leslie F., III, and Crown, David A., 2010, Geologic map of MTM –40277, –45277, –40272, and –45272 quadrangles, eastern Hellas Planitia region of Mars: U.S. Geological Survey Scientific Investigations Map 3096.*
   http://pubs.usgs.gov/sim/3096/

2. *Tanaka, K. L., and Golombek, M. P., 1989, Martian tension fractures and the formation of grabens and collapse features at Valles Marineris: Proceedings of the 19th Lunar and Planetary Science Conference, p. 383-396.*

3. *The Hitchhikers Guide to the Galaxy by Douglas Adams, Pan Books, UK, 1979.*

11 Meet the Architect

Would you believe it if I told you that I may have discovered who masterminded the entire Martian architectural project? Would you believe that there is a record of this dating from some 4 billion years ago and that this record is available to us today? All right, I don't really know the person's name or anything much about this individual or his life. But it does appear as though the person was male and he was obviously a great genius. This is one of those rare discoveries that send shivers up and down your spine. So let me explain how I came to know this and then you can judge for yourself.

One day as I was calculating the sizes and coordinates of craters between 30° N and 65° N on the map where Alba Mons is located, I came across a region of elevated rough terrain. When I looked at a crater-like feature approximately 1200 km east of Alba Mons, a ridiculous thought crossed my mind - this feature has the exact shape of a human ear! I then refocused my gaze to take in the larger region and suddenly an amazing figure arranged itself before my eyes. I could not believe what I was seeing. I was looking at a giant sculpture of what looked very much like a human head (Fig. 11.1)! The western edge of the region had the shape of a face in profile with a clearly visible nose and chin. The eye and mouth were missing in the face, and the upper back of the head was also missing or obscured. All of this suggests that the sculpture is very ancient and that it has eroded over time or has been damaged. The whole region

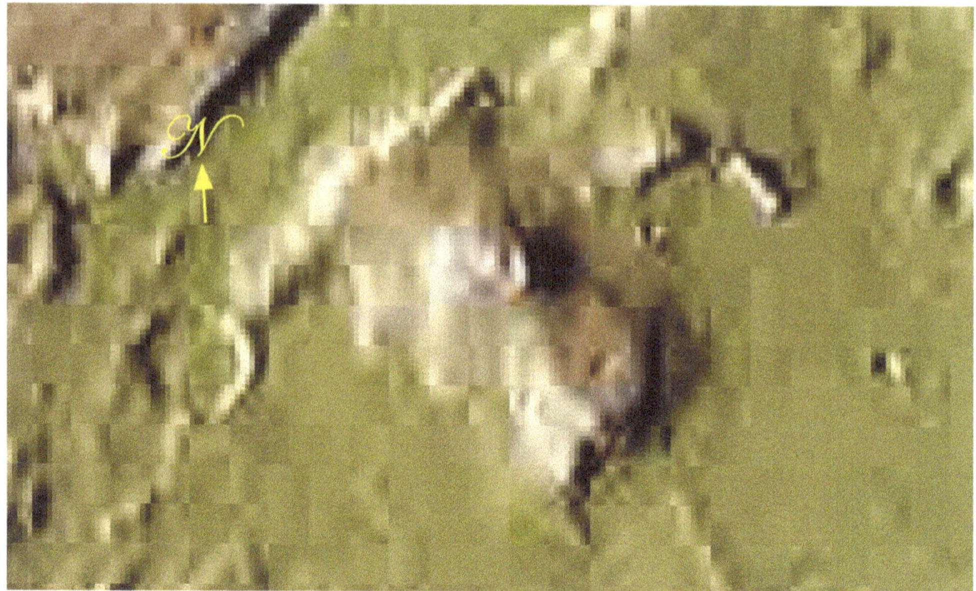

Fig. 11.1: *Landform in the shape of a human bust, complete with large hat which extends well beyond the face. The nose, chin, ear and shoulders are easily distinguishable. The bust is located at 275.82° E 41.33° N and is about 38 km in height. USGS Astrogeology.*

looked like a human bust in profile from the shoulders up, complete with a very large hat which had a brim that reached well beyond the front of the face. Surely this must be the image of a very important person in the history of Mars, something like a Martian Napoleon Bonaparte. My first impression was that the hat was an item that you would expect a high ranking military officer to be wearing rather than a civilian cap.

My enthusiasm was severely shaken when I looked up a THEMIS image of the same region (Fig. 11.2). This was not what I had expected! Yes, the ear was still very prominent and there was still a hat and shoulders. And now there was a very well formed mouth complete with a receding chin. But the nose! Almost like a short beak rather than a nose! And still no obvious structure for the eye. I was in a quandary. It seemed like there were 3 major possibilities for what I had taken to be a human bust:

(1) *The THEMIS image had been deliberately altered to try to make the bust look like a natural formation.*

(2) *The face shows a member of an alien species which has a very different shape of head than that of humans.*

(3) *This is a natural formation which simply appears somewhat like a human head in profile, just like clouds sometimes take on shapes which resemble animals or familiar objects.*

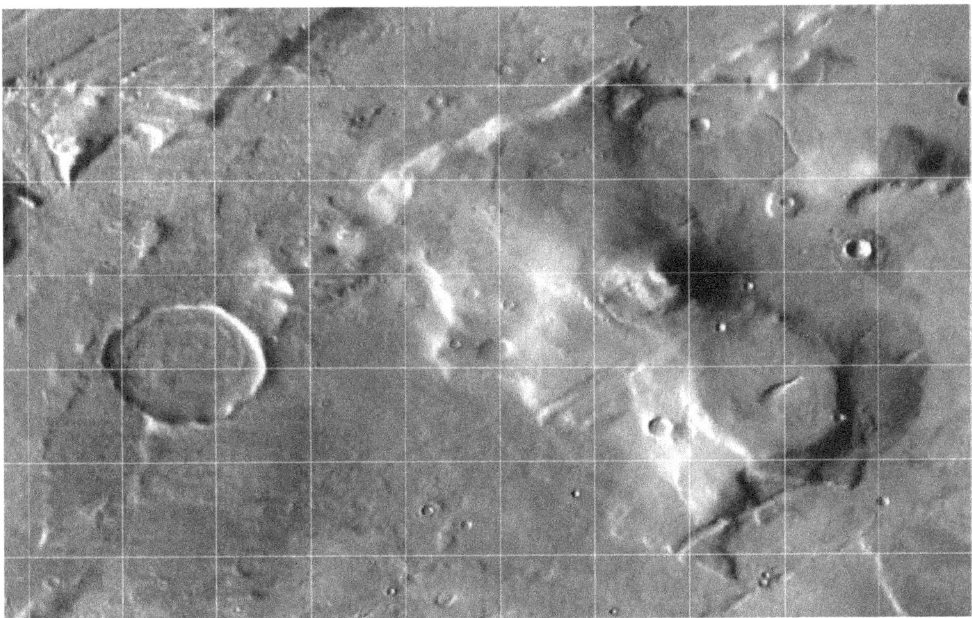

Fig. 11.2: *THEMIS image of the bust in Fig. 11.1. A well-defined mouth and chin are seen as well as the ear and hat. However the nose is shaped like a beak, being notched in front and protruding sharply outwards rather than sloping gently downwards. USGS Astrogeology.*

I decided to test possibility #1 by seeing if key components of the bust had the same coordinates in both the MOLA map and in the THEMIS image. If they were significantly different, this would reveal a deliberate distortion of the bust in the THEMIS image as has been observed for the Pentagram Pyramid and a face just north of the Pentagram Pyramid (see *Intelligent Mars I*). The image in Fig. 11.3 shows substantial deviations in the coordinates of landmarks on the bust and even in a small crater just east of the bust. The crosses in Fig. 11.3 show the locations of features found in the MOLA map. The centre of a large crater just west of the bust is exactly the same for both the MOLA map and in the THEMIS image (see left yellow cross). However, the centre of the ear canal in the THEMIS image is well to the east of the centre of the ear canal in the MOLA image (see red cross and the dark area representing the THEMIS ear canal to the right of the cross). As well, the ear structure is rotated counterclockwise by about 25° in the THEMIS image from its orientation in the MOLA map. An east-west groove in the shoulder of the bust as determined in the MOLA map (yellow cross lower right) lies significantly more to the north in the THEMIS image. Finally, the centre of a small crater to the east of the bust in the THEMIS image is northeast of its proper position (yellow cross

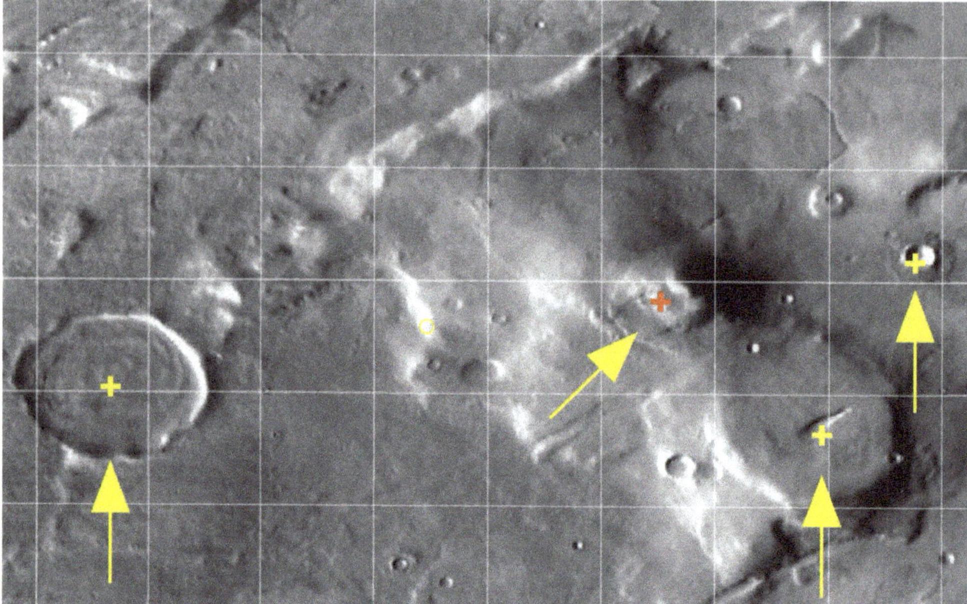

Fig. 11.3: *THEMIS image of the bust is found to be distorted. The western crater (left arrow) is perfectly centred. The ear canal is east of its proper position (red cross) and the ear is rotated counterclockwise by about 25 degrees. The east-west gash in the shoulder is north of where it should be (yellow cross lower right). The eastern crater is northeast of its true centre (yellow cross middle right). USGS Astrogeology.*

middle right). All of this suggests that the THEMIS image cannot be relied upon. Even more importantly, the fact that NASA went to the trouble to distort the bust by photoshopping it lends much credence to the interpretation of this topographical feature as a bust having a hat rather than a natural formation (i.e., eliminates possibility #3).

Fit of the Bust to a Human Head

My next step was to see how well the outline of a human head would fit the MOLA bust. I merged an outline of the human head in profile with a marker for the eye and an outline for the ear in their proper positions, and placed the composite image over the MOLA bust. In order to get it to fit properly, I found that it had to be rotated 27° in the counterclockwise direction from a north-south orientation. I then sized and positioned the head outline so that with the outline of the ear corresponding to the position of the MOLA ear, the outline of the nose lined up with what appeared to be a nose in the MOLA bust. The results of this exercise are shown in Fig. 11.4. The fit to the MOLA bust is very good except for the

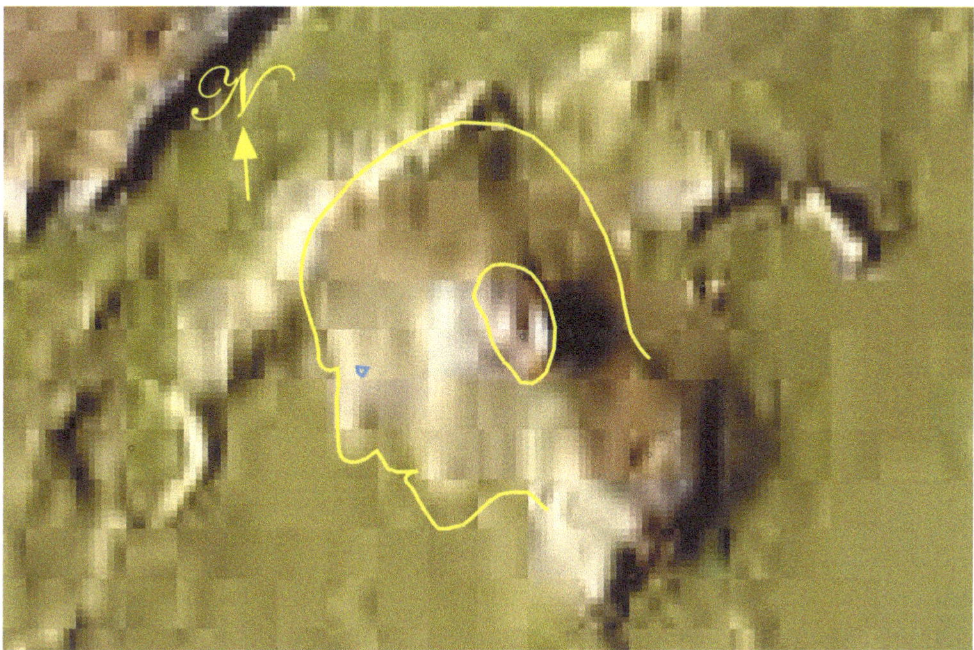

Fig. 11.4: *Outline of the human head superimposed on the bust. To fit the bust, the outline had to be rotated 27° counterclockwise from a north-south orientation. Although the lower jaw of the outline exceeds the edge of the bust by a small amount, the ear of the outline fits the ear of the bust perfectly. The eye of the outline is indicated in blue. USGS Astrogeology.*

front of the face below the nose where the upper and lower jaw of the MOLA image does not extend outward quite enough to reach the outline of the face. This might be accounted for at least in part by normal variation in the human jaw such as a receding chin. The upper back of the head appears to be missing in the MOLA bust. This could be the result of damage from a military attack which seems to have destroyed the eastern part of several structures such as the Pentagram Pyramid and the Pentagon Pyramid. The outline of the ear fits the MOLA "ear" perfectly and the top of the head goes up into the "hat". In summary, it would appear that there is no need to invoke possibility #2 that the bust is alien since the bust fits the human profile quite adequately. Also, the 27° rotation from due north supports the conclusion that this construct is a sculpture of a bust since 27° is 1/4 the size of the angle between the star points of a pentagram.

The Field of View of the Eye of the Bust

The close proximity to the Ayacucho Crater and Issedon Tholus suggests

Table 11.1: *Distances and clockwise bearing angles of Issedon Tholus features and the Ayacucho Crater to the eye of the bust.*

	Distance to Eye of Bust			
		Theoretical	Actual	Difference
Site	Formula	(km)	(km)	(km)
Issedon survey centre	$\varphi R'/(3\pi)$	579.62	579.61	-0.01
Issedon Caldera	$\sqrt{3}R'/10$	584.77	585.04	0.27
Issedon Caldera arrow tip	$eR'/(5\pi)$	584.26	584.52	0.26
Issedon centre of square	$e\sqrt{5}R/36$	573.41	573.70	0.29
Ayacucho Crater	$e\pi R'/72$	400.44	400.82	0.38
	Bearing Angle to Eye of Bust			
		Theoretical	**Actual**	**Difference**
Site	**Formula**	**(°)**	**(°)**	**(°)**
Issedon survey centre	$20e$	54.366	54.229	-0.137
Issedon Caldera	$\mathrm{atan}(\pi/\sqrt{5})$	54.558	54.589	0.031
Issedon Caldera arrow tip	$\mathrm{atan}(\pi/\sqrt{5})$	54.558	54.548	-0.010
Issedon centre of square	$21\varphi^2$	54.979	54.968	-0.011
Ayacucho Crater	$\mathrm{atan}(\varphi)$	58.283	58.447	0.164

that these sites might be related somehow to the bust. The bust is looking straight at them and they are less than 600 km away. I took the position of the eye to be the middle of the cornea of the fitted eye (middle of the western curvature of the blue eye shape in Fig. 11.4). I then calculated the distances and bearing angles of Issedon Tholus sites and the Ayacucho Crater to this point. Issedon Tholus was discovered in *Intelligent Mars II* to have been placed inside an enclosure which has the dimensions of a golden rectangle. The caldera and survey centre of Issedon Tholus lie in the northeast square section of the golden rectangle. There is an arrow shape inside the caldera whose tip has the coordinates of 36° W [SToPM] 36° N. The sacred distance formulae and bearing angles to the eye of the bust from Issedon Tholus elements and the Ayacucho Crater are listed in Table 11.1. All of the sacred distances formulae make some reference to the pentagram which is related to the golden mean. This creates a parallelism due to the presence of the golden rectangle surrounding Issedon Tholus. In their distance formulae to the eye, the Issedon survey centre has φ in the numerator, and the Issedon Caldera and its arrow tip have 5 as a factor in their denominators, the number of star points in a pentagram. The centre of the Issedon square not only has √5 in its numerator but also has the number 36 as the divisor in its sacred distance formula to the eye of the bust. This is very amazing not only because 36 is the number of degrees in the angle of a star point in a pentagram, but also because the survey centre, caldera and caldera arrow tip of Issedon Tholus are at 36° N, and the Issedon Caldera arrow tip is 36° W of the Sharonov Tower PM. With its sacred distance to the eye of the bust, the centre of the square is also associated with the number 36. The Ayacucho Crater has 72 in its sacred distance formula which is the number of degrees in the angles at the base of a star point in a pentagram. The association with the pentagram and golden ratio is also present in the formulae for bearing angles from all of these sites to the eye of the bust. The formula for the Issedon survey centre bearing angle has 5 as a factor, and the formulae for the other Issedon sites contain either √5 or φ. Note that √5 = 2φ - 1. For the Ayacucho Crater, the formula of atan(φ) is a simple way to represent the golden ratio.

By this time, I was beginning to suspect that the bust might be looking at other sites as well. I decided that I had better check out the distances and bearing angles to the eye of the bust for all sites that could possibly be "seen" by the bust (Fig. 11.5). The most easterly site that could be "viewed" was approximately the western edge of the Hebes Chasma. The line of sight to the eye from this location has a bearing angle of 7.2166° in the counterclockwise direction. The most northerly site that could be "viewed" from the eye of the bust was the Hecates Tholus Caldera. The line of sight to the eye from the centre of this caldera has a bearing angle of 84.3077° in the clockwise direction. Fig. 11.5 shows that the line of sight for the Hecates Tholus Caldera (top yellow line) runs slightly beneath the brim of the hat so it is just visible to the bust. The line of sight for the western edge of the Hebes Chasma runs just west of the edge of the upper lip of the bust. Unlike the Sharonov Crater eye, the eye of the bust also sees Alba Mons and, of course, Issedon Tholus and the Ayacucho Crater. The line of sight for Alba Mons is slightly below the top line which goes to the Hecates Tholus Caldera. Only some of the sites in the field of view are shown with lines of sight in Fig. 11.5 so as not to crowd the figure. The ones shown are listed in the caption to the figure. They include the Pentagon Pyramid, the arrow tip in the Issedon Tholus Caldera, the octagon centre of the Aum Crater, all the major mountains and the Fesenkov Crater. The dashed red line goes to the "cornea" of the Sharonov Crater, showing that the Sharonov Crater is not visible to the bust's eye nor is it aligned to the bust's eye. Hence these 2 eyes would not create binocular vision. Also the eye of the bust, like the eye of the

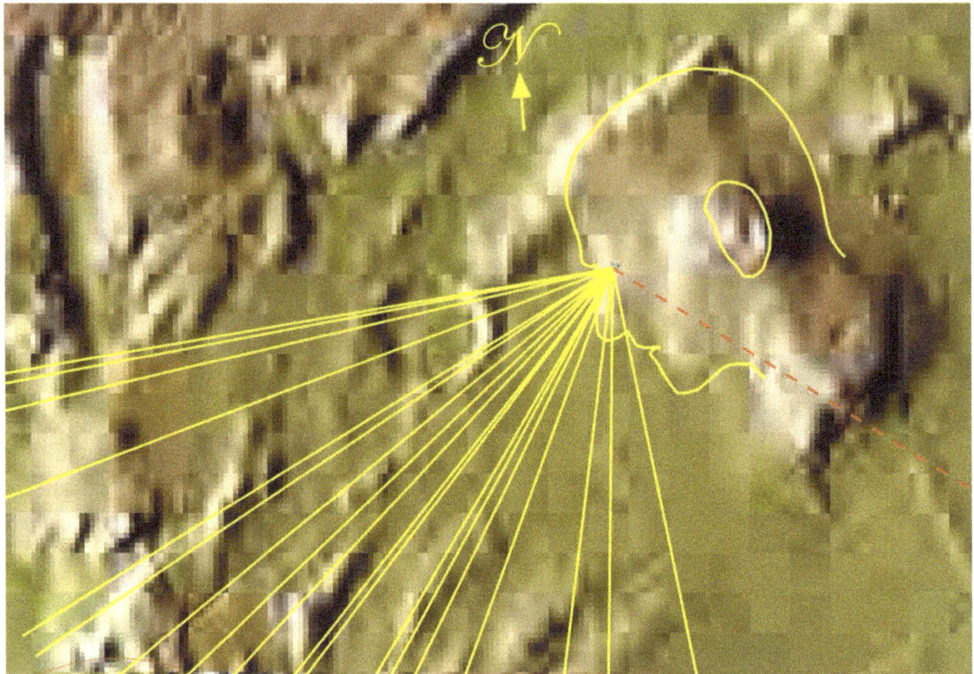

Fig. 11.5: *Lines of sight to various landmarks. From the top line, in the counterclockwise direction, the bust "views" the Hecates Tholus Caldera, Alba Mons, the Elysium Mons Caldera, the east-west midpoint of the Pentagon Pyramid, Apollinaris Mons, the Central Caldera of Olympus Mons, the arrow tip in the Issedon Tholus Caldera, the western centre of Jovis Tholus, the octagon centre of the Aum Crater, the northern centre of the Biblis Tholus Caldera, the northern crater of Ulysses Tholus, Arsia Mons, Pavonis Mons, Ascraeus Mons, Uranius Mons, the Tharsis Tholus Caldera, the Fesenkov Crater and the western edge of the Hebes Chasma. The dashed red line goes to the "cornea" of the Sharonov Crater and is not "seen" by the bust. USGS Astrogeology.*

Sharonov Crater, "sees" only symbolically since none of the sites could be viewed even with a powerful telescope due to the curvature of the planet.

Distances and bearing angles from various sites are shown in Tables 11.2 and 11.3. Although I measured a total of 64 sites, I show only some of the more interesting of these in the tables to conserve space. The sacred formulae in Table 11.2 use both the equatorial radius (R = 3396.19 km) and the northern polar radius (R' = 3376.20 km) of Mars. The first thing that I noticed is that a large number of the distance formulae are of the form i^2r/n where i is an irrational number, r is either the equatorial or the northern polar radius of Mars, and n is an integer. In the first 5 formulae, i is equal to π and the actual distances from these sites are all within 2 km of the theoretical distances. In the next 2 formulae, i is equal to e, and in the final formula of this type, i is equal to φ. The actual distance in these 3

Table 11.2: *Sacred distance formulae and distances of sites to the eye of the bust.*

	Distance to Eye of Bust			
	Theoretical		Actual	Difference
Site	Formula	(km)	(km)	(km)
Ulysses Tholus Caldera	$\pi^2R'/11$	3029.25	3027.27	-1.98
Olympus Mons Central Caldera	$\pi^2R/12$	2793.25	2792.02	-1.23
Ascraeus Mons	$\pi^2R'/16$	2082.61	2082.84	0.23
Tharsis Tholus S centre	$\pi^2R'/19$	1753.78	1752.38	-1.40
Uranius Tholus Caldera	$\pi^2R/30$	1117.30	1117.60	0.30
Biblis Tholus Caldera N centre	$e^2R/8$	3136.83	3137.61	0.78
Fesenkov Crater	$e^2R/21$	1194.98	1194.99	0.01
Uranius Tholus N Crater	$\varphi^2R'/8$	1104.88	1104.18	-0.70
Uranius Mons	$eR'/9$	1019.72	1019.63	-0.09
Alba Mons	$\sqrt{3}R'/5$	1169.55	1170.02	0.47
Aum Crater centre of octagon	$3R'/2$	5064.30	5070.78	6.48
Ascraeus Mons Caldera	R'/φ	2086.61	2086.19	-0.42
Apollinaris Mons	$5R/e$	6246.94	6246.11	-0.83
Hellas Basin northern tower	$10R'/e$	12420.35	12420.56	0.21
Pavonis Mons Caldera	$\varphi\pi R/6$	2877.25	2877.63	0.38
Pentagon Pyramid	$8R/(e\varphi)$	6177.32	6177.18	-0.14
Pentagram Pyramid	$\pi R'/(2\varphi)$	3277.63	3274.51	-3.12
Arsia Mons Caldera	$5R/(e\sqrt{3})$	3606.67	3607.05	0.38
Hecates Tholus	$15R/(e\pi)$	5965.39	5966.43	1.04

formulae is less than 1 km from the theoretical distance.

For Olympus Mons and the Tharsis Montes, I found that, with the exception of Ascraeus Mons, distance formulae were much closer to actual values with the calderas of these mountains than with their survey centres. Hence, I listed the calderas of these mountains in Table 11.2 rather than their survey centres, except for Ascraeus Mons for which 1 listed both. Note that the sacred distance formula from the Issedon Tholus Caldera ($\sqrt{3}R'/10$ - see Table 11.1) is exactly 1/2 that of the distance formula from Alba Mons ($\sqrt{3}R'/5$) and that the denominators both have 5 as a factor. For the octagon centre of the Aum Crater, the distance formula is simply 3/2 R′ which is a rather large 6.48 km less than the actual distance. Nevertheless, you will see in the next table, that the bust does consider this to be an important site since the bearing angle from the octagon centre to the eye of the bust is an integer.

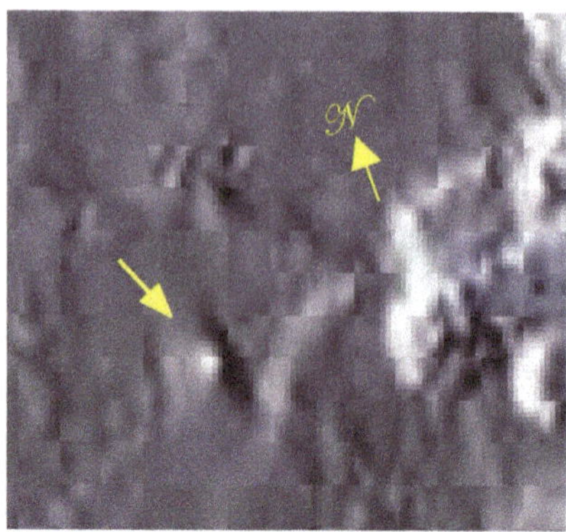

Fig. 11.6: *Tower (yellow arrow) in the northern region of the Hellas Basin at 63.1448° E 31.6714° S. The latitude coordinate is within 35 seconds of a degree of* $9(\varphi^2)^\circ$ *S [AMPL],* $8(\varphi^2)^{\circ\circ}$ *S [AMPL] and* $6\pi^{\circ\circ\circ}$ *S [AMPL], and is 2.77 minutes of a degree less than* $\text{atan}(\Phi)^\circ$ *S. The longitude coordinate is within 50 seconds of a degree of* $38\sqrt{5}^\circ$ *W [DPPM],* $42\varphi^{\circ\circ\circ}$ *W [DPPM] and* $25e^{\circ\circ\circ}$ *W [DPPM]. The tower may be honouring the equation of* $\varphi = (25/42)e$ *which differs from* φ *by only 0.000009. USGS Astrogeology.*

The sacred distance formula (10R'/e) to a tower-like structure found inside the northern part of the Hellas Basin (Fig. 11.6) would be exactly twice that (5R/e) to Apollinaris Mons except that the former uses the northern polar radius rather than the equatorial radius. The bust must have considered the Hellas Basin tower to be a very important site since it is only 220 meters further than its sacred formula of 10R′/e km, a distance of more than 12,000 km! Also (see Table 11.3) its clockwise bearing angle to the eye of the bust is only 8 seconds of a degree less than 69.5°. For such a distant site to have both a distance and bearing angle with so little discrepancy from meaningful values is truly amazing. The coordinates of this tower are remarkable in that the latitude coordinate is within 35 seconds of a degree of $9(\varphi^2)^\circ$ S [AMPL], $8(\varphi^2)^{\circ\circ}$ S [AMPL] and $6\pi^{\circ\circ\circ}$ S [AMPL]. It is also only 2.77 minutes of a degree less than $\text{atan}(\Phi)^\circ$ S. The longitude coordinate is within 50 seconds of a degree of $38\sqrt{5}^\circ$ W [DPPM], $42\varphi^{\circ\circ\circ}$ W [DPPM] and $25e^{\circ\circ\circ}$ W [DPPM].

The Pavonis Mons Caldera, the Pentagram Pyramid and the Pentagon Pyramid are all sites which are associated with the golden mean. It is therefore remarkable that the sacred distance formulae from their locations all contain the value φ. It should be noted that the formula from the Pavonis Mons Caldera contains π as well as φ which seems to reflect the formula of $(\varphi/\pi)^\circ$ N for the latitude of the centre of the caldera. The value of π also appears in the sacred distance formula of $\pi R'/(2\varphi)$ km for the Pentagram Pyramid which honours both the planetographic and planetocentric latitudes of π° N (see Chapter 9).

This brings us now to a discussion of the bearing angles of sites to the eye of the bust which are listed in Table 11.3. Here there is an unusually large number of bearing angles which are almost pure integers. There is also another bearing angle which is almost a pure integer and one-half value. For a maximum deviation of ± 0.0551° from being an integer, the presence of 13 integers among 64 bearing angles is statistically significant with a probability of 0.021 or about 1 chance in 48 that this could be due

Table 11.3: *Bearing angles of sites to the eye of the bust. Note the number of integer bearing angles. These bearing angles are all in the clockwise direction.*

	Bearing Angle to Eye of Bust			
	Theoretical		Actual	Difference
Site	Formula	(°)	(°)	(°)
Tharsis Tholus Caldera	11	11.0000	11.0026	0.0026
Uranius Mons	24	24.0000	24.0146	0.0146
Ascraeus Mons	30	30.0000	29.9688	-0.0312
Arsia Mons Caldera	33	33.0000	33.0014	0.0014
Uranius Tholus N Crater	35	35.0000	34.9548	-0.0452
Poynting Crater	37	37.0000	36.9449	-0.0551
Ulysses Tholus SE Crater	40	40.0000	40.0544	0.0544
Ulysses Tholus N Crater	41	41.0000	40.9944	-0.0056
Biblis Tholus Caldera N centre	42	42.0000	42.0463	0.0463
Biblis Tholus Caldera S centre	42	42.0000	41.9842	-0.0158
Aum Crater Octagon Centre	47	47.0000	46.9981	-0.0019
Issedon Tholus centre of square	55	55.0000	54.9682	-0.0318
Issedon Tholus N vertex of square	58	58.0000	58.0020	0.0020
Hellas Basin northern tower	69.5	69.5000	69.4977	-0.0023
Fesenkov Crater	3φ	4.8541	4.8645	0.0104
Ceraunius Tholus	19φ	30.7427	30.7109	-0.0318
Biblis Tholus Caldera N centre	26φ	42.0689	42.0463	-0.0226
Jovis Tholus East	31φ	50.1591	50.1263	-0.0327
Elysium Mons Caldera	50φ	80.9017	80.9352	0.0335
Pavonis Mons	12e	32.6194	32.6074	-0.0120
Pentagon Pyramid EW midpoint	27e	73.3936	73.4256	0.0320
Hecates Tholus Caldera	31e	84.2667	84.3077	0.0410
Jovis Tholus West	16π	50.2655	50.2922	0.0267
Tharsis Tholus South	$7\sqrt{2}$	9.8995	9.8931	-0.0064
Apollinaris Mons	$44\sqrt{2}$	62.2254	62.2210	-0.0044
Issedon centre of square	$21\varphi^2$	54.9787	54.9682	-0.0105
Olympus Mons Central Caldera	$23\varphi^2$	60.2148	60.2147	-0.0001
Nicholson Crater	$23\varphi^2$	60.2148	60.2086	-0.0062
Poynting Crater	$5e^2$	36.9453	36.9449	-0.0004
Tharsis Tholus South	π^2	9.8696	9.8931	0.0235

to randomness. This finding suggests that my positioning of the eye for the bust is very close to the intended value. The next group of bearing angles in Table 11.3 has sacred formulae which consist of a single irrational number multiplied by an integer. Most of these bearing angles seem to relate to the pentagram with the use of the irrational number φ and the integer of 27 which is 1/4 of the size of the angle between the star points of a pentagram. The final 5 bearing angles in Table 11.3 have sacred formulae which consist of the square of an irrational number multiplied by an integer. The simplest of these is the bearing angle of π^2 degrees from the southern peak of Tharsis Tholus. The bearing angle of the Olympus Mons Central Caldera (60.2147°) is only 22 seconds of a degree more than that for the Nicholson Crater (60.2086°) so these 2 sites are almost perfectly aligned from the viewpoint of the eye.

Conclusion

The bust appears to be looking at a very large number of important architectural achievements on Mars. The estimated location of the eye of the bust is associated with many meaningful distance and bearing angle formulae from these sites. These findings made me reconsider my initial guess that the bust might represent a military figure or ruler. The more likely possibility is that this person is the chief architect instead! What better way to honour his incredible achievements and genius than a public acknowledgement of his work in the form of a huge bust visible from outer space. Very likely he even designed and located his own bust. We are probably looking at the Slartibartfast of Mars! But did this individual design just the mountains, craters and pyramids in view, or did he design all of the topography of the planet including the Valles Marineris and other landforms which are at the back of the bust? Actually, since the eye can "see" the Hecates Tholus Caldera, it is in a position to see all of the Valles Marineris by looking right around the planet. It doesn't need eyes in the back of its head. When the line of sight to the Hecates Tholus Caldera reaches around the planet to intersect with the line of sight to the western edge of the Hebes Chasma, it is at a latitude of slightly more than 10° N. Hence, the eye is in position to see all important sites south of 10° N which are located at the back of the head of the bust, including the Argyre Basin. A large part of the northern hemisphere cannot be seen, however.

The fact that the entire Valles Marineris is actually in view of the bust is so similar to what Douglas Adams described about Slartibartfast that this author almost certainly must have had privileged knowledge about Mars. Douglas Adams had his character receive an award for his design of the Norwegian fjords. The award given to the Martian Slartibartfast for designing the Valles Marineris was probably a bust in his own likeness. With all of this in mind I have chosen henceforth to call the bust "The Architect".

The bust is close enough to the human form to be credible as an ancestor from which the human race evolved. Or should I say devolved since the Martian civilization was obviously light years ahead in terms of technology and spirituality. The rather humanoid appearance of this figure certainly brings into question the current theory of evolution of humanity from the apes. We could be looking at a picture of what our own biological ancestors looked like billions of years before our Darwinian evolution was supposed to have happened. We not only have to rethink our origins, but of the origins of all life in this universe. Our current theories can no longer stand up against these findings. It's now a wide open field, and we really have to start over from scratch.

Planetary Scale Arrangement of Artifacts

Having established the artificiality of enormous topographical features on Mars such as the Hellas Basin and the Valles Marineris, it is now time to examine if these and other features are simply discrete items, each serving a separate function, or whether they are somehow connected to one another. For instance, we have seen in Chapters 2 and 3 that the Aum Crater not only portrays the primordial sound depicted by a series of expanding squares and octagons, but also links this sound to the creation of both the planet and the Martian civilization. We have also seen how a plateau in the shape of a bust oversees a huge number of important sites including Elysium Mons, the giant mountains of the Tharsis Rise and the Valles Marineris. A similar overview of many of these sites was also found in *Intelligent Mars II* for the Sharonov Crater.

Planitiae in the Northern Hemisphere

Much of the northern terrain on Mars is occupied by 6 very large plains which are called planitiae (Fig. 12.1). Their origin is still controversial although at least some of them are thought to have been caused by impacts several billion years ago. There is some disagreement as to the

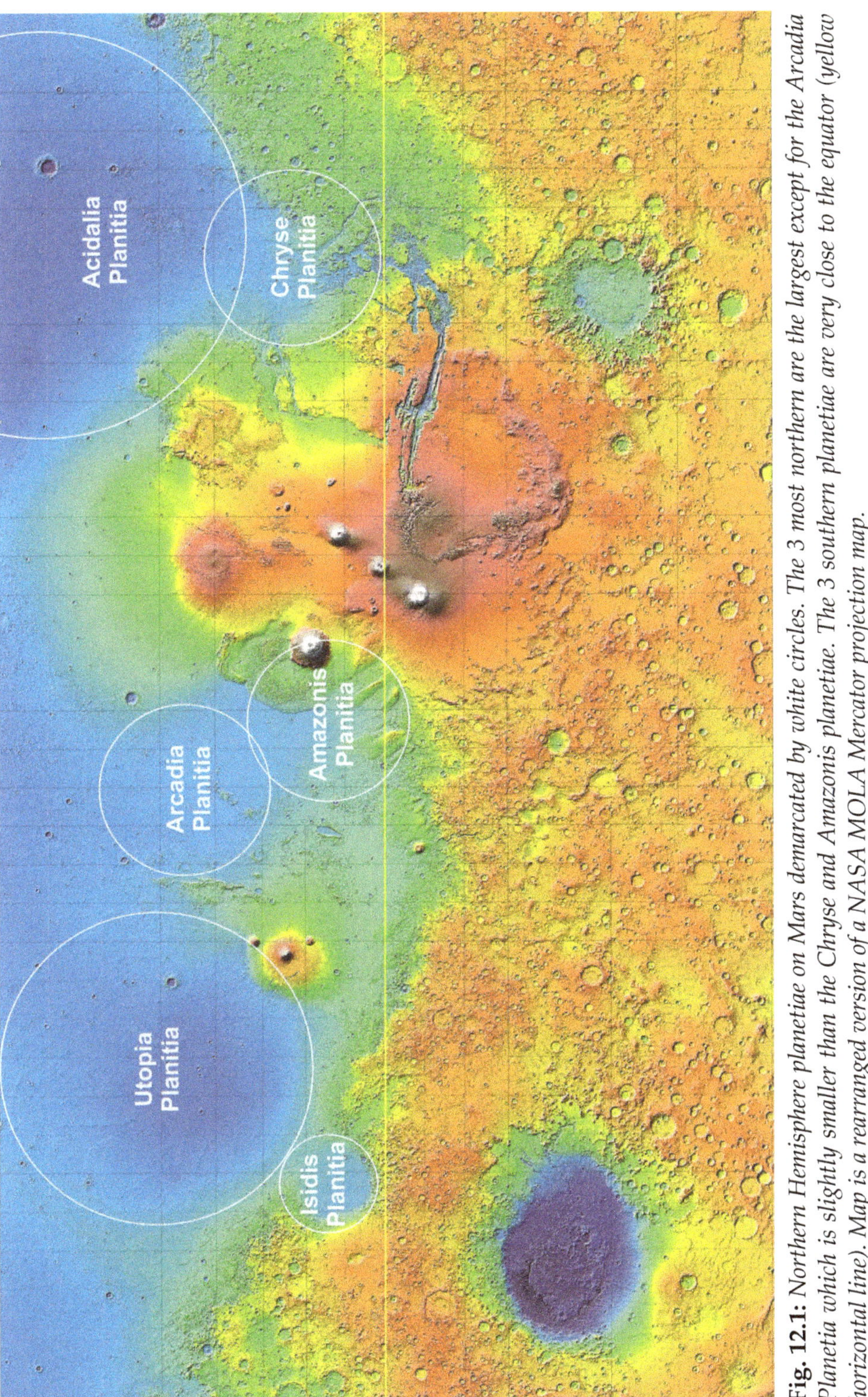

Fig. 12.1: *Northern Hemisphere planetiae on Mars demarcated by white circles. The 3 most northern are the largest except for the Arcadia Planetia which is slightly smaller than the Chryse and Amazonis planetiae. The 3 southern planetiae are very close to the equator (yellow horizontal line). Map is a rearranged version of a NASA MOLA Mercator projection map.*

sizes and exact coordinates of the centres of these plains. However, there is unanimity on the number and their general location. Assuming that the planitiae are essentially circular, I based my size and location estimates of the southern 3 planitiae on the curvature of the border between green and yellow areas on a MOLA map which represent 2 very different elevations of terrain. Similarly, for the 3 northern planitiae, I based my size and location estimates on the curved border between blue and green areas since the blue areas represent much lower terrain then the green areas. Although my estimated perimeter for the Amazonis Planitia includes Olympus Mons, it can be assumed that this mountain would have been constructed well after the planitia had been laid down. It should be noted that the circles in Fig. 12.1 are distorted since they are placed on a Mercator projection map which distorts the curvature of the planet.

As this chapter is about arrangement of artifacts, I should first establish the artificial nature of the planitiae. Ideally this should have been done in Chapter 10 which describes enormous scale artifacts since these are indeed enormous structures, ranging from about 1450 to 3000 km in diameter by my estimates. However, that would have made the chapter too long, so I elected instead to give a brief analysis of their artificiality here. I will start with examples of artificial sites within each of the planitiae. To conserve space, I will show only 1 example for each planitia.

Starting with the 3 northern planitiae, the perimeter of the Nier Crater within the Utopia Planitia has several linear segments, each of which has a very meaningful bearing angle (Fig. 12.2, image a). There are 2 segments with a clockwise bearing angle of 12° and 2 segments with a counterclockwise bearing angle of 36°. There is also a segment with a bearing angle of 0° and another with a clockwise bearing angle of 48°. A linear section of the western edge of the Erebus Montes in the Arcadia Planitia has a clockwise meaningful bearing angle of 33° which suggests artificiality (image b). Note that 33 is a very important number in Freemasonry. Its total length is about 70 km. There are 3 linear trenches in the Acidalia Planitia which are in close association to each other thus forming an enclosure (image c). The west and east sides of the enclosure have counterclockwise bearing angles of 39° and 24°. The southern side of the enclosure has a clockwise bearing angle of 54°. It is continuous with the northern edge of a crater which has a blunted arrow shape. The southern side of the arrowhead has a counterclockwise bearing angle of 54°, and the eastern side of the arrowhead has a counterclockwise bearing angle of 6°. Despite the angle of the back side of the arrowhead, the arrow points due west since the north and south sides of the arrowhead have symmetrical bearing angles. When the southern side is extrapolated to meet the northern side, the tip of the arrowhead is found to have the

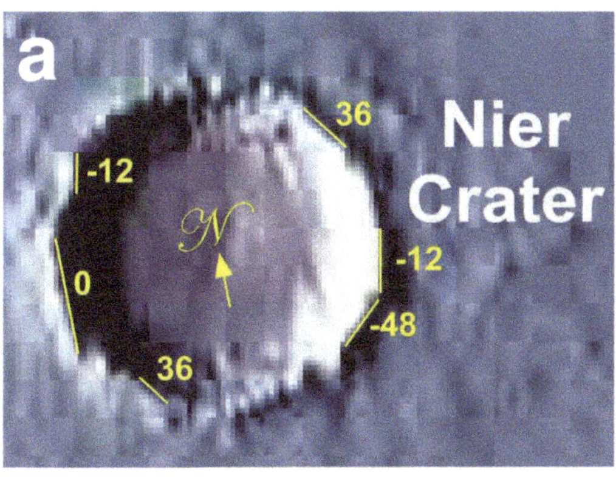

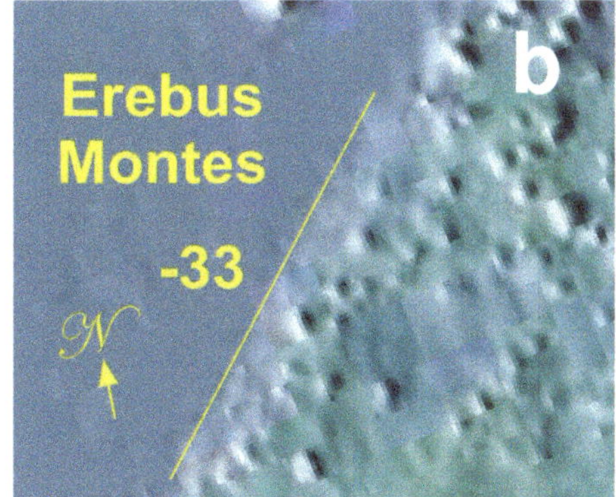

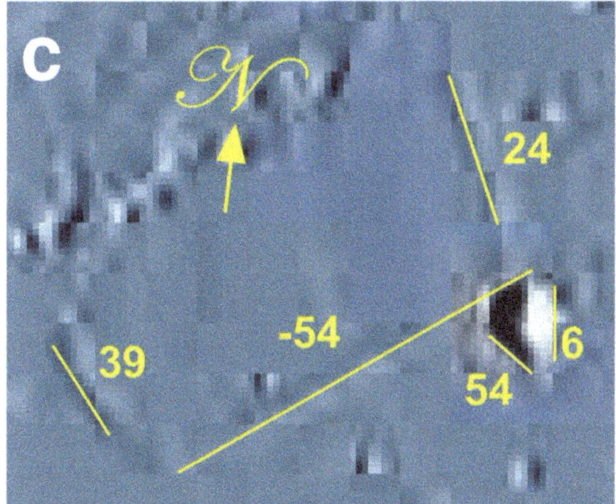

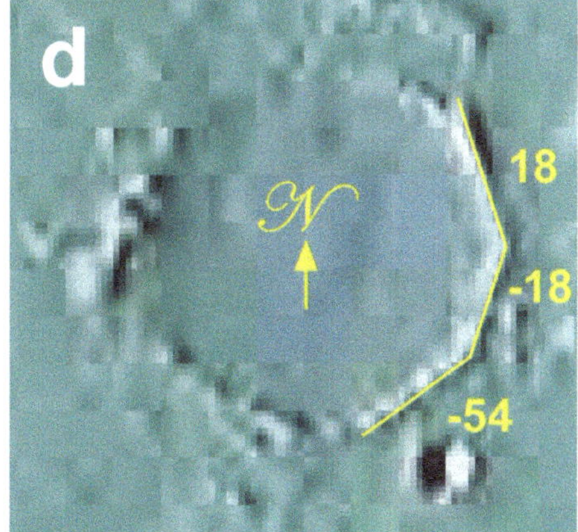

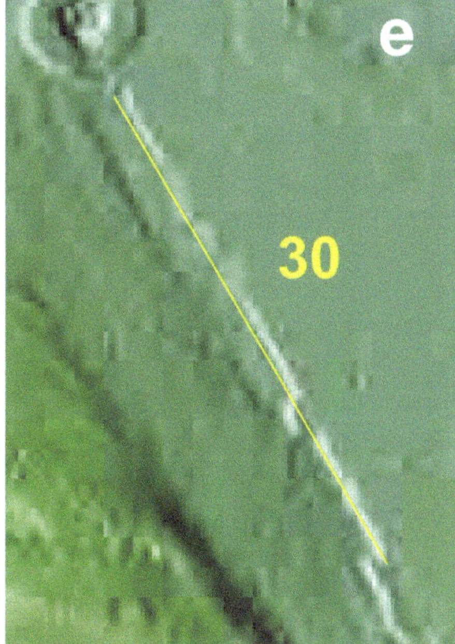

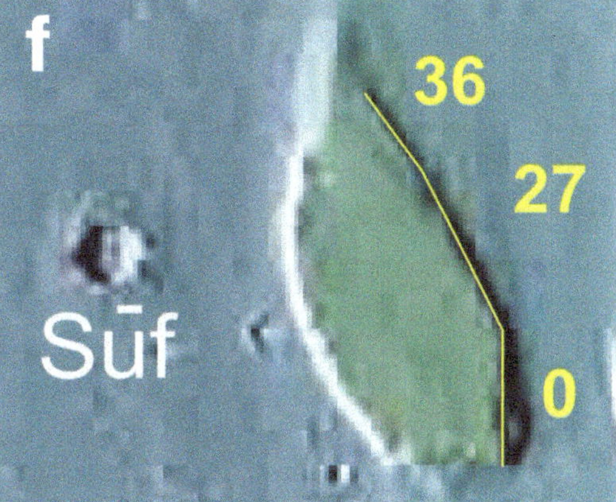

Fig. 12.2: *Artificial sites in the (a) Utopia Planetia, (b) Arcadia Planetia, (c) Acidalia Planetia, (d) Isidis Planetia, (e) Amazonis Planetia and the (f) Chryse Planetia. Note the square root sign middle left in image d. USGS Astrogeology.*

coordinates of 322.1035° E 36.8104° N. These translate into 60°°° E [PMPM] and 17√3°°° N with errors of less than 13 seconds of a degree.

Moving on to the 3 southern planitia, a very curious crater is located in the northeastern part of the Isidis Planitia (Fig, 12.2, image d). It could be considered to be an arrow crater since its 2 linear eastern sides have symmetrical bearing angles of ±18° which means it points due east. It also has a linear southern side which has a clockwise bearing angle of 54°. The numbers 18 and 54 are the half sizes of 2 important angles in the pentagram. But look at the western side of the crater. Do you see a very well formed square root sign? And just east of this is what looks to be a lower case 'f' letter whose base is connected to an oval-shaped mound. It looks very similar to the lower case digamma symbol, an archaic Greek letter which was also used for the number 6. Note that √6 = √2 x √3. In the southern part of the Amazonis Planitia there is a linear eastern edge of a depression in the floor of the planitia (image e). It has a bearing angle of +30° which suggests artificiality, and is over 95 km long. It is located just east of the Gordii Dorsum region. Although I have already presented some artificial sites found in the Simud Vallis and the Tiu Valles which are within the Chryse Planitia (Chapter 10), there are some sites which I overlooked. One of these is shown in image f which is an "island" located at the northeast edge of the Simud Vallis. It has 3 linear edges, one with a bearing angle of 0° and the other 2 with counterclockwise bearing angles of 27° and 36°. The number 27 is 1/4 the size of the angle between the star points and 36 is the number of degrees in a star point of a pentagram.

Having shown that artificiality occurs within all of the planitiae, it is time to look at how the planitiae relate to one another. One of the most remarkable arrangements is the placement of the 3 southern planitiae of Isidis, Amazonis and Chryse. The amazing aspect of these smaller planitiae is that their north-south midlines are located 120° of longitude apart from one another (Fig. 12.3). To test the likelihood of this happening naturally, I divided the planet into 3 separate regions of 120 degrees of longitude. I then further divided each region into 11 subregions of 10.91 degrees width (I chose the odd number of 11 to allow for a central subregion). Using the binomial expansion equation I tested for the probability of 3 sites, each lying within a separate 120 degree region, of falling into their respective central subregion. It turns out that, under random conditions, 2 out of the 3 planitiae would fall into their central subregions with a probability of 0.023 (a chance of 1 in 43). This result is considered to be statistically significant, i.e., it is unlikely due to chance. The probability for all 3 planitiae to fall into their central subregions is only 0.00075 (a chance of 1 in 1331). This result is considered to be highly statistically significant. Thus the situation shown in Fig. 12.3 is very unlikely to have occurred by chance alone.

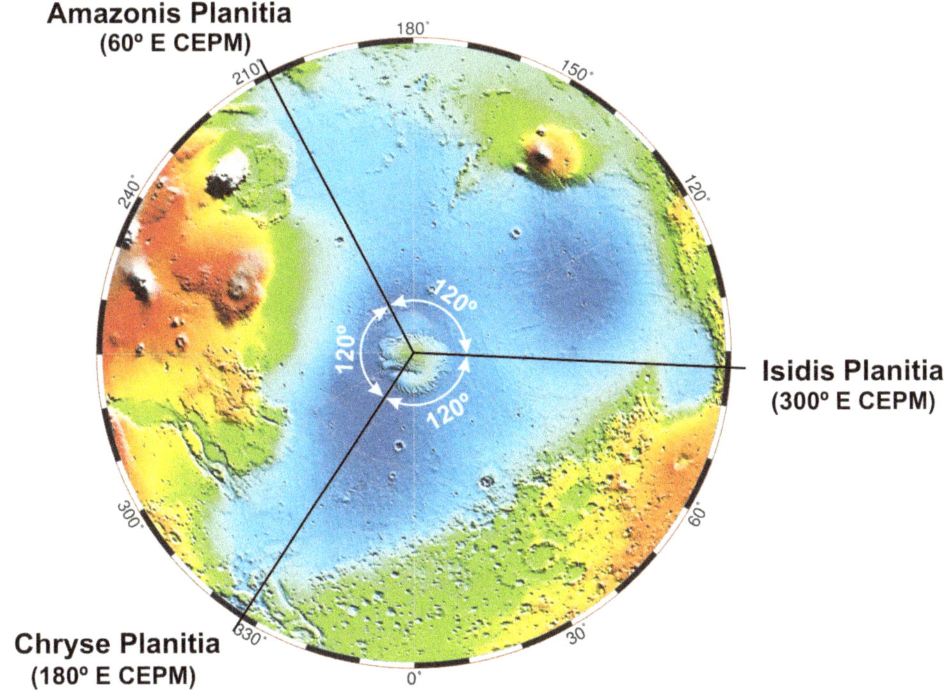

Fig. 12.3: *Amazonis, Chryse and Isidis planitiae are centred on longitudes which are 120° apart. The longitudes are very meaningful numbers when referenced to the Crater Edge Prime Meridian, all being multiples of 60. Image PIA02036.tif, courtesy of NASA.*

Another finding which strengthens the case for artificiality is that when my estimates of the longitudes of the midlines are referenced with respect to the Crater Edge Prime Meridian they work out to the very meaningful numbers of 60, 180 and 300 degrees E.

The 3 northern planitiae are also centred on longitudes which are separated by very meaningful numbers of degrees (Fig. 12.4). Arcadia Planitia is 72° east of Utopia Planitia, and Acidalia Planitia is 144° east of Arcadia Planitia and 144° west of Utopia Planitia. The value of 72° covers 1/5 of the circumference of the planet. Thus the 3 northern planitiae are separated by 1/5, 2/5 and 2/5 of planetary longitude. If their midline longitudes are extended to the equator as in Fig. 12.4, they would fit 3 of the star points of a hemispherical pentagram centred on the north pole.

Thus not only do the planitiae show artificiality within their boundaries, but also they appear to be artificially located with respect to their longitudes. Hence, the amazing conclusion is that these massive landforms must have been artificially created rather than having resulted from the impacts of huge objects from outer space.

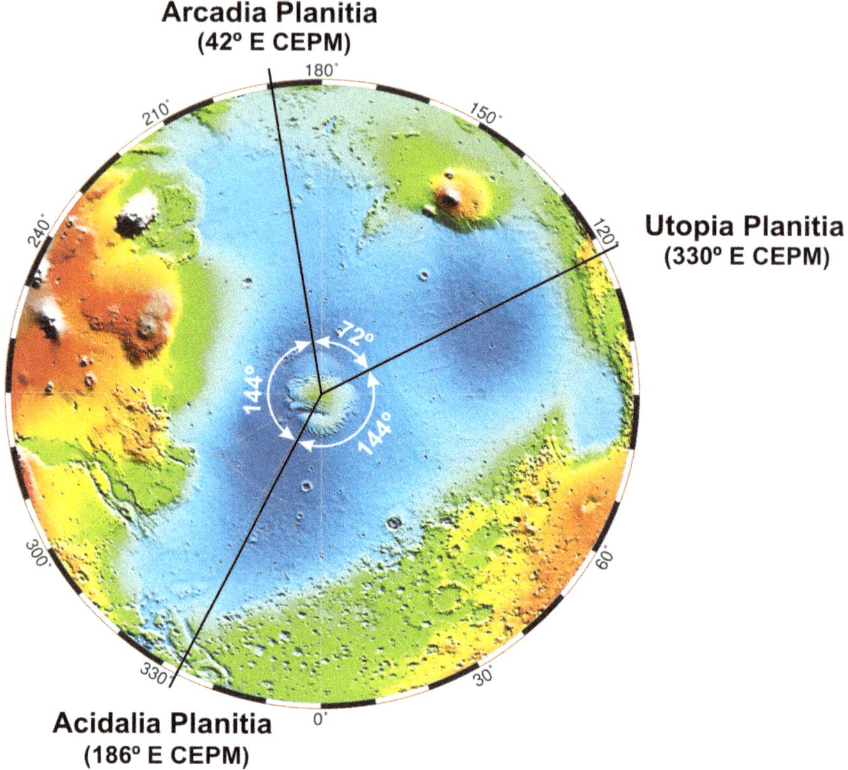

Fig. 12.4: *The 3 northern planetiae of Utopia, Arcadia and Acidalia planitiae are centred on longitudes which are 72°, 144° and 144° apart. The longitudes are referenced to the Crater Edge Prime Meridian (CEPM). Image PIA02036.tif, courtesy of NASA.*

Relationship of Alba Mons to the Hellas Basin

The next arrangement which I will discuss is the relationship between Alba Mons and the Hellas Basin. It is well known that Alba Mons is antipodal to the Hellas Basin. But how precise is this? It is very difficult to determine the exact location of the centres of either of these enormous structures. In *Intelligent Mars I*, I was unable to locate any survey craters to establish a centre for Alba Mons. I was forced to use the centres of Olympus Mons and Ascraeus Mons to survey for its centre.

When I examined the Hellas Basin for a plausible centre, I came across a crater with a diameter of about 25 km which has a central peak (Fig. 12.5). I calculated the coordinates of the centre of the crater (67.8506° E 41.2258° S) and then located the exact antipode of this location on the map of Alba Mons. To my amazement, I found 2 craters on Alba Mons which were exactly at the sacred distance of R/(7φ) = 299.8516 km (Fig.

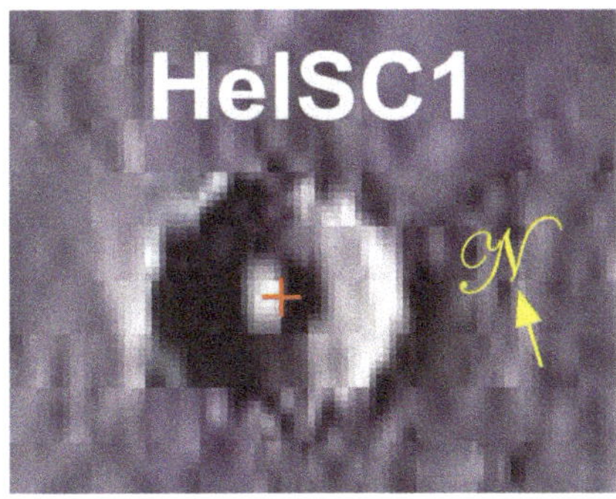

Fig. 12.5: *Crater HelSC1 in the central region of the Hellas Basin. The crater centre is marked by the red cross. It occurs exactly in the middle of the central peak which is indicated by white pixels to represent its upslope and dark pixels to represent its downslope. The centre of the crater (and central peak) is exactly antipodal to the survey site on Alba Mons (see Fig. 12.6). USGS Astrogeology.*

12.6) from this antipodal location. This sacred distance has the same format as a number of sacred distances from the survey craters of other mountains, e.g., R/(2π) km for the Ascraeus Mons survey craters and R/(3e) km for the Olympus Mons survey craters. This strongly suggests the Hellas crater marks the centre of the Hellas Basin and that its antipode marks the centre of Alba Mons. The 25 km wide crater can be considered to be the epitome of a survey crater since it is located on the exact spot that it surveys for (i.e., the centre of the Hellas Basin). I assigned it the name of HelSC1 since it lacks an official name. I assigned the names of AlmSC1 and AlmSC2 to the 2 craters which survey the antipodal location on Alba Mons since they also lack official names.

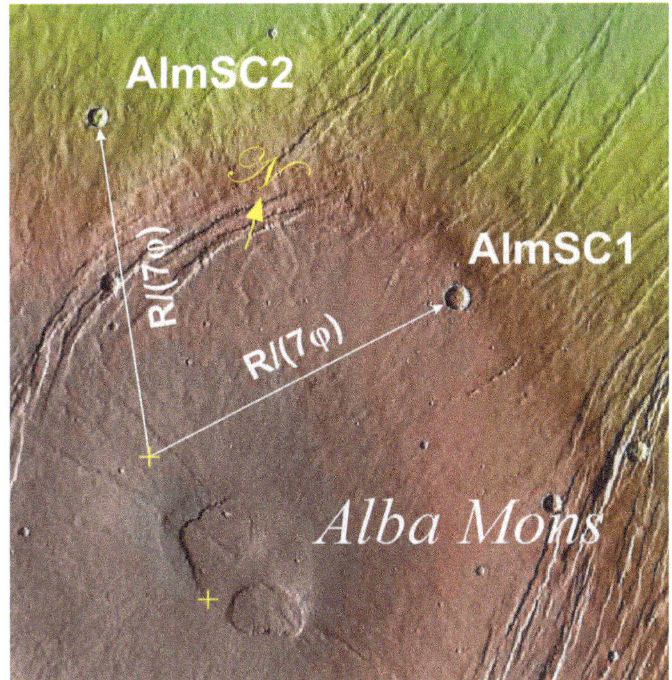

Fig. 12.6: *The AlmSC1 and AlmSC2 craters survey the location on Alba Mons which is exactly antipodal to the centre of the HelSC1 crater in the Hellas Basin. This location is marked by the higher yellow cross and is R/(7φ) km from the survey craters. The lower yellow cross marks the location of the point equidistant from the survey centres of Olympus Mons and Ascraeus Mons. Meaningful distances from other sites suggest it has much validity as an alternate centre of Alba Mons. USGS Astrogeology.*

Normally I use the first 3 letters of the name of the surveyed location followed by SC (for survey crater) and an integer, but since Alba Mons has the same first 3 letters as Albor Tholus, I needed to modify the name to distinguish these craters from the survey craters for Albor Tholus. Hence, I used the first 2 letters followed by 'm' for Mons.

The latitude of the newly surveyed centre of Alba Mons (247.8506° E 41.2258° N) is only 26 seconds of a degree less than a latitude of $14(\varphi^2)$ = 36.6452°° N. It is also only 1.16 minutes of a degree less than 33.0000°°° N. Its longitude (0.6782° E [PCPM]) is only 5 seconds of a degree less than e/4 = 0.6796. It is also 88.6630°° E [DPPM] which is 20 seconds of a degree less than $12e^2$ = 88.6687. The antipodal latitude of the Hellas Basin centre (67.8506° E 41.2258° S) is close to $14(\varphi^2)$°° S and 33.0000°°° S. Note that 33 is a very important number in Freemasonry. The longitude of the Hellas Basin centre is only 28 seconds of a degree less than 114Φ = 70.4559°° W [CEPM]. The rhumb distance between the 2 antipodal sites is 10,865.42 km which is only 2.39 km less than 16R/5 = 10,867.81 km. The bearing angle from the Hellas Basin to Alba Mons is 63.2690° (clockwise or counterclockwise) which is 1.37 minutes of a degree more than $\text{acos}(\sqrt{2}/\pi)$ = 63.2462°.

When I measured the sacred distances from important sites to the newly established centre of Alba Mons, I discovered that the distance to 8 of the sites could be expressed as an integer ratio multiplied by the equatorial (R) or the northern polar (R′) radius of Mars (Table 12.1). The integer ratios for the first 2 entries in Table 12.1 are the simplest, being 1/2 and 1/3. Some of the other integer ratios are either music intervals (3/2) or the inverse of music intervals (32/45, 45/64) from the chromatic scale. The distance of R/2 km now fits the Central Caldera of Olympus Mons instead of its survey centre as was the case for my original coordinates of Alba Mons surveyed from Olympus Mons and Ascraeus Mons in *Intelligent Mars I*. Another 10 sites are simple integer or inverse integer functions of R or R′ and of the square of π or e. All 4 of the giant mountains have meaningful sacred distance formulae to Alba Mons. The 3 Tharsis Montes have distance formulae containing φ and Olympus Mons has a distance formula containing π. The distance formula from the Pentagram Pyramid [$2R/(\varphi\sqrt{3})$ km] is almost exactly 1/2 of the value of the distance formula from the Pentagon Pyramid [$4R'/(\varphi\sqrt{3})$ km] except that the latter uses the northern polar radius rather than the equatorial radius. It is fitting that both formulae use φ since the dimensions for both pyramids contain this constant. The distance formula [$5R/(\pi\sqrt{5})$ km] from the peak of the ramp associated with the pentagram pyramid contains the numbers 5, √5 and π. The number 5 references the 5 star points of the pentagram, and √5 references the golden mean since $\varphi = (\sqrt{5} + 1)/2$. The number π reflects the latitude of the peak of the ramp tower which is π° N.

Table 12.1: *Sacred distance formulae from planetary sites to the centre of Alba Mons surveyed by craters AlmSC1 and AlmSC2.*

	Distance to Alba Mons			
	Theoretical		Actual	Difference
Site	Formula	(km)	(km)	(km)
Olympus Mons Central Caldera	R/2	1698.10	1700.54	2.44
Uranius Tholus N Crater	R/3	1132.06	1132.90	0.84
Elysium Mons Caldera	3R/2	5094.29	5091.09	-3.20
Ceraunius Tholus Caldera	3R′/8	1266.08	1266.40	0.32
Pavonis Mons Caldera	32R/45	2415.07	2413.52	-1.55
Biblis Tholus	45R′/64	2373.89	2372.69	-1.20
Biblis Tholus Caldera S centre	45R/64	2387.95	2389.23	1.28
Hellas Basin	16R/5	10867.81	10865.42	-2.39
Jovis Tholus East	π^2R′/24	1388.41	1388.26	-0.15
Jovis Tholus West	π^2R′/24	1388.41	1388.34	-0.07
Paros Crater	π^2R/25	1340.76	1338.60	-2.16
Issedon Tholus arrow tip	π^2R/39	859.46	859.38	-0.08
Olympus Mons	5R′/π^2	1710.40	1711.04	0.64
Pentagon Pyramid EW midpoint	14R/π^2	4817.48	4817.26	-0.22
Uranius Tholus	e^2R/22	1140.67	1141.04	0.37
Issedon Tholus centre of square	e^2R′/29	860.24	860.62	0.38
AscSC2 Crater	3R/e^2	1378.87	1380.08	1.21
Ascraeus Mons Caldera	4R′/e^2	1827.68	1827.53	-0.15
Ascraeus Mons	φR′/3	1820.94	1819.84	-1.10
Pavonis Mons	√5R/(2φ)	2346.71	2349.19	2.48
Arsia Mons	2R/(φ√2)	2968.38	2966.51	-1.87
Pentagon Pyramid	4R′/(φ√3)	4818.81	4818.91	0.10
Pentagram Pyramid	2R/(φ√3)	2423.67	2420.65	-3.02
Pentagram Pyramid ramp peak	5R/(π√5)	2417.28	2417.49	0.21

The most interesting bearing angles of sites to Alba Mons are listed in Table 12.2. Here we see 5 sites which use an integer multiplied by √2 and another 5 sites which use an integer multiplied by φ. Note that the western centre of Jovis Tholus has a bearing angle which belongs to both of these groups. All bearing angles of sites listed in Table 12.2 differ from theoretical values by less than 2 minutes of a degree. Note also that both the Pentagram Pyramid and the peak of its ramp use φ in their formulae for the bearing angle to Alba Mons.

Table 12.2: *Bearing angle formulae to the centre of Alba Mons antipodal to the Hellas Basin centre. Clockwise bearing angles are negative. Bearing angles with * are counterclockwise.*

	Bearing Angle to Alba Mons			
	Theoretical		Actual	Difference
Site	Formula	(°)	(°)	(°)
Jovis Tholus West	$8\sqrt{2}$	-11.3137	-11.3254	-0.0117
Ascraeus Mons*	$9\sqrt{2}$	12.7279	12.7332	0.0053
AscSC1*	$18\sqrt{2}$	25.4558	25.4548	-0.0010
Olympus Mons NE Caldera	$27\sqrt{2}$	-38.1838	-38.2159	-0.0321
Albor Tholus	$53\sqrt{2}$	-74.9533	-74.9714	-0.0181
Tharsis Tholus North*	$15\sqrt{5}$	33.5410	33.5437	0.0027
Nicholson Crater	$22\sqrt{5}$	-49.1935	-49.2234	-0.0299
Pavonis Mons Caldera	$\sqrt{3}/2$	-0.8660	-0.8673	-0.0013
Arsia Mons	6φ	-9.7082	-9.7399	-0.0317
Jovis Tholus West	7φ	-11.3262	-11.3254	0.0008
Pentagram Pyramid	13φ	-21.0344	-21.0160	0.0184
Uranius Tholus Round Top*	24φ	38.8328	38.8137	-0.0191
Issedon Tholus centre of square*	43φ	69.5755	69.5956	0.0201
Uranius Mons*	17e	46.2108	46.2028	-0.0080
Pentagram Pyramid ramp peak	$8\varphi^2$	-20.9443	-20.9665	-0.0222

When we look at the distances of important sites to the Hellas Basin we have to take into account that the sites are on the other side of the planet and hence their distances to the Hellas Basin can be measured either in the westward or the eastward direction. From the results that I obtained, it would appear that both directions were intended by the architects since the most meaningful sacred distance formulae were in the western direction for some sites and in the eastern direction for other sites. Both directions showed meaningful distance formulae for some sites. In Table 12.3, the distances listed are in the westward direction from each site to the Hellas Basin except when the site name has an asterisk (*) beside it to indicate the eastward direction. The coordinates used for the Hellas Basin are those of the centre of the HelSC1 Crater (Fig. 12.5). It can be seen in Table 12.3 that 8 sacred distance formulae consist of an integer multiplied by R (equatorial planetary radius) or R′ (northern polar radius) and divided by π^2. Another 6 sacred distance formulae use e^2 instead of π^2. Note that the sacred distance formula from the AscSC1a Crater in the westward direction is in the first group whereas its formula in the eastward direction is in the second group. Note also that the Alba Mons

Table 12.3: *Sacred distance formulae from planetary sites to the Hellas Basin. The distance is measured in the westward direction from the site to the Hellas Basin except when the site name has an asterisk(*), in which case, the direction is eastward.*

	Distance to Hellas Basin			
	Theoretical		Actual	Difference
Site	Formula	(km)	(km)	(km)
Elysium Mons Caldera	$17R/\pi^2$	5849.80	5849.13	-0.67
Issedon Tholus Caldera arrow tip*	$29R/\pi^2$	9979.05	9980.97	1.92
Jovis Tholus West	$30R'/\pi^2$	10262.42	10261.09	-1.33
Alba Mons (surveyed from mtns.)	$32R'/\pi^2$	10946.58	10944.46	-2.12
AscSC1a Crater	$33R'/\pi^2$	11288.66	11283.17	-5.49
Olympus Mons NE Crater*	$34R'/\pi^2$	11630.74	11629.80	-0.94
Issedon Tholus centre of square	$34R/\pi^2$	11699.60	11698.77	-0.83
Hecates Tholus*	$46R/\pi^2$	15828.88	15829.91	1.03
Nicholson Crater	$16R'/e^2$	7310.70	7311.72	1.02
Aum Crater centre of squares	$16R'/e^2$	7310.70	7313.07	2.37
AscSC1a Crater*	$21R'/e^2$	9595.30	9595.05	-0.25
Pentagram Pyramid*	$24R'/e^2$	10966.06	10965.30	-0.76
Fesenkov Crater	$26R/e^2$	11950.23	11947.14	-3.09
Elysium Mons Caldera*	$35R'/e^2$	15992.16	15991.34	-0.82
Hellas Basin northern tower	$2\pi R/35$	609.68	608.96	-0.72
Olympus Mons NE Crater	$2\pi R'/\sqrt{5}$	9486.87	9485.11	-1.76
Biblis Tholus	$2\pi R'/\sqrt{5}$	9486.87	9484.11	-2.76
Pentagram Pyramid	$12R/(e\varphi)$	9265.97	9266.13	0.16
Pentagon Pyramid	$4R'/\sqrt{5}$	6039.53	6039.61	0.08
Ascraeus Mons*	$15R'/(\varphi\pi)$	9962.81	9960.84	-1.97
Pavonis Mons	$15R/(\varphi\pi)$	10021.80	10023.42	1.62
Arsia Mons	eR	9231.80	9228.14	-3.66
Apollinaris Mons Caldera*	$4R'$	13504.80	13504.61	-0.19

site in the first group is the site surveyed from the Ascraeus Mons and Olympus Mons survey centres rather than from the AlmSC1 and AlmSC2 craters. The Nicholson Crater forms an isosceles triangle with the centre of squares in the Aum Crater. Both have the sacred distance formula of $16R'/e^2$ km to the Hellas Basin. Following these 2 groups of sacred distance formulae is the formula of $2\pi R/35$ km (i.e., the planetary circumference divided by 35) from the tower that is located in the Hellas Basin (see Fig. 11.6). An isosceles triangle is formed by the Olympus Mons NE Crater and Biblis Tholus with a common distance formula of

$2\pi R'/\sqrt{5}$ km in the westward direction to the Hellas Basin. The sacred distance formulae from the Pentagram and Pentagon pyramids both contain an irrational number (φ and √5 respectively) which references the geometry of these 2 structures. Note that the √5 is equal to 2φ – 1. The sacred distance formula from Ascraeus Mons [$15R'/(\varphi\pi)$ km] in the eastward direction matches that from Pavonis Mons [$15R/(\varphi\pi)$ km] in the westward direction except that the former uses R′ instead of R. The sacred distance formula from Arsia Mons is simply eR km, and the sacred distance formula from the Apollinaris Mons Caldera (in the eastward direction) is simply 4R′ km. The differences between actual and theoretical distances are quite small for all these sites when you consider that many of the distances exceed 10,000 km.

The bearing angles from the Hellas Basin to several sites are listed in Table 12.4 and are measured from the Hellas Basin in the eastward direction to the sites except those marked with an asterisk (*). Those in the eastward direction are clockwise and are listed as negative angle values. Those in the westward direction are counterclockwise and are listed as positive values. All of the bearing angles differ from theoretical formulae by less than 2 minutes of a degree (0.0333 degrees). All of the bearing angle formulae use a single irrational number (or the inverse of a single irrational number) multiplied by an integer, except for 3 of the formulae which use the square of φ. Note that the Paros Crater is aligned with the AscSC2 Crater since both have a bearing angle of $27\varphi^2$ degrees in the clockwise direction from the Hellas Basin. The bearing angle to the Aum Crater octagon centre is 25π° in the clockwise direction from the Hellas Basin. The number 25 is the square of 5, the number of star points in a pentagram. If we go in the counterclockwise direction, the bearing angle works out to 82.6199° which is not close to an interesting sacred formula. However, if we measure the bearing angle from the Aum Crater octagon centre to the Hellas Basin in the clockwise direction instead of in the counterclockwise direction from the Hellas Basin, the bearing angle works out to 31π° which is an amazing parallel to the 25π° obtained in the clockwise direction from the Hellas Basin.

Because Alba Mons is antipodal to the Hellas Basin, some researchers have speculated that the original formation of Alba Mons may have been causally related to the impact that was supposed to have produced the Hellas Basin. However, the finding of bearing angles with standard sacred geometry values for linear segments of the interior perimeter of the Hellas Basin point to the artificiality of this gigantic landform (see Chapter 10). If the Hellas Basin is artificial, then what about Alba Mons? There are very few straight lines on Alba Mons that permit measurement of their bearing angles but I did find 2 of these. Fig. 12.7 shows that the

Table 12.4: *Bearing angle formulae from the centre of the Hellas Basin in the eastward direction to other sites. These angles are clockwise (-ve values). The asterisk (*) denotes the westward direction from the Hellas Basin (bearing angles are counterclockwise and +ve). The superscript (1) denotes that the bearing angle is measured from the site to the Hellas Basin.*

	Bearing Angle From Hellas Basin			
	Theoretical		Actual	Difference
Site	Formula	(°)	(°)	(°)
Alba Mons (surveyed from mtns.)*	$45\sqrt{2}$	63.6396	63.6324	-0.0072
Paros Crater	$50\sqrt{2}$	-70.7107	-70.7163	-0.0056
Pentagram Pyramid	$52\sqrt{2}$	-73.5391	-73.5196	0.0195
Arsia Mons Caldera*	$56\sqrt{2}$	79.1960	79.1851	-0.0109
Albor Tholus	$30\sqrt{3}$	-51.9615	-51.9841	-0.0226
Alba Mons (surveyed from mtns.)	$37\sqrt{3}$	-64.0859	-64.0917	-0.0058
Uranius Mons*	$38\sqrt{3}$	65.8179	65.8020	-0.0159
Issedon Tholus centre of square*	$28\sqrt{5}$	62.6099	62.6268	0.0169
Hecates Tholus	$75/\varphi$	-46.3526	-46.3582	-0.0056
Olympus Mons NE caldera	$110/\varphi$	-67.9837	-67.9869	-0.0032
Olympus Mons*	$117/\varphi$	72.3099	72.2974	-0.0125
Pentagram Pyramid ramp peak	$119/\varphi$	-73.5460	-73.5196	0.0265
Biblis Tholus N centre	$120/\varphi$	-74.1641	-74.1685	-0.0044
Hecates Tholus Caldera*	$120/\varphi$	74.1641	74.1580	0.0061
Aum Crater centre of squares	$127/\varphi$	-78.4903	-78.4985	-0.0082
Uranius Tholus	43φ	-69.5755	-69.5627	0.0128
Paros Crater	$27\varphi^2$	-70.6869	-70.7163	-0.0294
AscSC2 Crater	$27\varphi^2$	-70.6869	-70.6669	0.0200
Poynting Crater	$28\varphi^2$	-73.3050	-73.3171	-0.0121
Hecates Tholus Caldera	$17e$	-46.2108	-46.2250	-0.0142
AscSC2 Crater	$26e$	-70.6753	-70.6669	0.0084
Poynting Crater*	$27e$	73.3936	73.4197	0.0261
Pentagram Pyramid*	$28e$	76.1119	76.1350	0.0231
Pentagram Pyramid ramp peak*	$28e$	76.1119	76.1210	0.0091
Olympus Mons central caldera*	23π	72.2566	72.2742	0.0176
Aum Crater octagon centre	25π	-78.5398	-78.5284	0.0114
Aum Crater octagon centre*[1]	31π	-97.3894	-97.3801	0.0093

southwest side of the main caldera is linear and has the remarkable bearing angle of 36° in the counterclockwise direction. This is the angle value of a star point in a pentagram. Another straight line structure was

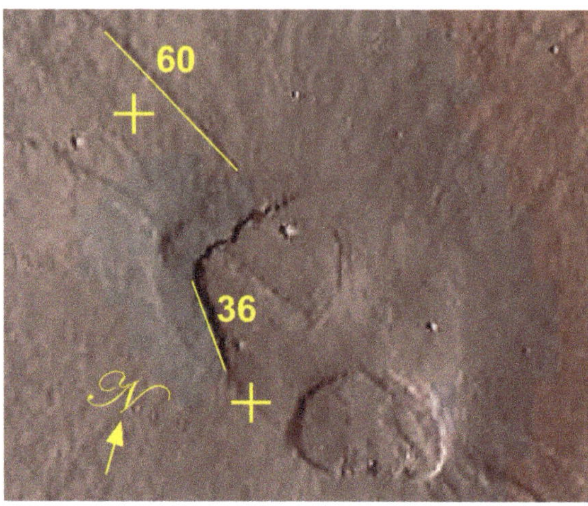

Fig. 12.7: *Linear structures on Alba Mons. The southwestern perimeter of the western caldera has a counterclockwise bearing angle of 36°. The northern edge of what looks like a lava flow has a counterclockwise bearing angle of 60°. The upper yellow cross marks the centre of Alba Mons surveyed from the AlmSC1 and AlmSC2 craters. The lower yellow cross is the Alba Mons centre surveyed from Ascraeus Mons and Olympus Mons. USGS Astrogeology.*

located just north of this location and it seems to be the edge of a lava flow. It has a bearing angle of 60° in the counterclockwise direction. This is the angle value of the 3 angles in an equilateral triangle.

A large scale view of the mountain reveals that the ratio of its east-west dimension (up to 3000 km) to its north-south dimension (about 2000 km) is close to 3:2. This is the same ratio that you would get with a vesica pisces, a geometric shape formed by 2 circles of the same size passing through each other's centre. The vesica pisces is the oval part of the shape that is created. I tried fitting a vesica pisces to Alba Mons and found that I could include the upper edifice of the mountain in the vesica pisces oval when it was centred on the site antipodal to the Hellas Basin (Fig. 12.8). The implications of this are enormous. The vesica pisces on Earth has been used

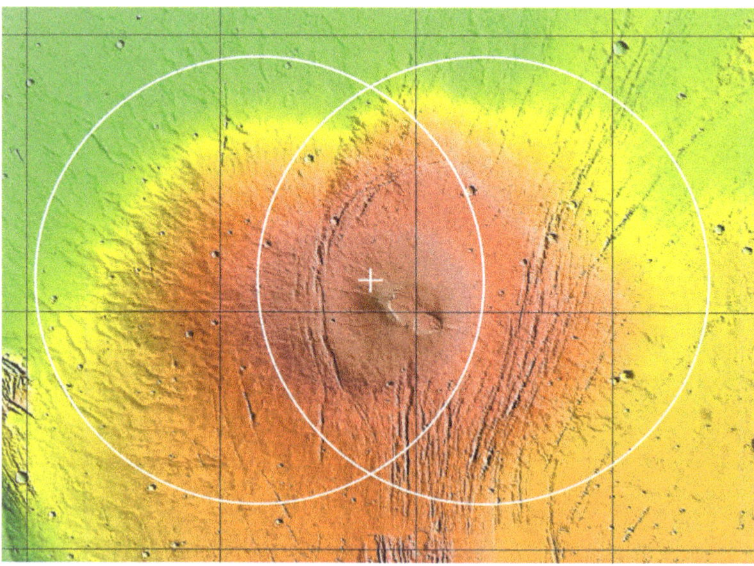

Fig. 12.8: *The upper edifice of Alba Mons sits nicely in the oval shape of a vesica pisces centred on the white cross which marks the location of the centre of Alba Mons antipodal to the centre of the Hellas Basin. MOLA Science Team, NASA Goddard Space Flight Center.*

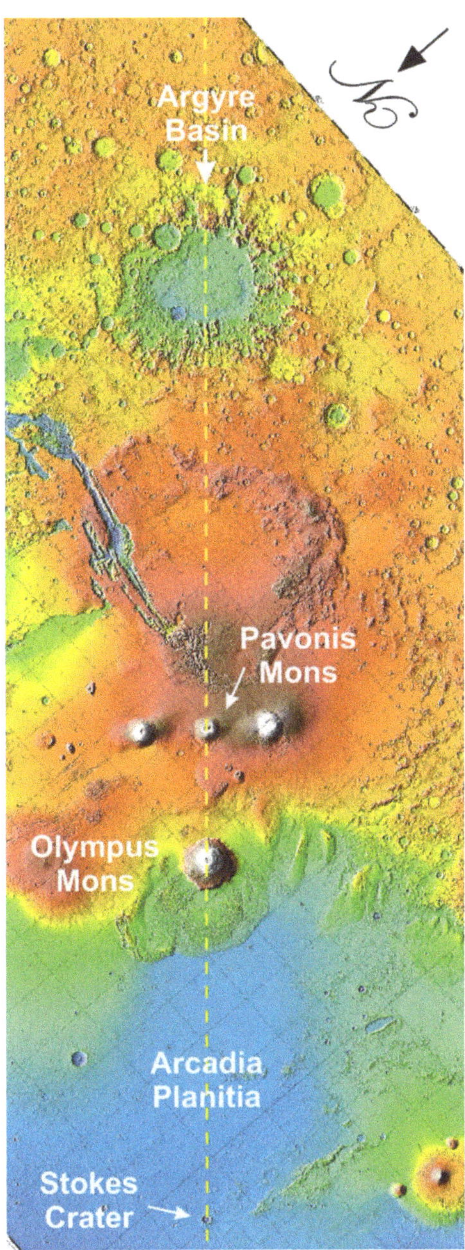

Fig. 12.9: *Alignment of the Stokes Crater, Arcadia Planitia, Olympus Mons, Pavonis Mons and the Argyre Basin. A straight line going through the centres of these 5 sites has a counterclockwise bearing angle of atan(π/e) = 49.1318°. MOLA Science Team, NASA Goddard Space Flight Center.*

to represent the primordial womb of the universe and is used in constructing sacred geometry patterns such as the Flower of Life and the Tree of Life. The vesica pisces can also be used to create measures of √2, √3, √5 and geometric shapes such as the equilateral triangle, the pentagram and the hexagram.

Hence, the evidence is against any causal relationship of the Hellas Basin to the creation of Alba Mons. But there very likely is a spiritual or metaphysical relationship, or why else would the architects go to such lengths to tie the two together with antipodal centres? One possibility is that the architects are revealing a physics beyond our comprehension. Perhaps the Hellas Basin – Alba Mons pair were intended to represent a black hole and a white hole respectively. That would then possibly make the interior of the planet a representation of a wormhole. However, the existence of white holes is still hypothetical as are wormholes. This is a wide open field for speculation.

Argyre Basin Alignment

A very striking alignment occurs with the Stokes Crater, Arcadia Planitia, Olympus Mons, Pavonis Mons and the Argyre Basin. Their centres all occur on the same straight rhumb line which has a counterclockwise bearing angle of atan(π/e)° = 49.1318°. This whole arrangement can best be seen by rotating the map in a counterclockwise direction by 130.8682° so that the Argyre Basin sits at the top of the figure (Fig. 12.9). What is most interesting about the Argyre Basin is that it has no clear perimeter which is what you would expect if this landform were caused by a tremendous impact. For this reason, together with the alignment of the Argyre Basin with 2 mountains, a large planitia and the Stokes Crater, I decided to examine this enormous landform more closely to see if I could detect any signs of artificiality.

The first thing that I noticed is that a very long intermittent groove line passes through the eastern side of the Argyre Basin (Fig. 12.10). The groove line was found to have a constant bearing angle of 7.5° in a clockwise direction and it extends for a distance of more than 1350 km. It appears slightly curved in Fig. 12.10 (top left image) since the map is a Lambert conformal conic projection map rather than a Mercator projection map. The groove is more than 2 km wide in some locations so it is more than a simple scratch on the surface.

The presence of a long groove line in the Argyre Basin suggests that the basin is artificial. Groove and ridge lines with a bearing angle of ±7.5° seem to form some sort of grid which was used for the placement of various landforms such as craters (Chapter 6). A closer look at straight line edges in the Argyre Basin area confirms the suspicion of artificiality. In the Charitum Montes region, located near the southern edge of the Argyre Basin, there is a long ridge which has a clockwise bearing angle of 60° (Fig. 12.11a). Just southeast of this is another linear ridge with a clockwise bearing angle of 12°. Immediately to the northeast of this location lies Oceanidum Mons with a linear summit having a clockwise bearing angle of 27° (image b). Moving on to the western region of the Argyre Basin, there is a linear ridge just southeast of the Salaga Crater which has the very meaningful clockwise bearing angle of $\operatorname{atan}(1/(\varphi\sqrt{5})) = 15.4504°$ (image c). This is the same angle that was found for the bearing angle of the northwest star point of the Pentagram Pyramid which points to the survey centre of Olympus Mons (*Intelligent Mars I*). It is virtually indistinguishable from the angle of $25\Phi = 15.4508°$ where Φ is the inverse of the golden ratio φ. I put both values next to the line in image c to indicate its bearing angle. To the northeast of this region, just east of the Cypress Crater is a depression in the floor of the Argyre Basin whose eastern side is linear and has a clockwise bearing angle of 12° (image d).

A few degrees north of the previous 2 sites lies the Luga Crater (Fig. 12.12a). A long trench whose east side is linear occurs on the west side of the crater. This trench extends beyond the north side of the crater, and its east side has a counterclockwise bearing angle of 9°. Just west of this trench is a linear ridge which has a clockwise bearing angle that is close to 4 very meaningful values, all within 12 minutes of a degree of each other. The values are $7\varphi = 11.3264°$, $8\sqrt{2} = 11.3137°$, $5\sqrt{5} = 11.1803°$ and $18\Phi =$

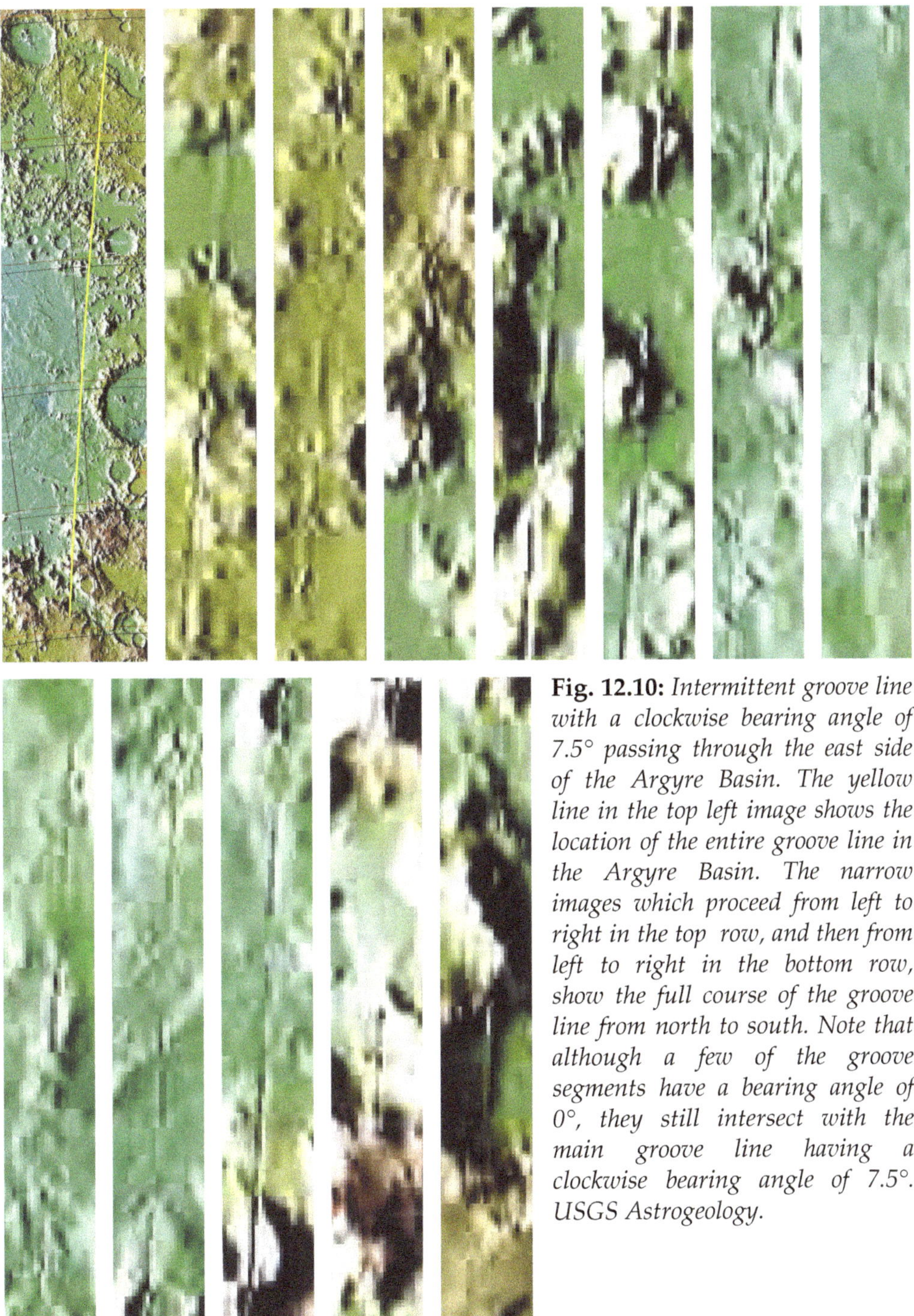

Fig. 12.10: *Intermittent groove line with a clockwise bearing angle of 7.5° passing through the east side of the Argyre Basin. The yellow line in the top left image shows the location of the entire groove line in the Argyre Basin. The narrow images which proceed from left to right in the top row, and then from left to right in the bottom row, show the full course of the groove line from north to south. Note that although a few of the groove segments have a bearing angle of 0°, they still intersect with the main groove line having a clockwise bearing angle of 7.5°. USGS Astrogeology.*

11.1246°. I put all 4 of these values next to the line fitting the ridge. About 5 degrees north of here in the Nereidum Montes there are 2 linear ridges,

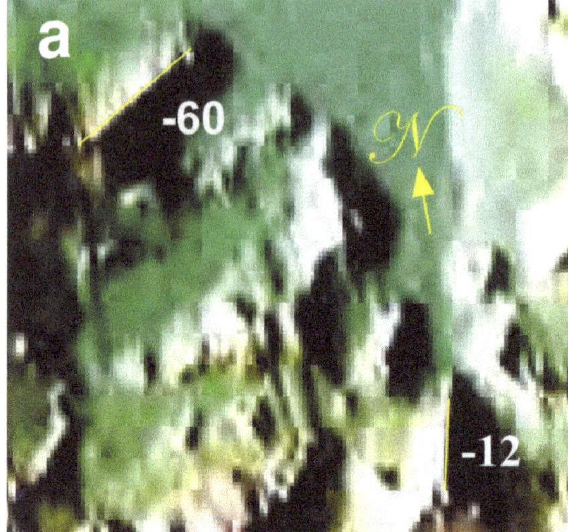

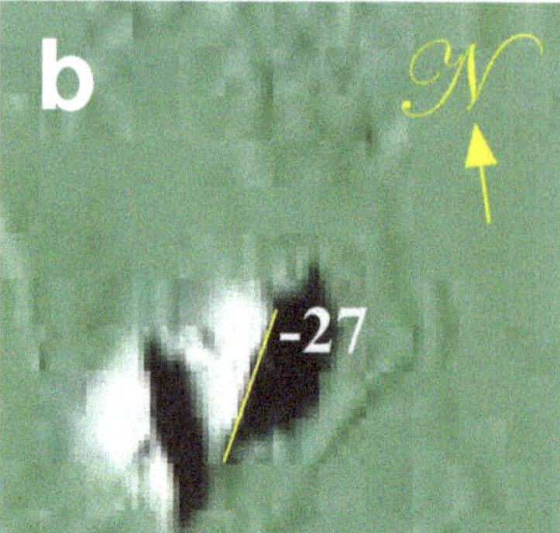

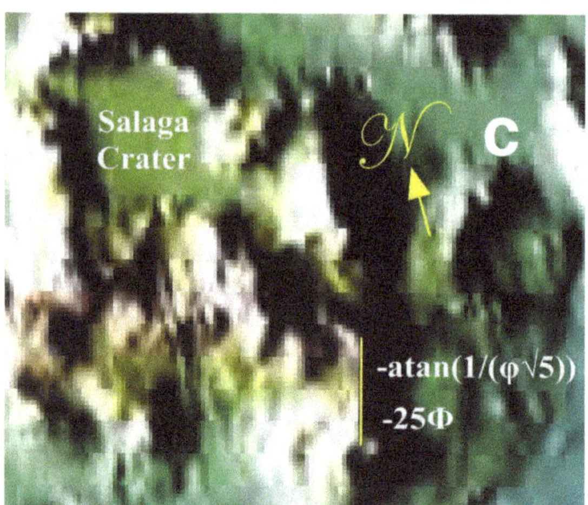

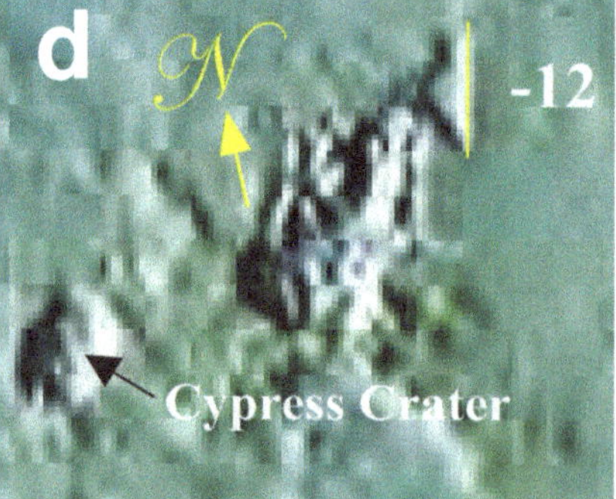

Fig. 12.11: *Linear ridges found in the Argyre Basin region. Image a is in the Charitum Montes. The ridge in image b is Oceanidum Mons. A linear edge of a depression is shown in image d. Negative numbers indicate bearing angle sizes in the clockwise direction. USGS Astrogeology.*

one with a counterclockwise bearing angle of 36° and another with a counterclockwise bearing angle of 9° (Fig. 12.12b). Finally in the northeast region of the Argyre Basin lies the Bozkir Crater whose western perimeter has a linear section with a clockwise bearing angle of 24° (Fig. 12.12c).

So there is plenty of evidence for the artificiality of the Argyre Basin. All of the bearing angles of linear structures within the basin have meaningful sacred geometry values. The numbers 9, 18 and 27 are factors of the 3 angles of the pentagram (36, 72 and 108) obtained by dividing the angles by powers of 2. The numbers 5 and 25 = 5 x 5 refer to the 5 star points of a pentagram. The irrational numbers of φ and Φ are present in the structure of the pentagram. The irrational number √2 is a factor in the size of the diagonal of a square. The irrational number √5 is a factor in the size of the diagonal of a double square. The number 12 is a factor in all 3 angles of the

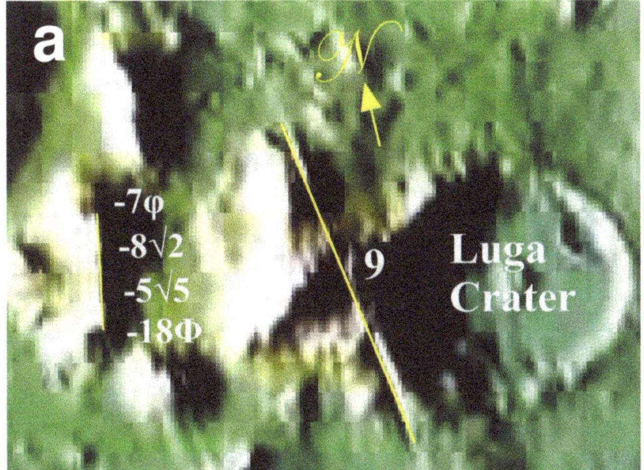

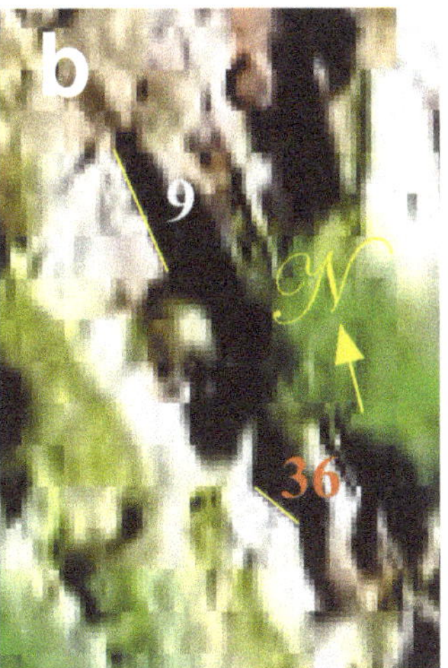

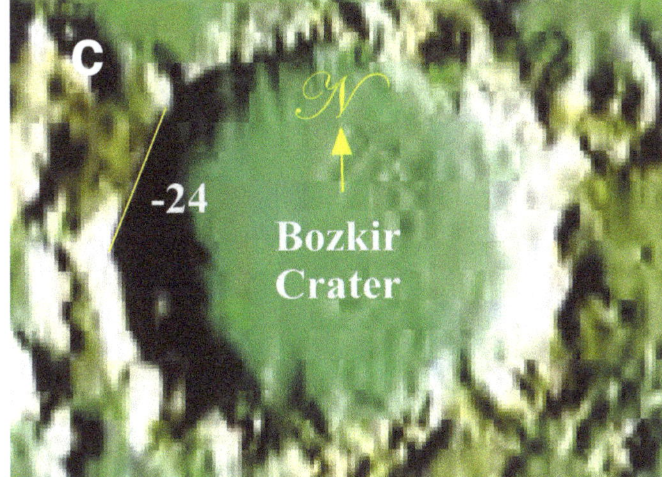

Fig. 12.12: *Image a shows a linear ridge and a linear trench. Image b shows 2 linear ridges in the Nereidum Montes. Image c shows a linear section of the Bozkir Crater perimeter. USGS Astrogeology.*

pentagram and 12 x 24 equals 288, the number of sacred degrees in a circle.

Since the Argyre Basin is aligned to Pavonis Mons and Olympus Mons, I searched for sites in the basin along the line of alignment. What I discovered is that in the Chalce Montes region located in the southeast corner of the basin, there is an irregularly-shaped crater which has a peak, not in the centre, but near the eastern edge of the crater (Fig. 12.13). The coordinates of the peak are 321.9829° E 53.2852° S. The longitude is 18.5584°° E [STrPM] which is very close to 18.5607 = asin(1/π) and is also 66.4982°° E [PCPM] which is very close to 66.5015 = $9e^2$. The latitude is 40.1649°° S [AMPL] which is close to 65Φ = 40.1722. If one draws a line with a counterclockwise bearing angle of atan(π/e)° from this peak, it runs parallel to a linear jagged ridge and passes through the survey centres of Pavonis Mons and Olympus Mons, and the centre of the Stokes Crater.

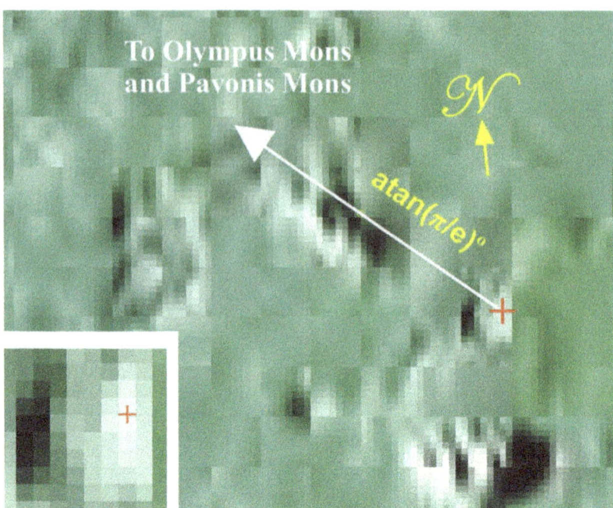

Fig. 12.13: *A straight line with a counterclockwise bearing angle of atan(π/e)° = 49.1318° originating from a peak in the east side of a crater (red cross) in the Chalce Montes region of the Argyre Basin. The line passes through the survey centres of Pavonis Mons and Olympus Mons, and through the centre of the Stokes Crater. The inset at bottom left shows an enlargement of the crater and its peak. USGS Astrogeology*

A check on the Stokes Crater using a high resolution MOLA map revealed that the centre of the crater is the alignment point with Pavonis Mons and Olympus Mons (Fig. 12.14). A straight line with a counterclockwise bearing angle of atan(π/e)° from the centre of the Stokes Crater passes through the survey centres of these mountains and ultimately to the peak in the crater in the Argyre Basin which was shown in the previous figure. Note that the bearing angle is measured for a line

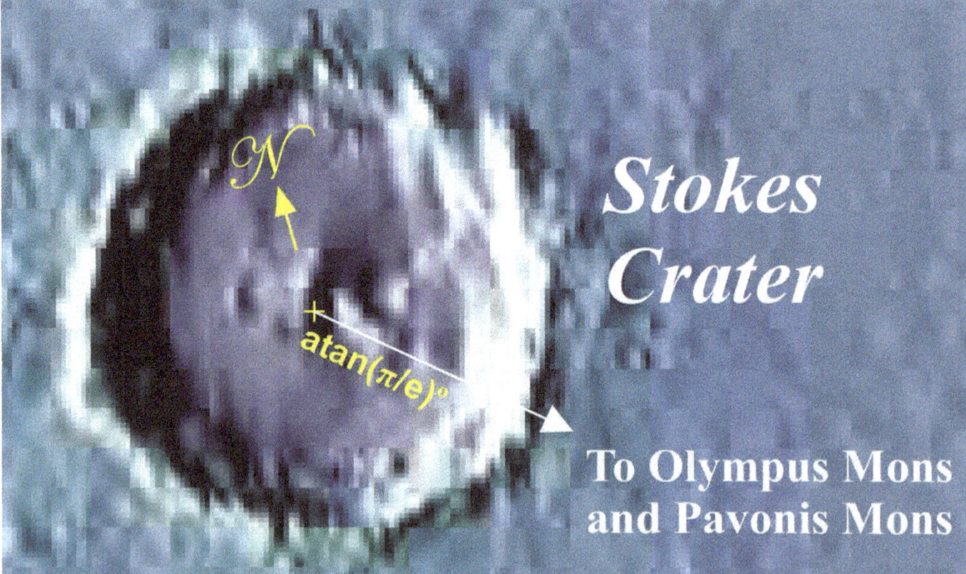

Fig. 12.14: *A straight line with a constant counterclockwise bearing angle of atan(π/e) ° = 49.1318° passes from the centre of the Stokes Crater (yellow cross) to the survey centres of Olympus Mons and Pavonis Mons. USGS Astrogeology.*

travelling in the opposite direction of the arrow in Fig. 12.14. The total distance from the centre of the Stokes Crater to the crater peak in the Argyre Basin is 9864.40 km which is not close to a meaningful sacred distance formula. The distance from the centre of the Stokes Crater to the survey centre of Olympus Mons is 3373.62 km which is only 2.58 km less than the length of the northern polar radius of Mars (3376.2 km).

Alignment of the Valles Marineris With Other Sites

Another striking alignment occurs when the caldera of Elysium Mons is connected to the HelSC1 centre of the Hellas Basin with a rhumb line having a constant bearing angle over its entire course (Fig. 12.15). This line passes through the central peak (169.8204° E 19.2740° N) of the Kotka Crater southeast of Elysium Mons (see inset, Fig. 12.15). The crater is about 38.5 km wide. The longitude of the peak is 22.6480° E [EMPM] which is very close to 22.6525 = 14φ. It is also 17.3184°°° E [DMPM] which is very close to 17.3205 = 10√3. The line of alignment passes just south of the Pavonis Mons Caldera and crosses the equator at 0.6758° E [PCPM] which is very close to 0.6796 = e/4. This occurs at the same longitude as the antipodal centre of Alba Mons which is the centre of the vesica pisces which can be fit to Alba Mons (Fig. 12.8). The line of alignment then passes along the length of the main canyon of the Valles Marineris system before finally reaching the centre of the Hellas Basin. The total length of the line is 15,991.34 km which is only 0.82 km less than $35R'/e^2$ km and is 74.94% or almost exactly 3/4 of the planetary circumference. The clockwise bearing angle of the line is 104.1338°.

The major canyon of the Valles Marineris also aligns with 2 other important sites when they are connected to the centre of the Hellas Basin with a rhumb line. These are Biblis Tholus and the Pentagram Pyramid. The latter is the more interesting since its rhumb distance to the Hellas Basin is 10,965.30 km which is only 0.75 km less than the value of $25R'/e^2$ km, a very interesting number not only because 25 is the square of 5 which is the number of star points in a pentagram, but also because it is a match to the $35R'/e^2$ km distance of the Elysium Mons Caldera to the Hellas Basin. The clockwise bearing angle of the line from the Pentagram Pyramid is very interesting as well since it is 103.8650° which is only about 3.5 minutes of a degree less than the value of 60√3°. Fig. 12.16 shows the course of the lines from the Elysium Mons Caldera, Biblis Tholus and the Pentagram Pyramid to the Hellas Basin as they pass through the Valles Marineris.

After showing that the main canyon of the Valles Marineris seems to be pointing to the Hellas Basin, I am now going to show how not just the main canyon, but also the entire Valles Marineris system of canyons is focused on the Hellas Basin. As demonstrated in *Intelligent Mars I*, Olympus Mons and the Tharsis Montes represent the Martian body shape similar to the Leonardo Da Vinci drawing of the Vitruvian Man. This body shape was shown to fit a circle, a square, 2 double squares, an equilateral triangle and a pentagram. If we take the pentagram which fits the Vitruvian Martian and draw lines from its northeast and southern star points to the centre of the Hellas Basin, we create a triangle which encompasses all of the canyons which make up the Valles Marineris system (Fig. 12.17). This would suggest that the Vitruvian Martian focuses on the Hellas Basin via the Valles Marineris. We will explore the possible significance of this arrangement in the following chapter.

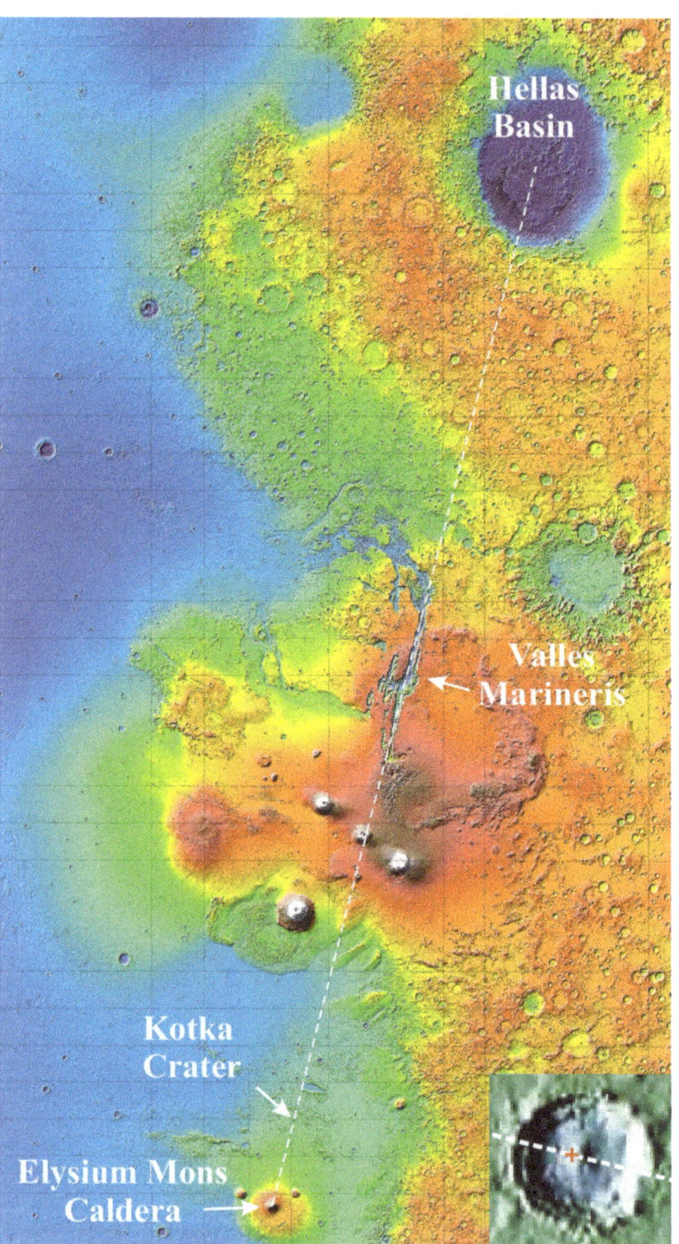

Fig. 12.15: *Alignment of the Elysium Mons Caldera, the peak in the Kotka Crater, the length of the major canyon in the Valles Marineris system, and the Hellas Basin centre. North is left side of image. Since the Kotka Crater is too small to be seen at this magnification, an enlargement is shown in the inset at bottom right. The red cross marks the location of the crater central peak. MOLA Science Team, NASA Goddard Space Flight Center. Inset: USGS Astrogeology.*

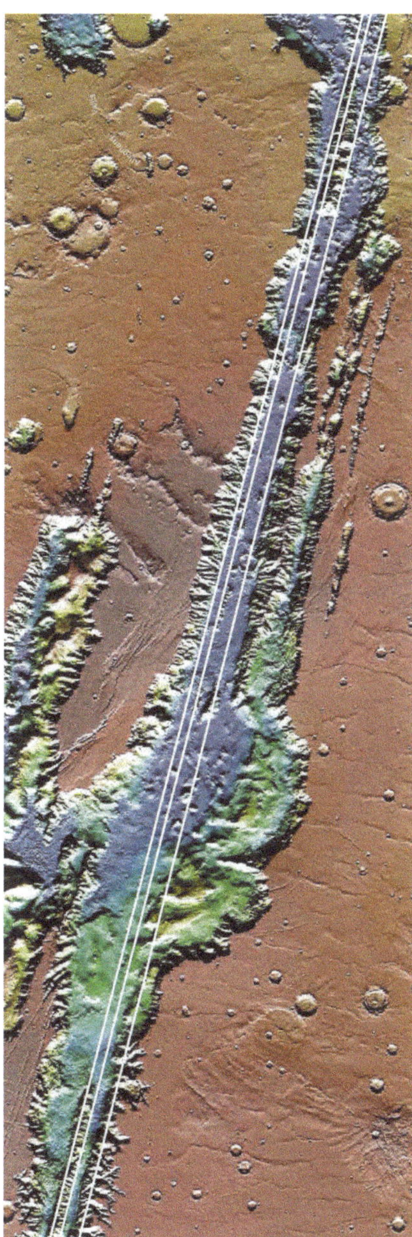

Fig. 12.16: *Course of lines through the Valles Marineris which join the Elysium Mons Caldera (left), Biblis Tholus (middle) and the Pentagram Pyramid (right) with the centre of the Hellas Basin. North is the left side of the image. USGS Astrogeology.*

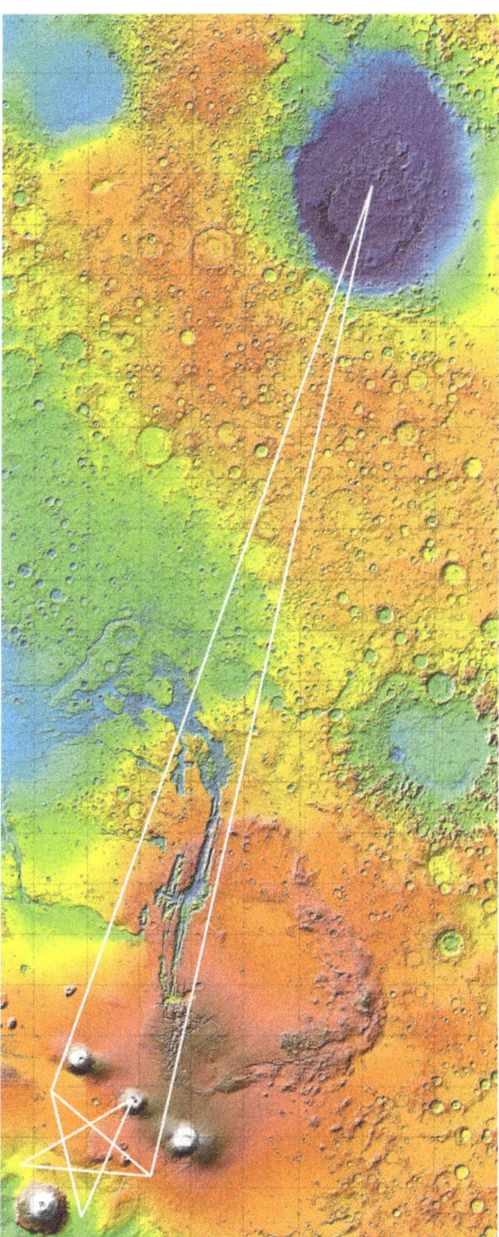

Fig. 12.17: *Lines to the Hellas Basin centre from the northeast and southern star points of the pentagram which fits the Vitruvian Martian. These lines encompass all the canyons which make up the Valles Marineris. North is the left side of the image. MOLA Science Team, NASA Goddard Space Flight Center.*

13 Sacred Scripture in the Form of a Planet

In *Intelligent Mars I*, it was shown how the Vitruvian Martian can be created from Olympus Mons and the 3 Tharsis Montes by using the Pentagram Pyramid as a key to unlock the size and position of the circle, equilateral triangle and pentagram which fit the Martian body shape. This present book has shown strong evidence for a series of concentric virtual squares and octagons emanating from the Aum Crater over thousands of kilometers. Data were presented which indicate that the Valles Marineris, the Hellas Basin, the Argyre Basin and 6 planitiae in the northern hemisphere are all artificial rather than creations of nature. Furthermore, it was discovered that Alba Mons and the Hellas Basin have centres which are antipodal to each other, and that Olympus Mons and Pavonis Mons are aligned with the Stokes Crater and the Argyre Basin. In addition, the Valles Marineris were shown to be a pathway connecting the Vitruvian Martian to the Hellas Basin. It is now time to try to integrate all of these structures into one unifying picture which covers the entire planet.

After studying all of these artificial constructs intensively, I have come to an incredible conclusion. When all of the individual pieces are assembled together, the entire planet of Mars forms a vast 3 dimensional sculpture that tells the story of creation, life, death, reincarnation and final ascent of the soul. In other words, Mars is sacred scripture encoded in the form of a planet! Instead of being in the form of a book written in the script of a spoken language, the sacred scripture of Mars is written in the language of planetary topography. It can only be read by those who understand sacred geometry, ancient spirituality, the Martian coordinate systems and the Martian mindset.

Like written sacred scriptures such as the books of the Bible and the book of Quran, the Martian planetary sacred scripture is composed of "chapters" placed in sequence. Each chapter tells a part of the overall story.

Chapter 1. A Story of Creation

The first chapter covers the story of creation. Here we see the Aum Crater producing "sound" waves in the shape of concentric squares and octagons. The Aum Crater represents the primordial sound bringing the planet and the Vitruvian Martian into existence. It does so not only with harmonics but also with sound waves (octagons and squares) whose interval ratios contain irrational numbers. For the creation of the planet, the most important square touches the equator with its northern vertex, and it has a diagonal size of $4.5e^2$ latitude degrees. The most important octagon for the creation of the planet is one whose northern side runs along the equator and has a radius size of 18 latitude degrees. For the creation of the Martian being, the most important octagon has an interval ratio containing the irrational number φ. It passes directly through the centre of the Pentagram Pyramid which plays an essential role in the construction of the Vitruvian Martian (*Intelligent Mars I*). Other octagons with interval ratios containing π, e, $\sqrt{2}$, or $\sqrt{3}$ come within 0.50 - 1.85 km of the Vitruvian Martian's navel, and a square with an interval ratio containing the irrational number $\sqrt{2}$ comes within 1.57 - 1.62 km of the Vitruvian Martian's square and pentagram centres.

The Aum sound is not the only primordial force creating the Vitruvian Martian. A second progenitor is the womb-phallus construct at the top of Pavonis Mons. This is a clear symbol of the female-male duality needed for the creation of Martians and many other living creatures. Both the phallus and the womb intersect at the latitude of $\varphi°$ N. Both contain π in their dimensions – the length of the phallus is π times its width, and the circumference of the womb is π times its diameter. The smaller circle which outlines the perimeter of the Pavonis Mons Caldera can be considered to be the progenitor of the larger circle forming the womb since the latter is exactly twice the size of the smaller. A close look at the pentagram fitting the Vitruvian Martian reveals that the tip of the pentagram star point which projects beyond the top of the Vitruvian Martian head lies right inside the phallus (Fig. 13.1). The phallus itself lies inside the womb. Part of the womb penetrates a small amount into the head of the Vitruvian Martian. All of this suggests that the womb-phallus

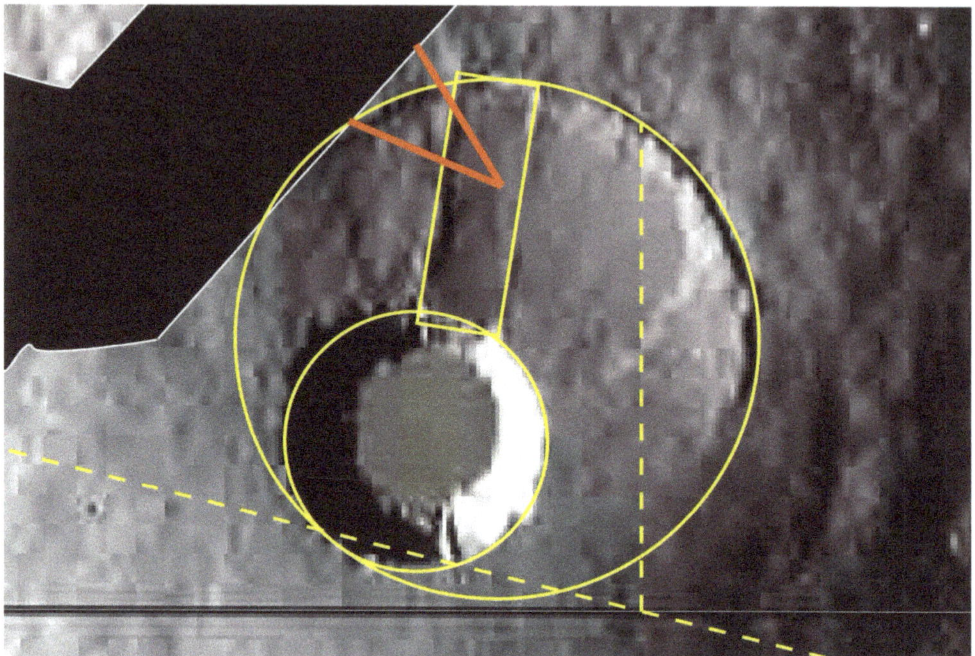

Fig.13.1: *Womb-phallus construct on Pavonis Mons showing that the tip of the pentagram star point located at the Vitruvian Martian's head (black band of material at top left) reaches into the phallus. The vertical dashed line marks the longitude of the antipodal site on Alba Mons. The slanted dashed line which passes through the equator (black horizontal line at bottom of image) is the line joining the Elysium Mons Caldera to the centre of the Hellas Basin. It passes through both calderas, and intersects the equator at the longitude of the antipodal site on Alba Mons. The ancient caldera (large circle) intrudes slightly into the top of the Vitruvian Martian's head. USGS Astrogeology.*

construct is somehow involved in the creation of the Vitruvian Martian.

There is yet a third progenitor which seems to be involved in the creation of the Vitruvian Martian and the planet. It is a huge vesica pisces formed by the massive Alba Mons located just north of the Vitruvian Martian. A vesica pisces traditionally represents the womb of the universe which creates all life in the form of geometric shapes. Since the centre of the Alba Mons vesica pisces is at the same longitude as the eastern part of the womb-phallus construct on Pavonis Mons, it is likely that the Alba Mons vesica pisces is intended by the Martian architects to depict a partnering of the vesica pisces with the womb-phallus to produce a zygote. It then directs the development of the fetus which will eventually be born to become the Vitruvian Martian. So what is the role of each of these partners of creation? Why do you need a vesica pisces when you already have a womb and a phallus which represent the female and male contributors which produce a fertilized egg or zygote?

It is interesting to note that our terrestrial version of sacred geometry uses the term "life" for the geometric patterns that are created from the vesica pisces. Thus we have the Seed of Life, the Egg of Life, the Flower of Life and the Tree of Life, terminology that may have been passed on to us from Mars (Fig. 13.2). What the architects may be telling us is that the womb-phallus cannot create a zygote without the infusion of a life principle. This is where the vesica pisces comes in. In this scenario, the Aum Crater produces the Aum sound which provides the material for creation. The vesica pisces directs the womb-phallus to create a zygote from the Aum sound into which it infuses a soul thus beginning the life of a new individual. The Alba Mons vesica pisces continues to use the material of the Aum sound to develop the zygote into a fetus which, when mature, is born from the Pavonis womb into the framework of the 4 giant mountains of Olympus Mons and the 3 Tharsis Montes. Still using the harmonic and irrational number vibrations of the Aum sound, the vesica pisces guides the growth of the child into an adult Vitruvian Martian according to the principles of sacred geometry. Note that the Vitruvian Martian body fits a circle, a square, an equilateral triangle and a pentagram, all constructs that can be obtained from the geometry of the vesica pisces (Fig. 13.3). The same process seems to be used for the construction of the planet since the rhumb line joining the Elysium Mons Caldera to the Hellas Basin centre passes through the equator at exactly the same longitude as the antipode site on Alba Mons and the eastern part of the womb on Pavonis Mons (Fig. 13.1). Rather than infusing a human soul into the planet, the vesica pisces may infuse a form of consciousness which interacts with living organisms to form a self-regulating biosphere.

Chapter 2. A Story of Sustenance

The bottom pair of feet of the Vitruvian Martian are located on Olympus Mons which itself is located right next to the Amazonis and Arcadia planitiae. This arrangement suggests that the Vitruvian Martian is being portrayed as standing next to water. This water would have been part of a vast northern ocean in ancient times and would have likely been salt water rather than fresh water. This probably symbolizes that the Vitruvian Martian is sustained by water, and its body is composed largely of salt water much like the human body. Far above the Vitruvian Martian is the Argyre Basin (Fig. 12.9) which probably represents the sun. In Fig. 12.9, note how the Argyre perimeter projects "rays". Whether this would be the present sun, or some radiating body that Mars might have orbited in earlier times, is open for speculation. At any rate, the sun, like water, is what sustains the Vitruvian Martian. All life on the planet would depend on the sun thus providing the Vitruvian Martian a supply of food.

Chapter 3. A Story of Life's Journey

The northeast and southern star points of the pentagram fitting the Vitruvian Martian project a triangular shape to the centre of the Hellas Basin (Fig. 12.17). This triangle completely encloses the Valles Marineris which suggests that the Vitruvian Martian is somehow connected to and focused on this immense system of canyons. It is as though the Vitruvian Martian has to traverse the Valles Marineris before arriving at Hellas. Could it be that the Valles represents life's challenges, a "vale of tears" that each of us must take during our lifetimes, full of diversions, but a main channel exists to keep us on the straight and narrow? Sacred scriptures warn us of the dangers of unethical behaviours and provide us with direction (e.g., the 10 commandments) to help us lead a good life. The pictorial representation of this with the Valles Marineris could be that, if we go astray, we end up trapped in canyons like the Hebes and Juventae chasmas which are cut off from the main system or we get diverted in a wrong turn to a dead-end canyon like the Ophir Chasma. Sticking to a righteous life allows us to stay on course and get to our destination with the least amount of disruption.

Chapter 4. A Story of Death and Rebirth

At the end of the Valles Marineris, 2 paths are available. One leads into the Aureum Chaos and then into high terrain before coming to the Hellas Basin. The other leads into the Chryse Planitia and eventually into the vast ocean in the northern hemisphere. Could this represent heaven (northern hemisphere) and hell (Hellas Basin)? Or is there another possible interpretation? The Hellas Basin is exactly antipodal to Alba Mons. In the previous chapter I speculated that this pair might actually represent a black hole (Hellas Basin) and a white hole (Alba Mons). If this were the case, then the Hellas Basin might represent a path for reincarnation after death. The liberated soul would enter the black hole represented by the Hellas Basin, and pass through a wormhole conceived to exist in the interior of the planet. The wormhole would connect the black hole to the white hole represented by Alba Mons. Since a white hole theoretically emits energy and matter, the soul would exit from Alba Mons through the vesica pisces. The vesica pisces would then work with the womb-phallus at the top of Pavonis Mons and take the Aum sound waves, shape them into 3-dimensional geometric patterns, and reincarnate the soul into a new zygote. The soul would once again become the Vitruvian Martian to experience a whole new lifetime.

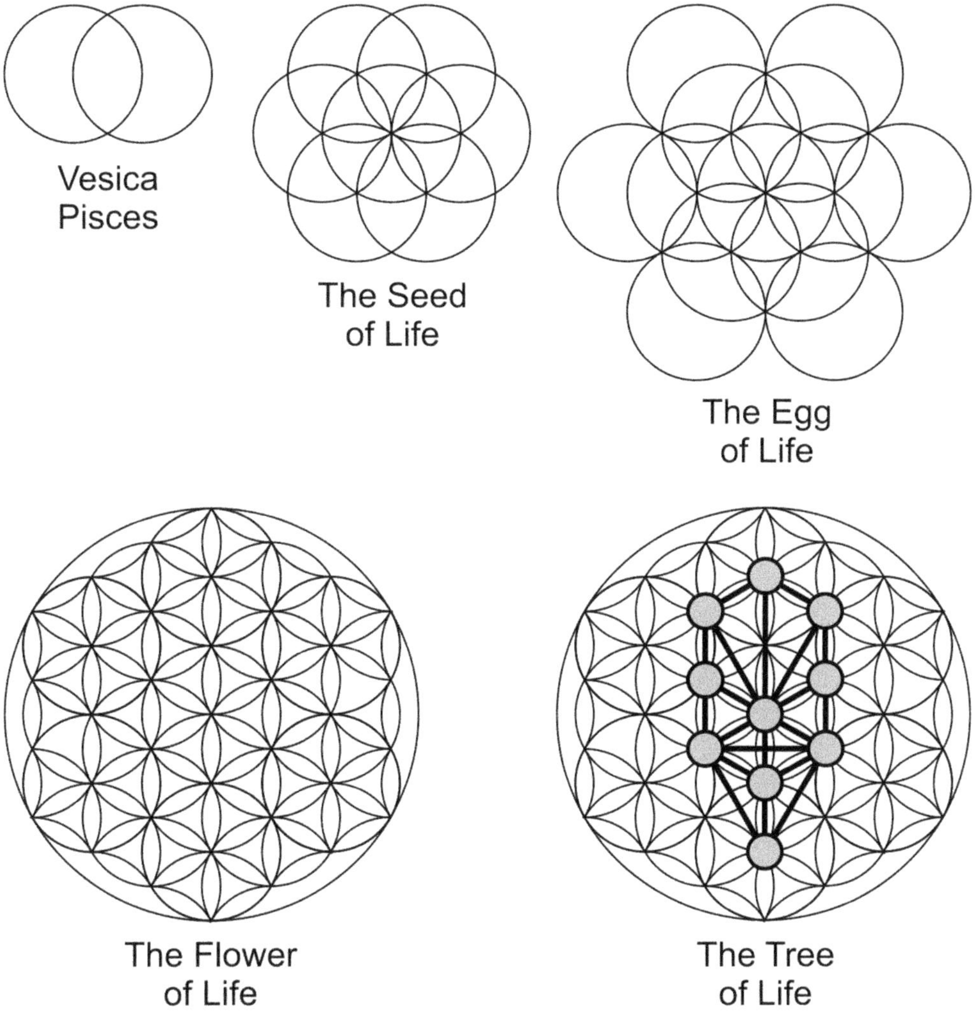

Vesica Pisces

The Seed of Life

The Egg of Life

The Flower of Life

The Tree of Life

Fig. 13.2: *Geometric patterns that can be derived from the vesica pisces.*

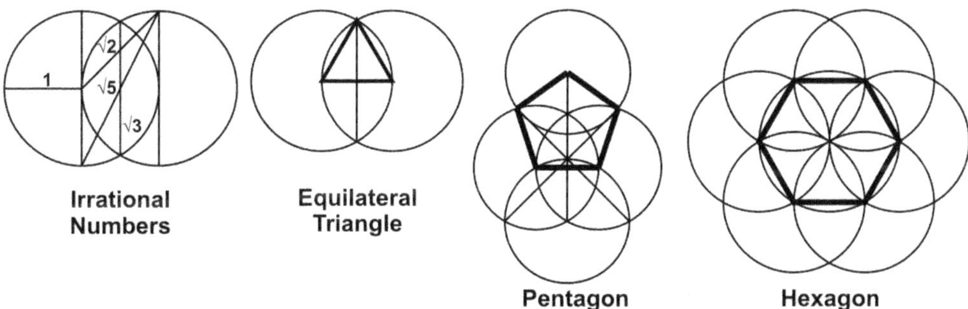

Fig. 13.3: *Irrational numbers and geometric shapes that can be derived from the vesica pisces. Note that the derived pentagon is a close approximation of a true pentagon. A pentagram can easily be created from the pentagon by extending its sides to meeting points.*

Chapter 5. The Final Ascent

In every lifetime, the soul would advance in wisdom and knowledge and eventually reach a culmination state where it would be worthy to escape the reincarnation cycle. When this is achieved, the soul would then exit to a state of nirvana where it no longer would have to go through a physical lifetime. This is probably represented by the alternative path that is available at the end of the Valles Marineris, the one that leads into the Chryse Planitia and eventually into the vast ocean in the northern hemisphere.

Summary and Conclusions

To summarize, I have assembled all of the major pieces of the Martian architecture in a single diagrammatic picture (Fig. 13.4). At the bottom right corner, we see a representation of the Aum Crater producing "sound" waves in the shape of concentric squares and octagons. These waves eventually reach the Vitruvian Martian to the left. I did not expand the squares and octagons into the Vitruvian Martian nor did I draw in the 4 mountains to which the Vitruvian Martian is fitted so as to keep the drawing less cluttered. Beneath the Vitruvian Martian are wavy blue lines representing the Amazonis and Arcadia Planitiae. Far above the Vitruvian Martian is the Argyre Basin depicted as a yellow sun. The right and left star points of the pentagram fitting the Vitruvian Martian are shown to project a triangular shape to the centre of a black oval at top left representing the Hellas Basin. Enclosed inside the triangle is a representation of the Valles Marineris in blue. At the bottom, left of the Vitruvian Martian, is a section of a MOLA map showing Alba Mons with a vesica pisces superimposed. A yellow cross marks the location of the site antipodal to the centre of the Hellas Basin. A black arrow is shown to symbolically depict a connection of the vesica pisces to the womb-phallus construct on Pavonis Mons. Finally, to the left of the top end of the Valles Marineris is the northern hemisphere of Mars. The Chryse Planitia is joined to the end of the Valles Marineris. The northern hemisphere is shaped like a crown with the equator as its headband and 3 lines of longitude passing through the middle of the Chryse, Isidis and Amazonis planitiae (as in Fig. 12.3) to connect the equator to the north pole where they form the top piece of the crown. The crown would then represent the reward which the soul receives for its accomplishment. The inset at the bottom left corner is a magnified image of the top of the head of the Vitruvian Martian showing that the tip of the star point of the pentagram which fits the Vitruvian Martian head is located in the phallus of the womb-phallus symbol at the top of Pavonis Mons (see also Fig. 13.1).

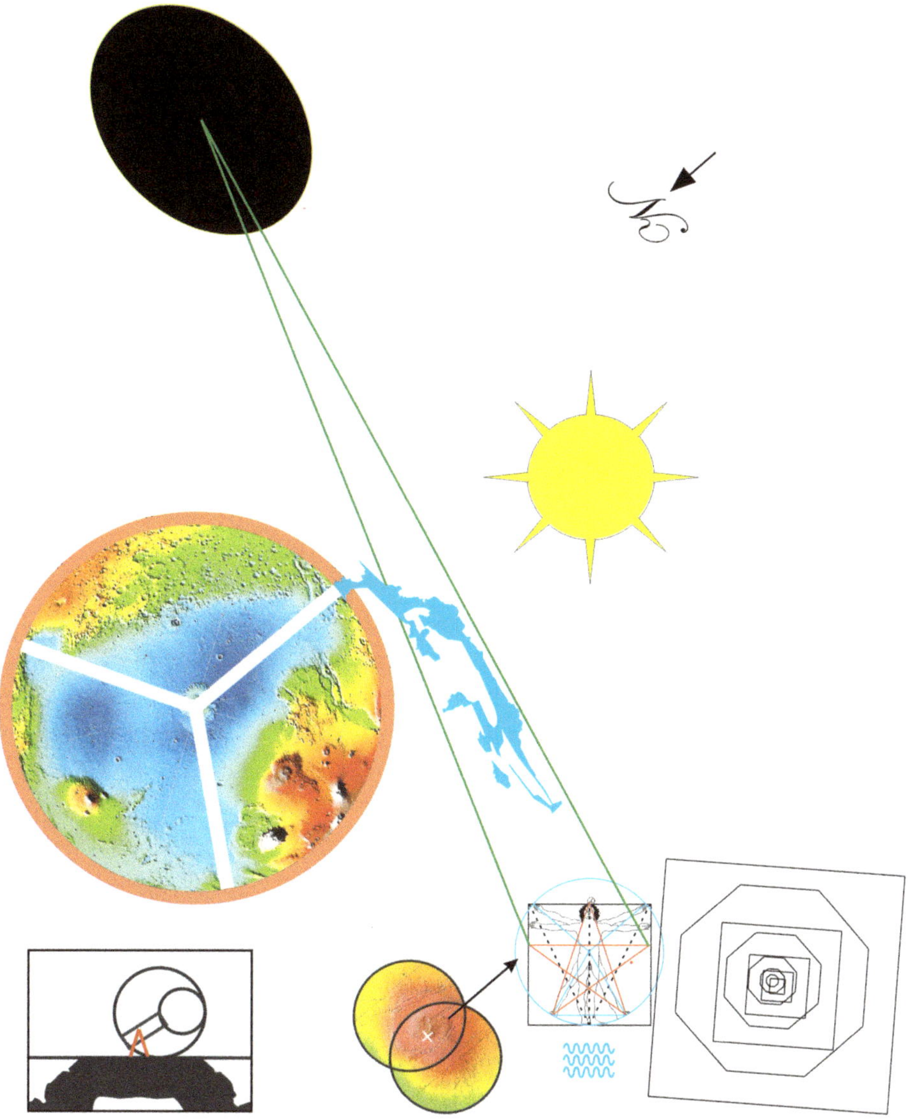

Fig. 13.4: *Diagram integrating the most important features of Martian topography. The Aum crater is depicted at bottom right as creating the Vitruvian Martian together with the womb-phallus construct on Pavonis Mons (see inset at bottom left) and the vesica pisces of Alba Mons (bottom middle). The Vitruvian Martian is sustained by water (Amazonis and Arcadia planetiae represented by wavy blue lines) and the sun (Argyre Basin). The right and left star points of the pentagram which fits the Vitruvian Martian project to the centre of the Hellas Basin (black oval) thereby enclosing the Valles Marineris. A black arrow symbolizes a connection of the vesica pisces to the womb-phallus construct on Pavonis Mons. The northern hemisphere of the planet comes off the end of the Valles Marineris. Images: PIA02036.tif, courtesy of NASA, and MOLA Science Team, NASA Goddard Space Flight Center.*

The entire planet of Mars seems, therefore, to be a massive spherical sculpture depicting the creation of the Martian life form, its nurturance by water and the sun, its passage through a lifelong arduous journey, its death and reincarnation, and finally its escape from the karmic cycle to enter into a final state of nirvana. The sculpture seems to be showing that the Alba Mons vesica pisces is the Divine Mother conducting the phenomenon of creation using the Aum sound and the womb-phallus as Her instruments. The Amazonis and Arcadia planitiae might represent the salt water composition of the body. The Argyre Basin likely represents the sun which was probably revered as an essential energy source for the growth of food and for the maintenance of a hospitable planetary temperature range compatible to human, plant and animal life. The Valles Marineris probably represents the obstacles that a soul encounters during its lifetime and a path for it to follow. The Hellas Basin could be either a region for punishment as is our current notion of hell, or it could be a black hole that results in the soul going through a wormhole and emerging through a white hole (Alba Mons) to reincarnate for a new lifetime. Heaven, represented by the northern hemisphere, could either be a place of bliss at the end of a worthy lifetime or it could represent the final ascent of the soul to a state of nirvana where it no longer needs to reincarnate. Hence, it is very possible that Martian spirituality included a belief in reincarnation and the immortality of the soul.

Underlying all of this is what seems to be a profound acceptance of sacred geometry as the basis of both physics and spirituality. They did not appear to put science and spirituality in separate categories as we do today. Sacred geometry was for them the method by which the Divine creates everything. Aum produces energy waves in the form of harmonics and irrational numbers. The vesica pisces and the womb-phallus use these energy waves to create geometrically shaped physical structures, instilling some with life. With Mars we have an extremely advanced textbook on the mysteries of creation and life itself. It might teach us how to tap into a boundless amount of both spiritual and physical energy by harmonizing with sacred geometry. The textbook of Mars has not yet been corrupted by self-serving agendas, but it seems that we are being blocked by NASA from seeing it in high resolution and unaltered photography.

So in conclusion, the *Intelligent Mars* series has demonstrated that a huge portion of the topography of Mars such as the major mountains, craters, planitiae, the Hellas Basin, the Argyre Basin and the Valles Marineris are very likely to have been artificially created rather than formed from natural causes. The extent of this artificiality leads one to suspect that the entire planet may have been constructed from the bottom up. This is in total contradiction to NASA's presentation of Mars as a natural planet arising from swirling gas and dust which first formed clumps that over time smashed into one another to create a larger and larger object until gravity shaped it into a sphere. The mountains are envisioned by NASA to have arisen by volcanic activity, and the craters, planitiae and basins from impacts. The evidence clearly shows otherwise.

Hence, Mars looks very much like it was constructed by a super race of beings which possessed technologies many orders of magnitude beyond our own. This is a complete paradigm shift which will be hard for many to accept. A host of further questions arise from this conclusion. If Mars is an artificial planet, how did it get to its present orbit around the sun? Does it possess a drive mechanism whereby it can be guided into place? Did it come from another star system or galaxy, and if so, how did it get to this solar system? Are other planets also artificial, notably Planet Earth and Venus? How far does this technology extend? Can it create stars such as our own sun? Can it create all the elements of the periodic table or does it require that the heavier elements be collected from the debris of supernovas? Where is the civilization that created Mars located now, or has it been vanquished and eliminated by dark forces or a natural disaster?

These are questions that will have to be tackled by current and future generations. Our real place in the universe has yet to be discovered. The infinity contained in the irrational numbers of π, φ, e, $\sqrt{2}$, $\sqrt{3}$ and $\sqrt{5}$ point to the existence of an Infinite Being. Everything else must be a manifestation of that Infinite Being or else It would not be infinite – it would be a contradiction to say that we are separate beings. We, our planet, all of Nature, the cosmos and extraterrestrials are One Being. I believe we will not find our true selves until we are prepared to incorporate the spiritual dimension into our physics and into our relationships with other beings in the universe. Learning all we can from the study of Mars will help us with this. But we ourselves need to be spiritually transformed or we will lack the necessary tools. We have to enter into love consciousness, our highest form of consciousness. Love has to transcend our reason, our emotions and our instincts. Otherwise we will blunder into territorialism, endless conflicts and wars. We cannot do this on our own. We need the help of the Divine Mother, the primordial creative power of the Divine represented by the Alba Mons vesica pisces. All we need to do is ask, and then go into a state of turiya (the 4th state of consciousness in Aum) for a few moments to not block Her energy. By repeating this once (preferably twice) daily, the Divine Mother will gradually transform us and bring us into the new era of unlimited harmony and creativity that awaits us.

***

www.ingramcontent.com/pod-product-compliance
Ingram Content Group UK Ltd.
Pitfield, Milton Keynes, MK11 3LW, UK
UKHW062304290726
14090UKWH00017B/870

9 780994 032126